Developments in Advanced Ceramics and Composites

T0316690

Developments in Advanced Ceramics and Composites

A collection of papers presented at the 29th International Conference on Advanced Ceramics and Composites, January 23-28, 2005, Cocoa Beach, Florida

Editors

Manuel E. Brito
Peter Filip
Charles Lewinsohn
Ali Sayir
Mark Opeka
William M. Mullins

General Editors

Dongming Zhu
Waltraud M. Kriven

The
American
Ceramic
Society
www.ceramics.org

Published by

The American Ceramic Society
735 Ceramic Place
Suite 100
Westerville, Ohio 43081
www.ceramics.org

Developments in Advanced Ceramics and Composites

For information on ordering titles published by The American Ceramic Society, or to request a publications catalog, please call 614-794-5890, or visit www.ceramics.org

ISSN 0196-6219

ISBN 1-57498-261-3

Contents

Multifunctional Material Systems Based on Ceramics

Carbon/Carbon and Ceramic Composite Materials in Friction

Reliability of Ceramic and Composite Components

Characterization Tools for Materials Under Extreme Environments

Functional Nanomaterial Systems Based on Ceramics

Preface

The 29th International Conference on Advanced Ceramics & Composites was held in Cocoa Beach Florida, January 24-28, 2005. This unique proceedings contains papers from scientists and engineers in government, industry, and academic organizations from around the world which were presented in the following symposia and focused sessions:

- *Ceramics in Environmental Applications*
- *Multifunctional Material Systems Based on Ceramics*
- *Carbon/Carbon and Ceramic Composite Materials in Friction*
- *Reliability of Ceramic and Composite Components*
- *Characterization Tools for Materials Under Extreme Environments*
- *Functional Nanomaterial Systems Based on Ceramics*

We would like to thank and express appreciation to those who attended and participated in the meeting, to the session chairs and organizers, and to those who helped us in the review of the manuscripts contained in this volume.

Manuel Brito
Peter Filip
Charles Lewinsohn
Ali Sayir
Mark Opeka
William Mullins

Ceramics in Environmental Applications

CHARACTERIZATION OF MnO-DOPED LANTHANUM HEXALUMINATE (LaMnAl$_{11}$O$_{19}$) IN TERMS OF SELECTIVE CATALYTIC REDUCTION OF NOx BY ADDITION OF HYDROCARBON REDUCTANT (HC-SCR)

M. Stranzenbach and B. Saruhan
German Aerospace Center (DLR), Institute for Materials Research
Linder Hoehe, Cologne, Germany, 51147

ABSTRACT

Environmental pollution by vehicles, turbines and aircrafts has enormously increased in the last decade. New generation lean-burn combustion engines which are more effective and require less fuel consumption will raise the emission of nitrogen oxides (NO$_x$) even more. Thus, especially in urban areas, more stringed environmental regulations are to release which can only be met by development of new catalytic materials and concepts.

Rare earth oxide catalysts are reported to offer highly effective conversion of NO$_x$ by methane (CH$_4$) in the terms of selective catalytic reduction (HC-SCR). Especially La$_2$O$_3$ is reported to have a high NO reduction by CH$_4$ to N$_2$, although its technical realization as catalyst material has not yet reported and can be challenging due to the hydroscopic property of La$_2$O$_3$.

Complex oxide compounds containing La$_2$O$_3$ however can be suitable alternatives and promising candidates for technical application as catalysts. In this study, LaMnAl$_{11}$O$_{19}$ is characterized in terms of HC-SCR with methane. Phase and morphological characterization of the powder synthesized by sol-gel route and coatings by electron-beam physical vapor deposition (EB-PVD) deposition was presented. LaMnAl$_{11}$O$_{19}$ crystallizes to the magnetoplumbite phase at about 1000°C and is then thermally stabile up to 1400°C.

FTIR spectra of pressed sol-gel powder showed that NO is adsorbed superficially and oxidized by the surface. The addition of oxygen led to changes of the spectrum in the nitrite-nitrate region and the formation of NO$^+$ and N$_2$O$_4$ species. The EB-PVD coated LaMnAl$_{11}$O$_{19}$ layer was catalytically characterized at 200°, 400° and 600 °C and showed catalytic activity towards NO depending on temperature.

INTRODUCTION

In the last decade, the need for low fuel consumption and high efficiency turbines and motors has promoted the use of lean-burn techniques leading to an increase in the combustion temperature. The consequent increase of cyclic process parameters raises the emission of NO$_x$ and leads to formation of oxygen rich exhaust gases. On the other hand, more stringent regulations have been released by legislators such as CEAP, LAER or EURO regarding emissions. As an intermediate precaution, NOx and hydrocarbon (HC) emissions are reduced to 0.3 g/km for diesel engines and almost to zero for petrol engines with the EURO4 regulation which is to apply from 2005. Further adjustment of the diesel and petrol engine regulations has been discussed for the future regulations [1]. From 2008, with the new CEAP6 regulation for aircraft turbines, NO$_x$ emissions are to reduce to a level far below the former CEAP2 regulation [2]. It is to expect that total emission further increases with the increase of the worldwide air traffic. European commission aims to achieve efficient and environment-friendly energy generation by targeting low to zero emission in power plant technologies and aircraft traffic.

Referring to these considerations more stringent regulations are anticipated in the future, particularly, for turbine engines operated in urban areas, emphasizing the necessity for employment of catalysts [3]. These catalysts must be able to reduce NO_x in net oxidizing conditions. Today's technologies allow various numbers of possibilities for NO_x reduction. In general, those can be divided into two main categories; primary and secondary catalytic systems. Primary system is mostly based on catalytic combustion, implying that a part of the air/fuel mixture is reacted heterogeneously over a catalyst before the actual combustion occurs [4,5]. For these types of catalysts, it is necessary to use premixed combustion which may bring enormous problems regarding liquid fuels, especially for aircraft turbines.

Secondary systems in turn achieve the reduction of NOx in the exhaust gas. There are two promising secondary systems for the reduction of NO_x-emission under lean-burn conditions. One of those is the storage-reduction technology (NO_x-SR). This technology allows the storage of NO_x by adsorption on the surface of the catalyst during the lean-burn phase and subsequently, desorption and reduction of the absorbed NO_x during the following fat-burn phase. Because the second phase is much faster than the storage phase, a significant NO_x reduction can be achieved in total [6].

The second promising candidate technology is the selective catalytic reduction of NO_x (SCR) with addition of ammonia or hydrocarbons as reductant [7]. For stationary systems ammonia is widely used, although for mobile systems hydrocarbons are preferred. As the reductant and NO_x react over the catalyst, NO_x is converted to N_2 and O_2 and thus, reduced. The problem regarding the systems working with ammonia addition is the production of non-reacted NH_3 which inflows the environment and/or forms corrosive products. In turn, the hydrocarbons (fuel, kerosene) are more easily applicable on mobile systems.

MnO-doped La-hexaluminate has been suggested as a promising material for the above mentioned catalytic systems. Earlier studies on rare earth lanthanide oxides report high catalytic activity as SCR-catalyst material. With La_2O_3 and Sr/La_2O_3, a high conversion of NO_x with methane is shown with or without oxygen [8,9]. Due to highly hydroscopic property of La_2O_3, its technical realization as catalyst material is difficult, making it necessary to direct the investigations towards finding out more stable chemical compounds of La-oxides for this application.

MATERIALS AND METHODS

Manganese (II) oxide doped lanthanum hexaluminate ($LaMnAl_{11}O_{19}$) has been produced by means of two processing routes. Figure 1 shows the procedure for sol-gel synthesis of lanthanum manganese hexaluminate powder. After drying at 200°C, the powder was calcined for 1 hour at various temperatures from 600°C to 1400°C. The detailed synthesis procedure is described elsewhere[10].

XRD-measurements of the calcined powder were performed with a step size of 0,020° at 2θ angles from 10° to 80° on a Siemens D5000 diffractometer using nickel-filtered CuKα radiation. The samples were characterized microstructurally and compositionally by Energy Dispersive X-ray spectroscopy (EDX) supported Scanning Electron Microscope (SEM Leitz LEO 982). X-ray diffractograms of the used $LaMnAl_{11}O_{19}$ powder showed that the crystallization to magnetoplumbite phase occurs above 1100°C. Therefore, the powder was pre-calcined for 1 h at 1000°C in order to carry out the FTIR measurements.

The FTIR spectra were recorded on a Bomen MB 102 FT-IR spectrometer equipped with a liquid-nitrogen cooled MCT detector at a resolution of $4cm^{-1}$. The detailed setup is described elsewhere[11].

Beside the sol-gel synthesized powder, EB-PVD (Electron-Beam Physical Vapor Deposition) manufactured LaMnAl$_{11}$O$_{19}$ coatings were analyzed. For deposition of coatings, a two-source 150 kW electronic beam coater was used consisting of separated chambers for loading, preheating, and deposition. Ingots which meet the special requirements of the EB-PVD deposition were bottom fed in the crucibles during evaporation to ensure continuous and constant deposition conditions. The substrates were rotated during the vacuum evaporation. SEM, EDX, and XRD-analysis of the EB-PVD coating layers were carried out as described above.

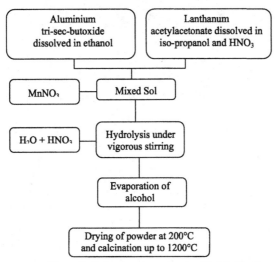

Figure 1: Sol-gel route synthesis of LaMnAl$_{11}$O$_{19}$

For the catalytic characterization of the hexaluminate coatings, a user-specific and computer controlled experimental setup consisting of a gas-mixer, a tube-furnace, a gas-tight, ceramic specimen holder, and two electrochemical gas sensors from Sensoric GmbH was employed.. As carrier gas synthetic air and as test gas nitrogen monooxide (NO) were used. In order to achieve a maximum surface area, a small honeycomb structured component was built by bonding pieces of the layers together. The size of this component was similar to that of the cross sectional area of the specimen holder being about 10 mm long.

RESULTS AND DISCUSSION

Production and Characterization of Materials
XRD-analysis of the sol-gel synthesized MnO-doped La-hexaluminate powder after calcining for one hour at various temperatures up to 1200°C showed that the powder was amorphous up to 900°C and crystallized to the magnetoplumbite phase at about 1000°C. Since the crystallization occurred without the formation of other phase(s) which is mostly encountered in the case of undoped La-hexaluminate, it can be attributed that the formation of the

magnetoplumbite phase at MnO-doped La-hexaluminate is due to a crystallisation process and not due to a reaction between $LaAlO_3$ and Al_2O_3.

Scanning electron microscopy (SEM) investigation of the sol-gel-synthesized and 1200°C calcined powder shows the formation of hexagonal, plate-like grains with random crystal growth, leading to an interlocking and high surface morphology (Fig.2). The high surface area of the morphology supports the catalytic effect of the material.

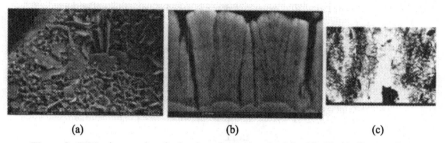

(a)   (b)   (c)

Figure 2: SEM micrographs of sol-gel synthesized and calcined $LaMnAl_{11}O_{19}$ powder morphology on the surface of a particle (a). SEM micrographs of 1100°C heat-treated EB-PVD manufactured MnO-doped La-hexaluminate coating showing the columnar morphology (b); Back scatter image detail of the interlocking microstructure after heat-treatment (small picture at c).

EB-PVD processing of $LaMnAl_{11}O_{19}$ with multiple beam patterns delivered a stoichiometric layer. Since the vapor pressures of individual oxides within $LaMnAl_{11}O_{19}$ composition, i.e. MnO, La_2O_3 and Al_2O_3 exhibit strong differences, EB-PVD- manufacturing of this composition requires less power than that of PYSZ evaporation and subsequently leads to lower substrate temperature. Therefore, the as-received EB-PVD coatings were amorphous and a heat-treatment process was needed for its crystallization. SEM micrographs shown in Figure 3 display the columnar EB-PVD-coating (Fig. 2b) and the formation of interlocking plate-like morphology of magnetoplumbite phase after heat-treatment as detected in back-scatter SEM-modus (Fig. 2c).

Magnetoplumbite phase of La-hexaluminate has a highly defective crystal structure, consisting of $[AlO_{16}]^+$ spinel blocks, intercalated by mirror planes of composition $[LaAlO_3]^0$. The introduction of Mn(II), a divalent ion, in the spinel block forms an electrically neutral structure. This promotes the formation of magnetoplumbite phase, with an interlocking needle-like morphology. Once formed, the phase is stable up to 1400 °C.

FTIR-Measurements

The specimen for the FTIR measurements was prepared as described above and activated by heating for 1 h in vacuum, followed by 100 Torr oxygen for 1h and again vacuum for 1 h. All steps were performed at 500 °C.

The FTIR spectrum of the activated sample was recorded at ambient temperatures and taken as a background reference. Each shown spectrum is subtracted by this background reference. On the surface of the activated samples only absorptions in the hydroxyl-stretching-region and the surface-carbonate-region were observed. The weak bands at 3862 cm^{-1} and 3675 cm^{-1} are attributed to Al^{3+}-OH [12] and La^{3+}-OH [13].

NO was added to the activated catalyst at room temperature for 20 minutes. The addition of NO resulted in an adsorption of NO on the surface which was disproportionate with time. Bands were observed at 1610 cm^{-1}, corresponds to the banding mode of adsorbed water, at 1398 cm^{-1}, identified as NO_2^- species, at 1208 cm^{-1}, characteristic for hyponitrite ($N_2O_2^{2-}$) and a defused adsorption at 1512 cm^{-1}, which is typical for surface nitrates.

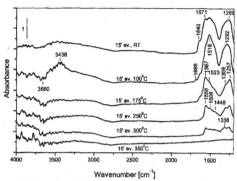

Figure 3: Temperature controlled vacuum diffraction of adsorbed surface species under NO/O$_2$

The addition of 60 Torr of oxygen into the reactor (containing 30 Torr of NO) led to great changes in the spectrum. The nitrosyl bands disappeared and the nitrite-nitrate region changed drastically. New bands at 2229 cm^{-1} and 1752 cm^{-1} belonging to NO$^+$ species and adsorbed N$_2$O$_4$ appeared indicating that oxidation of NO took place. The adsorption of NO$_2$ can occur by disproportionation under the participation of Lewis acid-base pairs and/ or surface hydroxyls [14].

Evacuation of the reactor for 15 min at ambient temperatures resulted only in a few changes in the spectrum, associated to the disappearance of NO$^+$ and N$_2$O$_4$ bands (Fig. 3). After evacuation temperature controlled diffraction was done. At 100 °C only changes in the shapes and intensities were observed, indicating that a rearrangement of the NO$_3^-$-species occurs.
Remarkable decomposition of the surface NO$_x$ species begins at 175 °C (Fig. 3). From the thermal stability and the magnitude of the v3 spectral splitting the nitrates bands, it was assigned that momo-, bi- and bridged species form (Fig. 3). The high temperature stable species can be identified as NO^{2-} species [13]. Detailed analyses of the desorption process of the surface nitro-nitrato structures are underway, in order to investigate the decomposition to NO and/or NO$_2$, or to N$_2$ and O$_2$.

Adsorption of NO/O$_2$ for 30 min was repeated and then evacuated for 10 min at room temperature. Than methane (CH$_4$) was added to the FTIR cell and heated for 15 min at various temperatures (Fig. 5). The heating to 250 °C led to a loss of the adsorbed nitrate species. The subtraction spectrum in Fig. 6B shows a parallel re-adsorption of the adsorbed H$_2$O (bands 3875 cm^{-1} and 3675 cm^{-1}). Heating to 350°C for 20 min showed a strong decrease in the nitrite-nitrates bands, accompanied by the adsorption of H$_2$O (band 1610 cm^{-1}). This indicates that the oxidation of CH$_4$ took place leading to the formation of adsorbed water. Further increase of temperature up to 450°C for 20 min raised the adsorption of water instead of decreasing. The subtraction spectra showed the formation of a new band at 1372 cm^{-1} corresponding to bidentate carbonates and negative bands due to the nitrite-nitrate species consumption (Fig. 5B). The changes of the

7

spectra in the Lanthanum-NO species region could not be observed, due to the spectral bandwidth of the employed FTIR.

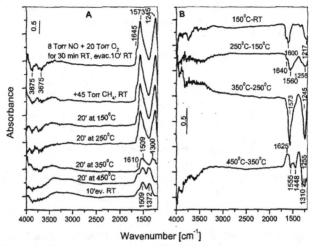

Figure 4: Spectra (left) and subtraction spectra (right) of an adsorbed NO/O_2 catalyst surface under CH_4 atmosphere (0,06 bar) at various temperatures

Catalytic Characterization

With the as-received EB-PVD MnO-doped La-hexaluminate layers catalytic experiments were preformed. Before performing the catalytic experiments, a dry run with no specimen was carried out in order to examine the operation of the experimental setup. A "background" graph was recorded at several temperatures up to 900°C and with NO concentrations of 25 and 50 ppm at a total flow rate of 400 ml/min. Synthetic air was used as carrier gas, thus there was an oxygen rich atmosphere of about 20% of excess oxygen and a humidity of 50 %. The results were reproducible. During background tests, no formation of nitrogen dioxide (NO_2) was detected at any temperature and with all NO concentrations. In turn, it was detected that the measured NO concentration decreases with increasing temperature, assuming that this decrease is either due to adsorption effects at walls of the ceramic holder or decomposition of NO to N_2 and O_2. Relying on these assumptions, all the measured results shown here were corrected by the pre-measured background values.

The experiments were carried out with as-received i.e. amorphous/crystalline layers of EB-PVD manufactured MnO-doped La-hexaluminate. NO reduction and conversion was observed at temperatures 200, 400 and 600 °C. Simultaneously an increase in NO_2-concentration has been detected which was almost compatible to the concentration reduction of NO at the corresponding temperature (Fig. 5). Percentage ratio of NO reduction to the theoretical expected NO concentration seems to depend on the NO concentration and on temperature. The temperature dependence is likely due to the catalytic activity temperature of $LaMnAl_{11}O_{19}$ whereas the concentration dependence is probably to the present surface area of the honeycomb structure. Comparison of the measured concentrations of NO and NO_2 with the theoretically expected concentrations indicates some negligible deficit which is presumably due to the conversion of the gas to another species, which can not be detected with the present setup.

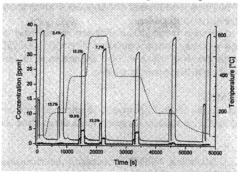

Figure 5: Catalytic activity of LaMnAl$_{11}$O$_{19}$ EB-PVD layer under NO+synthetic air atmosphere at different temperatures (NO: dotted; NO$_2$: continuous; temperature: dashed). The numbers show the NO reduction in percentage to the theoretical expected concentration

Considering these results and observed FTIR-spectra, it can be presumed that chemisorption of O$_2$ occurs on oxygen defect sites, leading to the formation of charged oxygen species which might promote the formation of NO$_2$ and the adsorption of NO$_x$ species on surface sites of LaMnAl$_{11}$O$_{19}$. Moreover, it must be noted that the FTIR measurements are carried out under low pressure atmospheres, whilst the catalytic measurements at atmospheric pressure. It is known that thermal stability of the adsorbed surface nitrates increases with pressure which may explain the observed discrepancy between two different experiments. The decomposition of the adsorbed (surface) nitrates to O$_2$ and N$_2$ upon increase of the temperature is not unlikely, although it can not be definitely disclosed. New measurements with a better equipped experimental setup are underway.

CONCLUSION

FTIR spectra show adsorption of NO under oxygen and oxygen free atmospheres. On adsorption of NO at room temperature, surface-nitrates form, most likely as anionic nitrosly and nitro-species. The adsorbed NO undergoes disproportionation with time involving surface OH⁻ groups. Oxygen addition changes the spectrum, mainly due to the oxidation of NO, NO$^+$ and N$_2$O$_4$ surface species. The evacuation leads to the disappearance of NO$^+$ and N$_2$O$_4$ and results in slight changes in the nitrate region of the spectrum. Under vacuum nitrates are stable up to 300°C.

Catalytic characterization of EB-PVD layer of LaMnAl$_{11}$O$_{19}$ confirms the promotion of NO-oxidation. The FTIR measurements were carried out under vacuum, It is likely that the decomposition and desorption temperatures under atmosphere pressure differ and are higher from those. Activity temperatures around 700°C under combustion conditions therefore seem to be feasible.

Due to the limitations in bandwidth of the employed FTIR equipment, the behavior of NO species at La-surface sites could not be characterized in order to determine HC-SCR mechanism. Regarding the methane at FTIR spectra in Fig. 5, it can be assumed that the Mn-surface sites were blocked by adsorbed NO-species which are likely non-reactive towards CH_4. Doping $LaMnAl_{11}O_{19}$ with single-oxidation-state ions, like Mg, it may force that the Mn-ion remains in 2^+-oxidation state resulting in adsorption of NO_x-species which are more reactive with methane or formation of free active surface-sites.

According to previous investigations, NO can be converted over rare earth metal oxide catalysts, especially over La_2O_3, with a high rate to N_2 and O_2 by CH_4 in the present of oxygen. Therefore further experiments which involve the detection of La-sites and modification of $LaMnAl_{11}O_{19}$ by doping with Mg may be eligible.

ACKNOWLEDGEMENT

The authors thank to Prof. Margarita Kantcheva for her support with the FTIR measurements and Prof. Paul Kaul for his support with the catalytic characterization.

REFERENCES

[1]Rat, E. P. u., "Richtlinie 98/69/EG", *Amtsblatt der Europäischen Gemeinschaft* L350/1 (Deutsche Fassung) (1998).

[2]ICAO, "Research to develop new technology targets aircraft engine emissions", *ICAO Journal* 59 (5) (2004).

[3]EPA, U. S., "EPA-454/R-00-002", in *National Air Pollutant Emission Trends, 1900-1998* Office of Air Quality, Planning and Standards, Research Triangle Park, NC 27711 (2000).

[4]Vatcha, S. R., "Low-emission gas turbines using catalytic combustion", *Energy Conversion and Management* 38 (10-13), 1327-1334 (1997).

[5]Carroni, R., Schmidt, V., and Griffin, T., "Catalytic combustion for power generation", *Catalysis Today* 75 (1-4), 287-295 (2002).

[6]Takahashi, N., Shinjoh, H., Iijima, T., Suzuki, T., Yamazaki, K., Yokota, K., Suzuki, H., Miyoshi, N., Matsumoto, S.-I., and Tanizawa, T., "The new concept 3-way catalyst for automotive lean-burn engine: NOx storage and reduction catalyst", *Catalysis Today* 27 (1-2), 63-69 (1996).

[7]Radojevic, M., "Reduction of nitrogen oxides in flue gases", *Environmental Pollution* 102 (1), 685-689 (1998).

[8]Zhang X. K., Walters A. B., and Vannice M. A., "NO Adsorption, Decomposition, and Reduction by Methane over Rare Earth Oxides", *Journal of Catalysis* 155 (2), 290-302 (1995).

[9]Vannice, M. A., Walters, A. B., and Zhang, X., "The Kinetics of NOx Decomposition and NO Reduction by CH4 over La_2O_3 and Sr/La_2O_3", *Journal of Catalysis* 159 (1), 119-126 (1996).

[10]B. Saruhan, L. Mayer, and Schneider, H., "Lanthanumhexaluminate as Interphase Material in Oxide Fiber-Reinforced Oxide-Matrix Composites", *Ceramic Transactions* 94 (ed. by N. P. Bansal and J. P. Singh) The American Ceramic Society, Westerville - OH), 215-226 (1999).

[11]Kantcheva, M., Saruhan, B., Stranzenbach, M., Agiral, A., and Samarskaya, O., "Characterization of $LaMnAl_{11}O_{19}$ by FT-IR Spectroscopy of Adsorbed NO and NO/O_2", *Applied Surface Science* to be published.

[12]Koenzinger, H. and Ratnasamy, P., *Catalytic Review-Science Engineering* 31 (17) (1978).

[13]Huang, S.-J., Walters, A. B., and Vannice, M. A., "Adsorption and Decomposition of NO on Lanthanum Oxide", *Journal of Catalysis* 192 (1), 29-47 (2000).

[14]Kantcheva, M. and Ciftlikli, E. Z., "FTIR Spectroscopic Characterization of NOx species adsorbed on ZrO_2 and ZrO_2-SO_4", *Journal of Physical Chemistry B- Condensed Phase* 106 (15), 3941 (2002).

HIGH POROSITY CORDIERITE FILTER DEVELOPMENT FOR Nox/PM REDUCTION

Isabelle Melscoet-Chauvel
Corning Incorporated
SP-DV-02-1
Corning, NY 14831
USA

Christophe Remy
Corning S.A.S
7bis, Avenue de Valvins
77210 Avon
FRANCE

Tinghong Tao
Corning Incorporated
SP-DV-02-1
Corning, NY 14831
USA

ABSTRACT

This paper presents the latest progress in high porosity filter product development for the 4-way application as well as the corresponding property and performance attributes. Two new compositions, Dev-HP1 and Dev-HP2, at 72% porosity with median pore sizes of 17 μm and 20 μm are identified and under development. From composition development, narrow pore size distribution and good pore connectivity were achieved in comparison to the previous version of high porosity filters, Dev-EC, and also the standard commercial product, DuraTrap® CO. As a result, the pressure drop performance has been significantly improved while a high filtration efficiency of more than 95% has been preserved (artificial soot lab test). The detailed physical property and performance data for these two new 72% porosity filters (vs. Dev-EC and DuraTrap® CO) are discussed. It is anticipated that significantly improved pore microstructure and high wall permeability will allow high catalyst loading in the wall for the 4-way catalyst application.

INTRODUCTION

Diesel engines are the most energy efficient powertrains among all types of internal combustion engines known today. This high efficiency translates to very good fuel economy and low greenhouse gas emissions (CO_2) which helps reduce the global climate effect. Other diesel engine advantages that have not been matched by competing energy conversion machines include durability, reliability and fuel safety. Nowadays, in Europe, about 50% of passenger cars are diesel powered, and this trend is increasing, with a diesel share of new passenger cars in Western Europe at 51.9% [1]. Due to the potentially harmful effects on health and on the environment from both NO_x and PM (particulate matter) emissions, there is a need for a reduction of these emissions, and the regulations are tightening in Europe, Japan, and the USA [2]. For example, US 2010 regulations require tailpipe emission to be 0.2g/bhp-hr NO_x and 0.01g/bhp-hr PM for heavy duty applications.

To meet US2010 regulations for both PM and NO_x emissions, there are several technologies proposed including diesel particulate filter (DPF), NO_x trap or selective catalytic reduction (SCR) for NO_x, and their combination. The combination of PM and NO_x is the most attractive due to expected cost and space saving, but it is also the most challenging technology. Toyota has introduced their first combined PM/NO_x emission control technology in late 2003 in Europe (Toyota Avensis 2.0L) and in Japan (Hino truck 4.0L), using the diesel particulate and NO_x reduction (DPNR) catalyst system with NO_x trap on cordierite filter developed jointly by Toyota, NGK, DENSO et al. [3,4]. With the worldwide tightening regulations, it seems that all diesel vehicles sold in these areas (Europe and Japan) will ultimately have integrated NO_x and PM functions. Alternatively, selective catalytic reduction (SCR) catalysts can be used to replace NO_x trap catalysts in a DPF to provide NO_x reduction. Because such a system is also able to reduce CO and HC emissions through catalytic oxidation, the system is also called 4-way catalyst system. Both NO_x trap and SCR 4-way systems have their own advantages and drawbacks in catalyst technology and system design. However, their NO_x reduction performance (efficiency and capacity) is dependent upon the total amount of catalyst loading in DPF filter. A high porosity filter is the leading approach to achieve high catalyst storage. How to maintain the delicate balance between high porosity and thermo-mechanical durability is a challenging problem, and this paper sheds some light on the latest progress in high porosity filter product development.

BACKGROUND

Catalytic converters based on cordierite have been developed and widely used over the past 35 years for the automotive market thanks to a long history of discovery and development [5]. Cordierite is a refractory ceramic with a melting temperature around 1450°C. Another key feature of cordierite is its low coefficient of thermal expansion (typically lower than $4 \times 10^{-7}/°C$), which explains its excellent thermal shock resistance, an attribute necessary for automotive applications. Similar attributes of cordierite have also been found desirable for diesel particulate filter applications.

MATERIALS AND PROPERTIES

Specific amounts of selected raw materials with carefully controlled particle size distributions are batched together, mulled and then extruded to form monoliths, which are then dried and fired with specific schedules, before finally being plugged to make a filter (adjacent channels alternatively plugged at each end in order to force the diesel exhaust through the porous walls acting as a mechanical filter). Figure 1 shows the different processing steps followed to obtain a cordierite ceramic monolith.

Mix Clay + Talc + SiO_2 + Al_2O_3 Raw Materials
With pore formers and organic binders
⇓
Mull, Extrude, Dry
⇓
Firing ≈ 1380-1450°C
Figure 1. Generic Processing of Cordierite Monolith

The porosity is controlled during the solid state reaction taking place during firing of the monolith. Higher and more controllable levels of total porosity can be achieved by adding pore formers to the batch. By careful material composition development and subsequent drying and firing process optimization, we are able to engineer and control filter pore microstructures and their characteristics in terms of total porosity, pore size and pore size distribution.

Table 1 summarizes the key physical properties for two newly fabricated filters in development, Dev-HP1 and Dev-HP2, in direct comparison to a previous version high porosity filter, Dev-EC, as well as DuraTrap® CO, a commercial cordierite filter product. The detailed methods used to measure physical properties of filter materials were described in a previous paper [6].

Table 1. Physical Properties for High Porosity and Reference Filters

Property/Composition	DuraTrap® CO	Dev-EC	Dev-HP1	Dev-HP2
Porosity (volume %)	50	64	72	72
Intrusion Volume (ml/g)	0.4	0.7	1.0	1.0
Fine Pore Size at 10% pore filling d_{10} (μm)	5	5	9	10
Mean Pore Size d_{50} (μm)	12	11	17	20
Coarse Pore Size at 90% pore filling d_{90} (μm)	35	20	32	37
d-factor $(d_{50}\text{-}d_{10})/d_{50}$	0.58	0.54	0.47	0.50
Coefficient of Thermal Expansion (CTE) - (RT to 1073 K) ($\times 10^{-7}$)	3.5	4.8	8.7	7.5
Modulus Of Rupture- MOR at RT, in MPa (cells per square inch/wall thickness in mil)	2.76 (200/12)	1.67 (200/12) 1.88 (300/13)	1.19 (300/13)	1.25 (300/13)
E-Modulus at RT, in MPa	5.52	2.36	1.08	0.98
Strain ($\times 10^{-4}$)	5.00	7.95	11.03	12.82
Thermal Shock Parameter (K)	1429	1657	1267	1709
Bulk Volumetric Heat Capacity (J/l K)	500	420	346	346

The parameters d_{10}, d_{50} and d_{90} are the pore diameters at respectively 10%, 50% and 90% volumes filled by mercury. The d-factor, $(d_{50}\text{-}d_{10})/d_{50}$, is used to describe the fine fraction of the pore size distribution (PSD). The coefficient of thermal expansion (CTE) of a material reflects how much this material expands (or contracts) during heating (or cooling). The cell per square inch (cpsi) and wall thickness both indicate the cell geometry of the filter. The mechanical strength is measured by using a four-point bending method, which provides the corresponding modulus of rupture (MOR), or fracture stress. The elastic modulus (E-Mod, or Young's modulus) is a way to measure the degree of stiffness of a material. It is commonly known that for brittle materials such as ceramics, the lower the elastic modulus, the lower the stress for a given strain (deformation/elongation).

Pore Characteristics and Microstructures

As shown in Table 1, two new filter compositions, Dev-HP1 and Dev-HP2, were identified that achieve ultra-high porosity of over 70%. This is a substantial increase in porosity compared to previous versions of high porosity cordierite filters which have porosities around 60% [3,4,7] and a commercial cordierite filter product of 50% porosity, DuraTrap® CO. In addition to increased porosity, two distinct median pore sizes at 17 μm and 20 μm were developed for these new filter compositions [7]. The reason why these two median pore sizes were selected was to gain a better understanding as well as a technical evaluation of the catalyst coating response. Moreover, they also enable a deeper understanding of the delicate trade off between pore size and mechanical properties, especially for porosity levels as high as 72%.

In order to refine the filter pressure drop and catalyst coatability, an effort has been made to improve the pore size distribution of these two new filter compositions since the material mechanical strength is limited by the largest pore size, not the median pore size. The d-factor is used to describe the fine fraction of the pore size distribution. The lower the d-factor, the lower the pressure drop and also the better the catalyst coatability expected due to a reduction in the fine fraction of pores. The fine pores are always coated first due to the capillary effect in catalyst coating process and are expected to have a negative impact on back pressure. As listed in Table 1, Dev-HP1 and Dev-HP2 filters have d-factor values of 0.47 and 0.50 respectively, compared to 0.58 for DuraTrap® CO and 0.54 for Dev-EC.

Another important aspect of these two new high porosity filters is the significant improvement in their mercury intrusion volumes. The intrusion volume in a filter is a direct measure of physical space or volume available in the wall for catalyst storage space. There is about a 40% increase in intrusion volume for both new compositions compared to previous Dev-EC, and about 150% increase vs. DuraTrap® CO.

The pore characteristics described above can also be seen in their corresponding SEM pictures shown in Figure 2. It is obvious that both new high porosity filters show better uniformity in pore microstructure and connectivity compared to Dev-EC and DuraTrap® CO filters. Furthermore, the increased median pore size from 17 μm to 20 μm can be observed in the microstructures as well, which will allow higher catalyst loading while preserving a reasonably low back pressure.

Thermal and Mechanical Properties

Both Dev-HP1 and Dev-HP2 are early stage experimental compositions as product concepts. From the preliminary process development to make these two new high porosity compositions into filters, their coefficients of thermal expansion are at $8.7 \times 10^{-7}/°C$ and $7.5 \times 10^{-7}/°C$ respectively, which are higher than the reference filters Dev-EC at $4.8 \times 10^{-7}/°C$ and DuraTrap® CO at $3.5 \times 10^{-7}/°C$. Clearly, the coefficients of thermal expansion for both Dev-HP1 and Dev-HP2 preferably need to be decreased to be comparable to DuraTrap® CO.

As the porosity continues to increase, the filter mechanical strength is impacted. As listed in Table 1, the moduli of rupture for the two new high porosity filters are significantly reduced

14

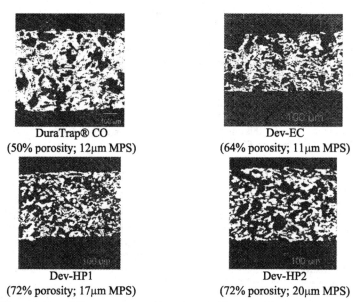

DuraTrap® CO
(50% porosity; 12μm MPS)

Dev-EC
(64% porosity; 11μm MPS)

Dev-HP1
(72% porosity; 17μm MPS)

Dev-HP2
(72% porosity; 20μm MPS)

Figure 2. Pore Microstructures of DuraTrap® CO, Dev-EC, Dev-HP1 and Dev-HP2

(at 172 and 182 psi respectively, even for the 300/13 geometry) compared to more than 240 psi for Dev-EC and 400 psi for DuraTrap® CO filters (200/12 geometry). It is interesting to note that Dev-HP1 at 17 μm median pore size has nearly identical strength to Dev-HP2 at 20 μm median pore size, presumably due to contributions of its pore size distribution. There is an expected strength gain from catalyst coating, especially for heavy washcoat loadings in the 4-way catalyst application.

The material elastic moduli for the two new filters are also directly related to their porosity and pore microstructures. With an 8% porosity increase from Dev-EC to both Dev-HP1 and Dev-HP2, the elastic moduli have been respectively decreased by 54% and 58%. A similar trend is observed from DuraTrap® CO to Dev-EC with a 57% elastic modulus reduction by increasing the porosity from 50% to 64%. These results show significant lower elastic modulus or reduction in rigidity in the ceramic body for our new high porosity filters and hence provide a better capability to withstand thermal and mechanical stress for the filter application.

Thermal Shock Parameter

In diesel particulate filter applications, the filter needs to be regenerated periodically by burning soot in presence of oxygen to reduce pressure drop resulting from soot accumulation. This regeneration event generates an exotherm temperature and related thermal gradient. Depending on the resulting stress level, this could cause cracking of the filter. Similar rapid heating and cooling cycles also occur in gasoline automotive catalytic converter applications where it is common to use the thermal shock parameter (shown below) to describe the tendency of the product to crack upon such thermal gradients:

$$TSP = \frac{ModulusOfRupture}{ElasticModulus \bullet CTE}$$

From this definition, the thermal shock capability of the filter depends upon the combination of mechanical strength, elastic modulus, and CTE. The lower the CTE and elastic modulus, the higher the thermal shock capability. As depicted in Table 1, both Dev-HP1 and Dev-HP2 show adequate thermal shock capability comparable to DuraTrap® CO and Dev-EC, despite higher CTE and lower mechanical strength. Dev-HP2 shows an even better thermal shock parameter than both DuraTrap® CO and Dev-EC filters. Clearly, this results from the extremely low elastic modulus for this filter. The thermal shock parameter for the Dev-HP1 filter is lower than that of the reference filters DuraTrap® CO and Dev-EC due to its higher CTE. Further improvement in mechanical strength and CTE in composition and process development benefits the thermal shock capability of both Dev-HP1 and Dev-HP2.

FILTER PERFORMANCE

In DPF applications, the most important performance attributes are pressure drop, filtration efficiency, and filter durability. Since the filter durability is a function of service life, and also the two new high porosity filters are in early stage of product concept feasibility demonstration, the durability for these two new filters is not included in this study.

Filter pressure drop performance was conducted in lab test device at ambient temperature with air and artificial soot (Printex-U from Degussa Corporation) [8]. All filters were 5.66" or 0.143 m in diameter and 6" or 0.152 mlong. The filter was loaded with artificial soot by aerating the fine soot powder into a compressed air stream. After a specific and controlled amount of artificial soot is loaded inside the filter, the filter is removed, weighed, installed on a cold flow pressure drop rig and tested for pressure drop as a function of flow rate. The filter is then incrementally loaded with greater amounts of soot and the overall process is iterated until reaching the desired soot loading level (generally up to 5g/l). Pressure drop is typically plotted as a function of soot loading for the highest measured flow rate ($356m^3$/hr). From pressure drop vs. gas flow rate curves, the filter wall permeability at clean and soot-loaded conditions can be determined.

Filtration efficiency was also measured on the same soot-loading device. Filters 2" or 0.05 m in diameter by 6" or 0.152 m long were used in this measurement. A microdiluter was coupled with a condensation particle counter. These instruments are used in conjunction to dilute the aerosol flow upstream and measure the particles present at upstream and downstream of a particulate filter [8]. The filter is weighed before and after the test. The filtration efficiency of the entire system is determined by comparing the upstream and downstream numbers of artificial soot particles while the filter is in a quasi stationary test operation. The ratio of soot particles that are trapped in the filter over the total soot particules injected leads to the filtration efficiency of the system. For a DPF, it is desired to have such number equal to or greater than 90%.

Soot Loaded Pressure Drop

Soot loaded pressure drop performance for two high porosity filters, Dev-HP1 and Dev-HP2, is shown in Figure 3 in direct comparison to DuraTrap® CO and Dev-EC filters (all of them in the bare state). It is obvious that both new high porosity filters have substantially improved soot loaded pressure drop performance, about 44% reduction vs. Dev-EC and about 60% reduction vs. DuraTrap® CO. This excellent pressure drop performance can be ascribed to the improvement of total porosity, pore characteristics and microstructures as discussed above.

16

As pore characteristics (total porosity, pore size and distribution, and pore connectivity) continue to evolve, it is interesting to observe that the deep bed filtration mechanism during initial soot loading (< 0.5 g/l) becomes less pronounced. This is counter-intuitive, especially for increased median pore size of Dev-HP1 and Dev-HP2, since enlarged pore size will increase the probability of soot penetrating into the wall. One possible explanation is that overall soot laden gas flow is well distributed through the wall porosity due to good pore connectivity and pore uniformity and hence leading to slower gas velocity. As a small amount of soot penetrates into the wall porosity, it will increase soot loaded wall flow resistance similar to that through surface soot cake layers. As a result, deep bed filtration becomes less pronounced.

Whereas these filters have significantly different soot loaded pressured drop, one can also observe that the difference in clean pressure drop is in comparison relatively small. This can be explained by the fact that the difference in pressure drop across the wall between these different filters in the clean state is somehow masked by other factors also contributing to the total pressure drop (such as channel friction or entrance and exit gas compression/expansion).

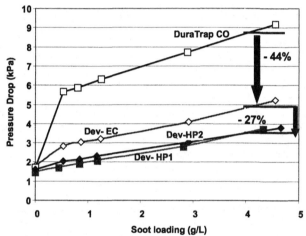

Figure 3: Soot Loaded Pressure Drop for Cordierite Filters

Filtration Efficiency

When both filter porosity and pore size are increased for better catalyst coatability and pressure drop performance, it is always necessary to verify that the filtration efficiency is not compromised. Figure 4 shows filtration efficiency as a function of time for both new high porosity filters vs. Dev-EC. An internal standard filter with 35 μm mean pore size was used as a reference to calibrate and qualify the measurement and test setup. As expected, the standard reference starts at 70-80% efficiency due to its initial large pore size of 35 μm. As soon as soot builds up, the filtration efficiency increases since soot shrinks pore opening. It is obvious that the two new high porosity filters along with Dev-EC filter show excellent filtration efficiency greater than 97%.

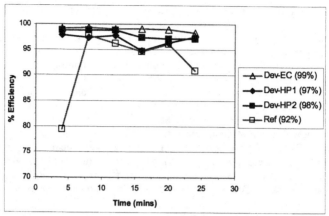

Figure 4. Filtration Efficiency for Different Families of Cordierite Filters

Filter Wall Permeability

From the pressure drop measurement, wall permeability can be determined for both clean and soot-loaded conditions. The pressure drop as a function of wall permeability is plotted in Figure 5. As discussed above, clean pressure drop for the filters is dominated by other factors than the filter wall permeability after the permeability reaches a threshold value of about 1×10^{-12} m^2. DuraTrap® CO is right at that threshold for clean wall permeability. Therefore although Dev-EC and Dev-HP2 are two- and six-times better in wall permeability, the overall clean pressure drop among these filters is not so different (as shown in Figure 4). On the contrary, as the wall permeability is significantly reduced by soot as shown in Figure 5 (data point from same set of filters for soot loaded and clean values), the pressure drop then becomes more sensitive to the difference in wall permeability among these different filters.

Wall permeability is one of the most relevant parameters used to predict pressure drop after catalyst coating. With about a factor of three improvement in wall permeability of Dev-HP2 compared to Dev-EC as well as additional improvement in porosity and pore connectivity, it is highly anticipated that these two new high porosity filters should be able to have high catalyst loadings with coated pressure drops comparable to DuraTrap® CO.

CONCLUSION

The new high porosity filters, Dev-HP1 and more particularly Dev-HP2, show a significant improvement over the existing cordierite product, DuraTrap® CO, and also the first generation of high porosity cordierite, Dev-EC. These filters have a total porosity of 72% with narrow pore size distribution and mean pore sizes of 17 μm (Dev-HP1) and 20 μm (Dev-HP2). These basic product attributes are highly desirable for high washcoat loading necessitated for the 4-way application while retaining low soot loaded pressure drop as well as a reasonably low CTE and an acceptable mechanical strength.

ACKNOWLEDGMENTS

We wish to thank our colleagues Adam Collier, Benjamin Stevens and Stanley Solsky for their continued help on this project. We would also like to recognize the help from our pressure drop testing team, our plugging team as well as our internal characterization team.

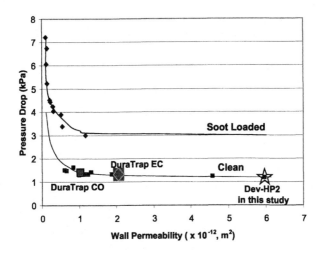

Figure 5. Pressure Drop as a Function of Filter Wall Permeability for
Clean and Soot Loaded (4g/l) States

REFERENCES

[1.] "Diesel share of new passenger cars in Western Europe at 51.9%", DieselNet Update December 2004.

[2.] Timothy V. Johnson, "Diesel Emission Control Technology –2003 in Review", SAE 2004-01-0070.

[3.] Akira Shoji, Shiji Kamoshita, Tetsu Watanabe and Toshiaki Tanaka, "Development of a Simultaneous Reduction System of NOx and Particulate Matter for Light-Duty Truck" SAE 2004-01-0579.

[4.] Yoshiyuki Kasai, Shinichi Miwa and Tatsuyuki Kuki, Kouji Senda and Yoshitsugu Ogura, "New Cordierite Diesel Particulate Filter Material for the Diesel Particulate – NOx Reduction System", SAE 2004-01-0953.

[5.] Martin Murtagh, Gregory Merkel, and Douglas Beall, "Cordierite as a Macro-Cellular Ceramic Material (Invited)", CB-S3-40-2004, The 28th International Conference & Exposition on Advanced Ceramics & Composites, Cocoa Beach, FL, Jan. 25-30, 2004.

[6.] Tinghong Tao, Yuming Xie, Steven Dawes and Isabelle Melscoet-Chauvel, Marcus Pfeifer and Paul Spurk, "Diesel SCR NOx Reduction and Performance on Washcoated SCR Catalysts", SAE 2004-01-1293.

[7.] Tinghong Tao, Willard Cutler, Kenneth Voss, and Qiang Wei, "New Catalyzed Cordierite Particulate Filters for Heavy Duty Engine Applications", SAE 2003-01-3166.

[8.] Robert J. Locker, Natarajan Gunasekaran and Constance Sawyer, "Diesel Particulate Filter Test Methods", SAE 2002-01-1009.

THERMAL STABILITY OF CORDIERITE SUPPORTED V_2O_5-WO_3-TiO_2 SCR CATALYST FOR DIESEL NOx REDUCTION

Yuming Xie
Corning Incorporated
SP-DV-01-9
Corning, NY 14831

Christophe Remy
Corning S.A.S.
7bis, Avenue de Valvins
77210 Avon, France

Isabelle Melscoet-Chauvel
Corning Incorporated
SP-DV-02-01
Corning, NY 14831

Tinghong Tao
Corning Incorporated
SP-DV-02-01
Corning, NY 14831

ABSTRACT

Thermal stability of cordierite supported and extruded SCR catalysts was evaluated and a mechanism was developed to explain the observed differences. These catalyst systems are used to reduce NOx in diesel engine exhaust to meet European and US requirements. The cordierite supported catalyst showed significantly better catalytic conversion efficiency after aging at 650°C than the extruded catalyst, although low temperature activity was compromised because of the lower catalyst loading of the cordierite supported catalyst. Characterization of the supported catalyst indicated that the catalyst had very little surface area loss upon aging and maintained most of its acidity. In contrast, the extruded catalyst showed dramatic surface area and acidity loss. The acidity correlates with catalytic activity, confirming that the reaction kinetics are controlled by the availability of V-OH sites at the temperature range.

INTRODUCTION

Diesel powered vehicles have been receiving significant attention due to their high energy conversion efficiency and associated improved fuel efficiency. Emissions standards for such vehicles are becoming more and more stringent, requiring substantial reductions of nitrous oxides (NO_x) and particulate matter in both Europe and the United States as these regions begin implementing Euro IV/V and US 2007/2010 standards, respectively. Numerous technologies have been studied in response to the upcoming NO_x emission standards, including lean NO_x catalysts, NO_x traps, and selective catalytic reduction (SCR) of NO_x. V-W-Ti based SCR catalysts have been adopted by the automotive market because of their effectiveness in reducing NO_x. Continued advancement of this technology has allowed it to become the technology of choice in Europe to meet Euro IV and possibly Euro V NO_x standards for heavy duty (class 8) and medium duty (class 4-7) truck applications.

Currently, two SCR technologies are being considered for diesel emission control: extruded catalysts and washcoated catalysts. Extruded SCR catalysts typically show good catalytic performance at low temperatures but have limited thermal durability, i.e., limited maximum use temperature and low thermal shock resistance. In contrast, washcoated SCR catalysts typically show adequate catalytic performance at the same low temperatures but have excellent thermal durability up to about 700°C. This work focuses on developing fundamental understanding of the differences of these catalysts under various conditions with surface characterization tools and catalytic performance testing.

SAMPLES AND CHARACTERIZATION

Samples

Two commercially available Corning Celcor® cordierite substrates and a commercially available extruded SCR catalyst were used in this study. A commercially available V_2O_5-WO_3-TiO_2 SCR catalyst was washcoated on the cordierite substrates at several loadings. This composition and chemistry was similar to that of the extruded catalyst. The Celcor® cordierite substrates had the following typical physical properties:

Table1. Physical properties for cordierite substrates

Property	Celcor® 600/4	Celcor® 400/7
Porosity (volume %)	35	35
Mean Pore Size, d50 (microns)	3.9	3.9
Open Frontal Area (%)	81	74
Geometrical Surface Area (in2/in3)	88	69
Substrate Bulk Density (g/mL)	0.32	0.44

Microstructure

SEM micrographs of cross sections of both the extruded and washcoated catalysts are shown in Figure 1. Silica glass fibers which provide mechanical reinforcement in the

22

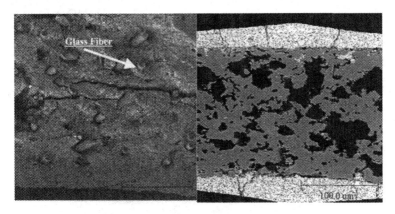

Figure 1. SEM micrographs of cross sections of the extruded catalyst (left)
and the washcoated cordierite catalyst (right).

extruded catalyst can be readily seen. The relatively low calcining temperatures (≤550°C) used in extruded catalysts preserves the surface area of the titania support. Even so, extruded catalysts often crack because of their high coefficient of thermal expansion (CTE). In contrast, washcoated catalysts typically retain most of the thermal shock resistance of the underlying cordierite material, since the washcoating is only about 30-40 µm thick. The washcoat layer usually does exhibit 'mud cracks' due to the drying / calcining processes used during the washcoat process, but these have minimal impact on the thermal durability of the composite structure. [1].

SCR Reaction and Its Mechanism

The SCR catalytic reaction and its mechanism have been well documented over the past 20 years. Although there is still some debate on the precise reaction mechanisms [2], one of the most widely accepted reaction schemes is shown in Figure 2 [3]:

23

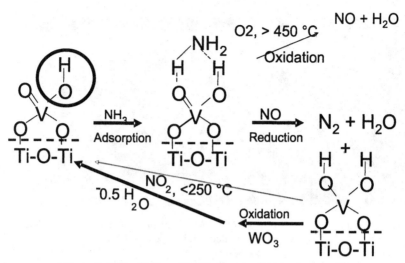

Figure 2. SCR of NOx with NH₃ on a V-W-Ti catalyst.

The active component in an SCR catalyst is dispersed vanadia on the surface of titania. The active V-OH site is a weak acid which adsorbs ammonia. The ammonia is oxidized on the site and becomes the active reducing agent NH_2, which reacts with NO to form benign N_2 and H_2O. The active site V(V)-OH is reduced and becomes non-active V(IV)-OH. If the temperature is above 250°C, the V(IV)-OH site can be oxidized to become V(V)-OH with the help of WO_3, therefore, completing the catalytic cycle. If the temperature of the catalyst is below 250°C, the reduced V(IV)-OH needs NO_2 for oxidation to become catalytically active V(V)-OH. Otherwise, the catalytic cycle cannot be completed and the reaction kinetics slow down. If the catalyst temperature is higher than 450°C, part of the adsorbed NH_2 can be oxidized directly by O_2 to become NOx, a highly undesirable result.

Catalyst Performance Test

Performance of the catalyst was characterized using the test set up shown in Figure 3. The main reactor consisted of a quartz tube 30 mm in diameter and about 600 mm long. The furnace temperature was linearly ramped from 150°C to 900°C in 90 minutes. The effluent gases were analyzed with an on-line FTIR analyzer (Nicolet Magna IR560, OMNIC QuantPad software). This setup is capable of the simultaneous detection and quantification of NO, NO_2, N_2O, NH_3, H_2O, HNO_3, CO, CO_2, and HC. All gas lines were heated to 170°C to prevent the accumulation of ammonium nitrate and condensation of water. The feed gas consisted only of NO (500ppm), NH_3 (500ppm), O_2 (5%) and vaporized water (5%), with the balance being N_2. Experimental samples of 1" diameter and 1" length were used. The space velocity was set at 30,000 l/hr, which is comparable to typical diesel engine exhaust flow. Fresh and aged (650°C) samples were studied, with aging being done by thermal exposure in air. 650°C was chosen in order to accelerate aging even though it is higher than the typical useful maximum temperature of 550°C recommended for extruded SCR catalysts.

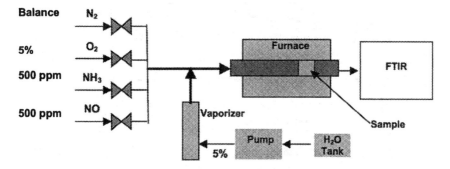

Figure 3. Schematic of the catalyst performance test set up.

Catalyst Surface Acidity Characterization

Temperature programmed desorption (TPD) of ammonia was employed to analyze surface acidity of the catalysts under various conditions. In a typical measurement, about 1.0 gram of ground catalyst was loaded into an Altamira-200 sample holder. After drying at 350°C in a helium flow of 50 cc/minute, the sample was soaked with 10% ammonia in He for over half an hour at room temperature. Helium was flushed across the sample surface to remove excess ammonia physically adsorbed on the catalyst until a stable baseline developed, as determined by the thermal conductive detector (TCD). The sample temperature then was raised at 10°C/minute to 550°C, while the effluent gases were monitored by TCD which detects the desorbed ammonia. The catalyst surface area was measured with nitrogen multiple layer adsorption (BET) at liquid nitrogen temperature.

RESULTS AND DISCUSSION

Catalyst Performance

Figure 4 shows the catalytic activities of the extruded and washcoated catalysts as a function of aging time at 650°C, normalized to their fresh catalytic activity, and measured at three temperatures. It is apparent that the washcoated catalyst preserved its original activity after aging while the extruded catalyst showed significantly reduced activity after aging. After 70 hours aging, the activity of the extruded catalyst approached zero.

These results indicate that the washcoated catalyst is capable of being used at temperatures as high as 650°C. This apparent good thermal stability of the washcoated catalyst is going to be extremely beneficial for Euro V and US2010 applications where diesel particulate filters will likely be inevitable and high temperatures are expected from the need for filter regeneration. This thermal stability also provides some extra margin of safety even for current Euro IV applications.

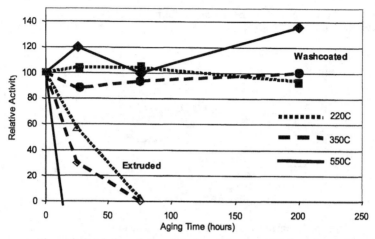

Figure 4. Catalytic activity of washcoated and extruded catalysts after aging.

Surface Area

Specific surface area and catalyst surface acidity were evaluated to try to understand the differences in catalyst activity upon aging between the extruded and washcoated approaches. Figure 5 shows specific surface area of the catalyst as a function of aging time at 550°C. This aging temperature was selected after learning of the rapid loss of catalyst activity of the extruded catalyst at 650°C. Indeed, the extruded catalyst surface area decreased monotonically and reached less than 10 m^2/g after 5 weeks at

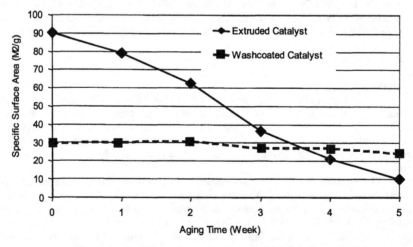

Figure 5. Specific surface area change with the aging at 550°C

26

550°C – more than a 90% loss of surface area. The washcoated sample was more stable, with less than a 20% loss of surface area under the same aging conditions. One explanation for this is that the support slows down the thermal sintering process of catalyst by enhancing heat transfer away from the catalyst, into the underlying cordierite material.

Catalyst Surface Acidity Characterization

As shown in the SCR reaction mechanism in Figure 2, surface acidity is an important performance parameter for SCR catalysts. Temperature programmed desorption (TPD) of ammonia was used to quantify catalyst surface acidity. Figure 6 shows the extruded catalyst surface acidity as a function of aging time at 550°C. As expected from changes in specific surface area upon thermal aging, the total surface acidity of the extruded catalyst decreased with aging time. This rate of reduction seemed substantially faster than that of the surface area, indicating there may be chemical changes in addition to the physical changes in the extruded catalyst. In the TPD spectrum, there are two groups of acid sites: the strong acid sites in the temperature range from 350 to 500°C and the weak acid sites from 200 to 350°C. The peak at around 100-150°C can be ascribed to NH_3 physisorption on the surface since it is close to the ammonium salt condensation temperature. It is obvious that the strong acid sites are extremely sensitive to the aging at 550°C. After 5 weeks, all of the strong acid was lost, along with the catalytic activity.

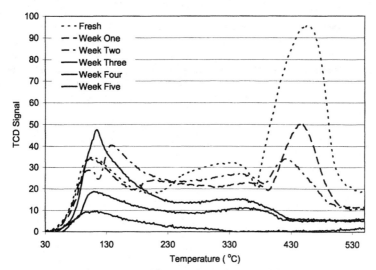

Figure 6. TPD of ammonia on the extruded SCR catalyst
as a function of aging time at 550°C.

As shown in Figure 7 and consistent with the specific surface area measurements, the washcoated catalyst appeared fairly stable, with very little change in surface acidity upon aging at 550°C. There is little change in either the strong or weak acid sites indicating virtually no chemistry change for the washcoated catalyst.

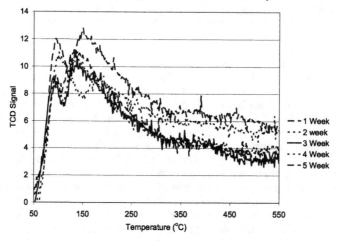

Figure 7. TPD of ammonia on the washcoated SCR catalyst
as a function of aging time at 550°C.

To further evaluate the thermal stability of the washcoated catalyst, samples were aged at 650°C. Figure 8 shows the TPD results, which indicate a shift of surface acidity from high temperature to low temperature or from strong acids to weak acids at this higher temperature aging. Total surface acidity, as indicated by the integrated area under the TPD curve, is virtually unchanged. The thermal stability of the washcoated catalyst activity as indicated by the TPD curves indicates that the cordierite plays an active role in the overall stability of the catalyst.

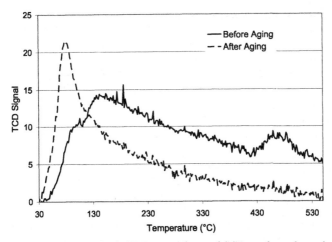

Figure 8. Ammonia TPD on washcoated SCR catalyst after aging
at 650°C for 200 hours.

From the SCR reaction mechanism shown in Figure 2, the amount of ammonia adsorbed on the catalyst surface depends on the amount of V-OH groups on the catalyst surface. At temperatures above 300°C, catalyst performance depends on gas diffusion to the catalyst surface kinetics. With similar catalyst chemistries, these kinetics, and associated NOx conversion efficiencies, are the same for extruded and coated catalysts. At lower temperatures, the reoxidation reduced catalytic V(IV)-OH sites become critical

In the absence of NO_2, the catalytic performance depends on the available V(V)-OH, which is proportional to the catalyst loading. Here the extruded catalyst showed advantages. At temperatures higher than 450°C, the SCR catalyst acts as both an oxidation catalyst and a reduction catalyst. Ammonia is going to be oxidized to NOx, and the higher the temperature, the more NOx is generated. Increasing the temperature or the time at temperature increases the deactivation of the extruded SCR catalyst, reducing the overall NOx conversion efficiency. Overall, the reaction scheme exhibited is shown in Figure 9.

CONCLUSION

Washcoated V_2O_5-WO_3-TiO_2 SCR catalysts have higher thermal stability than extruded V_2O_5-WO_3-TiO_2 SCR catalysts, as indicated by catalyst activity measurements done after aging at 650°C.

The extruded catalyst has lower thermal stability due to loss of surface active sites caused by surface area loss of the catalyst upon exposure to high temperature. The catalyst showed high NOx conversion efficiency at low temperature with high ammonia adsorption capacity, thereby confirming the reaction mechanism that conversion efficiency depends on catalyst loading at temperatures below 300 °C.

In contrast, the washcoated SCR catalyst has much better thermal stability. The high NOx conversion efficiency after exposure to high temperature is ascribed to high concentration of total surface acidity, which degraded very little upon thermal exposure.

However, the reason for the washcoated catalyst retaining its high surface area is worth further investigation.

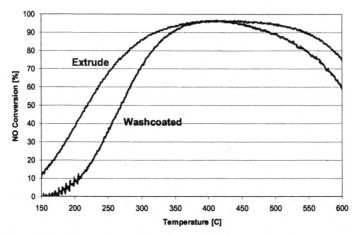

Figure 9. NOx conversion efficiency over extruded and washcoated SCR catalyst

REFERENCES
1. Tinghong Tao, Yuming Xie, Steven Dawes, Isabelle Melscoet-Chauvel, Marcus Pfeifer, and Paul C. Spurk, SAE Paper No. 2004-01-1293, 2004.
2. Guido Busca, Luca Lietti, Gianguido Ramis, Francesco Berti, Applied Catalysis B: Environmental, **18**(1998), p.1-36
3. H. Bosch, F.J.J.G. Janssen, F.M.G. van den Kerkhof, J. Oldenziel, J.G. van Ommen and J.R.H. Ross. *Appl. Catal.* **25** (1986), p. 239

A NEW FAMILY OF UNIFORMLY POROUS COMPOSITES WITH 3-D NETWORK STRUCTURE (UPC-3D): A POROUS Al$_2$O$_3$/LaPO$_4$ IN SITU COMPOSITE

Yoshikazu Suzuki
Institute of Advanced Energy, Kyoto University
Uji, Kyoto, 611-0011, Japan

Peter E. D. Morgan
Rockwell Scientific,
Thousand Oaks, CA, 91360, USA

Susumu Yoshikawa
Institute of Advanced Energy, Kyoto University
Uji, Kyoto, 611-0011, Japan

ABSTRACT

Uniformly porous composites with 3-D network structure (UPC-3D) have been recently developed via a pyrolytic reactive sintering process, which takes advantage of the evolved CO$_2$ gas from a decomposing carbonate source (e.g., dolomite CaMg(CO$_3$)$_2$, or calcite CaCO$_3$) and does not require any additional pore forming agent nor long-time burning-out process. In this paper, development of a new family of UPC-3D is examined for the Al$_2$O$_3$/LaPO$_4$ system. A porous Al$_2$O$_3$/20 vol.% LaPO$_4$ in situ composite (porosity: 42%) with very narrow pore-size distribution at ~200 nm has been successfully obtained by a reactive sintering process at 1100°C for 2 h.

INTRODUCTION

Porous ceramics are widely used as filters, lightweight structural components, electrodes, catalysts, catalyst supports, bioreactors, and so on. We have recently developed uniformly porous CaZrO$_3$/MgO and related composites with 3-D network structure (UPC-3D) via a pyrolytic reactive sintering process.[1-7] The pore-size distribution was very narrow (with pore size of ~ 1 μm), and the porosity was controllable (typically ~30-50%) by changing the sintering temperature. This process takes advantage of the evolved gas from a decomposing starting material and does not require any additional pore forming agent nor long-time burning-out process. Through liquid formation via mineralizer doping, strong necks are formed between constituent particles before completion of the pyrolysis, resulting in the formation of a strong 3-D network structure. Although the CaZrO$_3$-based systems are attractive due to their refractory nature etc., applications are limited for these compositions. Further expansion of UPC-3D families is desired in order to meet various demands for porous materials.

Al$_2$O$_3$/monazite composites (typically, Al$_2$O$_3$/La-monazite) are of interest because of their favorable characteristics, such as excellent chemical compatibility, machinability, H$_2$O vapor resistivity, as reported by Morgan and Marshall et al.[8-16] LaPO$_4$ is not only used as an interfacial coating material for Al$_2$O$_3$ fibers, but also as a second phase dispersoid. Konishi et al.[17] have prepared dense Al$_2$O$_3$/LaPO$_4$ by reactive hot-pressing in vacuum, using the following in situ reaction:

$$La_2(CO_3)_3 \cdot xH_2O + Al(H_2PO_4)_3 + Al_2O_3{}^* \rightarrow LaPO_4 + Al_2O_3 (+CO_2 +H_2O) \qquad (1)$$

($^*\alpha$-Al_2O_3 was added to the starting mixture in order to adjust the final composition)

In their study, vacuum hot-pressing at 1400°C was required to remove evolved gases and to obtain high density. From a different viewpoint, this reaction system can be applicable to design UPC-3D via the reactive sintering where evolved gases are used as "pore forming agents."

In this paper, as a new family of UPC-3D, development of a porous Al_2O_3/20 vol.% $LaPO_4$ *in situ* composite is examined. Preliminary results on microstructure and pore-size distribution of the porous Al_2O_3/$LaPO_4$ composite will be reported.

EXPERIMENTAL PROCEDURE

Commercially available $La_2(CO_3)_3 \cdot xH_2O$ (99.9% purity, Kojundo Chemical Laboratory Co. Ltd.), $Al(H_2PO_4)_3$ (Nacalai Tesque Inc,), α-Al_2O_3 (99.99%, 14m^2/g, Taimei Chemicals Co. Ltd.) and LiF (99.9%, Wako Pure Chemical Ind., Ltd) were used as the starting powders. In this preliminary work, x value in $La_2(CO_3)_3 \cdot xH_2O$ was expediently set as the given value by supplier (in this case, x = 2.6) for the compositional calculation. $La_2(CO_3)_3 \cdot xH_2O$, $Al(H_2PO_4)_3$ and Al_2O_3 powders with LiF (0.5 mass% for total starting powders) were wet-ball milled in ethanol for 4 h in a planetary ball-mill (acceleration: 4 G). The mixed slurry was dried and then sieved through a 150-mesh screen.

To obtain a bulk porous composite, the mixed powder was cold isostatically pressed at 200 MPa after mold-pressing. The green compacts with no binder, 15 mm in diameter and ~ 3 mm in thickness (cylinder) were sintered in air at 1100°C for 2 h to obtain the porous composite. The constituent phases of the bulk materials were analyzed by XRD (Cu-Ka, 40 kV and 40 mA), and the microstructure was characterized using a scanning electron microscope (SEM). The pore-size distribution was determined by the mercury porosimetry. The surface area was determined both by the mercury porosimetry and the N_2-absorption method. Figure 1 shows a schematic illustration of the reactive synthesis of the porous Al_2O_3/$LaPO_4$ composite.

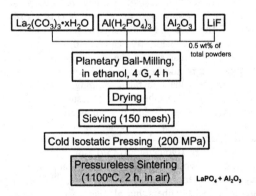

Fig. 1 Schematic illustration of the reactive synthesis of porous Al_2O_3/$LaPO_4$ composite.

RESULTS AND DISCUSSION

Sintered samples are crack free and almost the same size as green bodies, which implies the applicability of near net shaping. Figure 2 shows the XRD pattern for the porous composite sintered at 1100°C. After the sintering, the whole *in situ* reactions substantially completed at 1100°C for 2 h in the powder compacts, and the compact became composed of Al_2O_3 and $LaPO_4$ phases.

Figure 3 demonstrates a typical microstructure of the porous $Al_2O_3/LaPO_4$ composite. The microstructure was fine and rather homogeneous, which was composed of 100-200 nm of Al_2O_3 grains and 1-2 μm of $LaPO_4$ grains (confirmed by EDS and backscattered electron image). Formation of some larger grains (but still < 2 μm) can be attributable to some compositional deviation (i.e. slightly La-poor composition) from the ideal stoichiometry of *in situ* reactions, caused *e.g.* by the H_2O content in the starting lanthanum carbonate. La-poor composition yields other phases like $AlPO_4$,[12] and thus more precise *in situ* reaction design is currently in progress.[18]

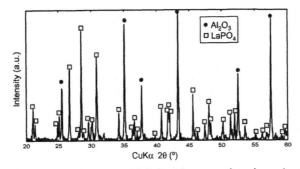

Fig. 2 XRD pattern of the porous $Al_2O_3/LaPO_4$ composites sintered at 1100°C.

Fig. 3 SEM image of the porous $Al_2O_3/LaPO_4$ composites sintered at 1100°C.

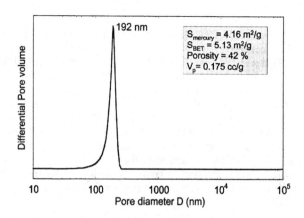

Fig. 4 Pore-size distribution of the porous $Al_2O_3/LaPO_4$ composites sintered at 1100°C.

Figure 4 shows the pore-size distribution of the porous $Al_2O_3/LaPO_4$ composites. Similarly to the other UPC-3Ds, this porous composite also had very narrow pore-size distribution. The peak-top position was about 200 nm, which was smaller than the $CaZrO_3$-based UPC-3Ds. Details of the porous $Al_2O_3/LaPO_4$ and $Al_2O_3/CePO_4$ *in situ* composites will be reported in a forthcoming paper.[18]

SUMMARY

In this study, preliminary results on the porous $Al_2O_3/LaPO_4$ composite via reactive synthesis were presented. $Al_2O_3/LaPO_4$ can be successfully synthesized by the simple one-step reactive sintering technique. The $Al_2O_3/LaPO_4$ UPC-3D had narrow pore-size distribution at about 200 nm.

ACKNOWLEDGEMENT

This study was supported by Grant-in-Aid for Science Research No. 16685019 (For Younger Researcher: Category A) by MEXT, Japan.

REFERENCES
[1]Y. Suzuki, P. E. D. Morgan and T. Ohji, "New Uniformly Porous $CaZrO_3$/MgO Composites with Three-Dimensional Network Structure from Natural Dolomite," *J. Am. Ceram. Soc.*, **83** [8] 2091-93 (2000).
[2]Y. Suzuki, M. Awano, N. Kondo and T. Ohji, "CH_4-Sensing and High-Temperature Mechanical Properties of Porous $CaZrO_3$/MgO Composites with Three-Dimensional Network Structure," *J. Ceram. Soc. Jpn.*, **109** [1] 79-81 (2001).

[3]Y. Suzuki, M. Awano, N. Kondo and T. Ohji, "Effect on In-Doping on Microstructure and CH₄-Sensing Property of Porous CaZrO₃/MgO Composites," *J. Eur. Ceram. Soc.*, **22** [7] 1177-82 (2002).

[4]Y. Suzuki, H. J. Hwang, N. Kondo and T. Ohji, "*In-Situ* Processing of a Porous Calcium Zirconate/Magnesia Composite with Platinum Nanodispersion and Its Influence on Nitric Oxide Decomposition," *J. Am. Ceram. Soc.*, **84** [11] 2713-15 (2001).

[5]Y. Suzuki, N. Kondo and T. Ohji, "*In-Situ* Synthesis and Microstructure of Porous CaAl₄O₇ Monolith and CaAl₄O₇/CaZrO₃ Composite," *J. Ceram. Soc. Jpn.*, **109** [3] 205-209 (2001).

[6]Y. Suzuki, N. Kondo and T. Ohji, "Reactive Synthesis of a Porous Calcium Zirconate/ Spinel Composite with Idiomorphic Spinel Grains," *J. Am. Ceram. Soc.*, **86** [7] 1128-31 (2003).

[7]Y. Suzuki, N. Kondo, T. Ohji and P. E. D. Morgan, "Uniformly Porous Composites with 3-D Network Structure (UPC-3D) for High-Temperature Filter Applications," *Int. J. Appl. Ceram.Tech.*, **1** [1] 76-85 (2004).

[8]P. E. D Morgan, D. B. Marshall, and R. M. Housley, " High-Temperature Stability of Monazite-Alumina Composites," *Mater. Sci. Eng. A,* **195** [1-2] 215-22 (1995).

[9]P. E. D Morgan, and D. B. Marshall, " Ceramic Composites of Monazite and Alumina," *J. Am. Ceram. Soc.*, **78** [6] 1553-63 (1995).

[10]D. B. Marshall, J. B. Davis, P. E. D. Morgan, and J. R. Porter, "Interface Materials for Damage-Tolerant Oxide Composites," *Key Eng. Mater.*, **127**, 27-36 (1997).

[11]S. M. Johnson, Y. Blum, C. Kanazawa, H. J. Wu, J. R. Porter, P. E. D. Morgan, D.B. Marshall, and D. Wilson, "Processing and Properties of an Oxide/Oxide Composite, " *Key Eng. Mater.*, **127**, 231-238 (1997).

[12]D. B. Marshall, P. E. D. Morgan, R. M. Housley, and J. T. Cheung, "High-Temperature Stability of the Al₂O₃-LaPO₄ System," *J. Am. Ceram. Soc.*, **81** [4] 951-56 (1998).

[13]J. B. Davis, D. B. Marshall, R. M. Housley, and P. E. D. Morgan, "Machinable Ceramics Containing Rare-Earth Phosphates," *J. Am. Ceram. Soc.*, **81** [8] 2169-75 (1998).

[14]J. B. Davis, D. B. Marshall, and P. E. D. Morgan, "Oxide Composites of Al₂O₃ and LaPO₄," *J. Eur. Ceram. Soc.*, **19** [13-14] 2421-24 (1999).

[15]J. B. Davis, D. B. Marshall, and P. E. D. Morgan, "Monazite-Containing Oxide/Oxide Composites," *J. Eur. Ceram. Soc.*, **20** [5] 583-87 (2000).

[16]D. B. Marshall, J. R. Waldrop, and P. E. D. Morgan, "Thermal Grooving at the Interface between Alumina and Monazite," *Acta Mater,* **48** [18-19] 4471-74 (2000).

[17]Y. Konishi, T. Kusunose, P. E. D. Morgan, T. Sekino and K. Niihara,"Fabrication and Mechanical Properties of Al₂O₃/LaPO₄ Composite," *Key Eng. Mater.*, **161-163**, 341-44 (1999).

[18]Y. Suzuki, P. E. D. Morgan, and S. Yoshikawa, "Uniformly Porous Al₂O₃/LaPO₄ and Al₂O₃/CePO₄ Composites with Narrow Pore-Size Distribution," *J. Am. Ceram. Soc.*, in preparation.

NOVEL, ALKALI-BONDED, CERAMIC FILTRATION MEMBRANES

S. Mallicoat, P. Sarin, W. M. Kriven
Department of Materials Science and Engineering
University of Illinois at Urbana-Champaign
Urbana, IL, 61801, USA

ABSTRACT

Ceramic filters allow operations at extreme pressures and pHs, high temperatures, and have longer operating lifetimes. However, costly ceramic filtration systems are often rejected in favor of polymeric filters. The primary cause for the high cost of ceramic filters is the expensive processing which includes powder synthesis and sintering cycles. The goal of our research is to develop a new type of ceramic filter using alkali bonded ceramics (ABCs). In this study, we present an update on our effort to create a ceramic membrane with interconnected porosity of controlled pore size using alkali bonded ceramics. Synthesis of ABC membranes can be separated into four main steps (a) synthesis of nanosize polymeric colloidal particles, (b) preparation of self-assembled templates using polystyrene (PS) nanospheres, (c) intrusion of the template with ABC slurry, which hardens at room temperature, and finally (d) dissolution of the template in a solvent. PS nanospheres of diameters 100 and 300 nanometers were synthesized and self-assembled to form templates of different shapes and sizes. Results on intrusion of ABC slurry into the template formed on a microporous nylon filter paper will be presented.

INTRODUCTION

Membranes used in filtration systems can be broadly classified into organic and inorganic filters. Organic filters, such as those made up of cellulose acetate, polyamides, and polysulfones, dominate the market today despite poorer performance in comparison to inorganic membranes in several aspects. In general, inorganic membranes, particularly those made of ceramics, offer superior chemical resistance, wider operational temperature limits, greater resistance to extreme pH conditions, higher pressure limits, longer operating lifetimes, and improved backflushing capabilities. Despite these benefits, inorganic membranes suffer from high fabrication costs primarily due to expensive powder processing and sintering at high temperatures.

Alkali bonded ceramics (ABCs) are amorphous or nano-crystalline aluminosilicate materials. They are formed by reacting inexpensive aluminosilicate raw materials, such as natural clays and industrial by-products, with alkali silicate solutions. Some of the commonly used sources of aluminosilicate materials are metakaolin[1-8], fly ash[9-13], and slag.[14-17] Upon mixing of the aluminosilicate powders with the alkali silicate solutions, a slurry is formed which hardens at ambient temperatures and forms a rock-like material. ABCs are also referred to as geopolymers in the literature.[18]

ABCs are comprised of a matrix of aluminosilicate material which encapsulates unreacted or filler particles that may be present in the system. The matrix is believed to be an amorphous analogue of zeolites and consists of a three-dimensional framework of linked SiO_4 and AlO_4^- tetrahedra. The excess negative charge, resulting from the AlO_4^- tetrahedra, is balanced by alkali cations. Porosimetry analysis (see Figure 1) has confirmed that the average pore size is less than four nanometers and that 95% of the internal surface area is present in pores of diameter less than ten nanometers. ABC's are, however, impermeable materials, with a measured permeability value of 10^{-9} cm/s.[19]

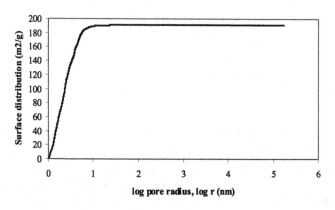

Fig. 1. Integral surface distribution as a function of pore radius obtained from porosimetry.

ABC's have a number of properties that make them favorable candidates for use in filtration applications. ABC's can have compressive strengths of ≥ 50 MPa,[3,8,14] which is considerably higher than that required by most high performance applications such as those utilizing reverse osmosis membranes. Such membranes operate at maximum pressures of 10 MPa.[20] The mechanical properties of ABCs are largely dependent on the processing parameters, most notably, the oxide molar ratios and the degree of reactivity of the aluminosilicate source. In addition, ABCs are thermally stable and maintain their strength at very high temperatures.[21] Acid resistance is another property of ABCs that has been studied extensively.[14,19,22] Metakaolin based ABC's immersed in solutions of sodium sulfate (4.4% wt), sea water (ASTM D 1141-90), and sulfuric acid (0.001M) respectively showed good structural stability even up to 270 days of testing.[6]

Unlike traditional ceramics, both the raw materials and processing of ABCs are inexpensive. Moreover, ABCs can be easily molded into any shape, since they do not stick to organics. ABCs offer excellent potential as a filtration material if a method can be developed to control the porosity. The main objective of this effort is to develop a low-cost ceramic screen filter for water filtration applications. This paper presents progress made towards this goal, and describes the use of polymeric templating to create the microporous structure required for filtration applications.

EXPERIMENTAL PROCEDURE
Raw Materials

Metastar 402 (Imerys Minerals, UK) was used as a metakaolin source, and contains a small amount of muscovite as impurity. The chemical composition of metakaolin, as determined by X-Ray Fluorescence (XRF), was $2.3SiO_2 \cdot Al_2O_3$. The Brunauer-Emmett-Teller (BET) surface area of the metakaolin powder, determined by nitrogen adsorption using Micromeritics ASAP2000 (Micromeritics, Norcross, GA), was 12.7 m^2/g with the mean particle size (d_{50}) of 1.58 μm.

Synthetic metakaolin was prepared using the organic steric entrapment method as described by Gordon et al.[23] The chemical composition of the synthetic metakaolin was

2.0SiO$_2$•Al$_2$O$_3$. The BET surface area of the synthetic metakaolin was 166 m^2/g, and the mean particle size (d_{50}) was 1.62 μm.

Potassium silicate solutions were prepared by dissolving silica fume (Cab-o-Sil, 99.98% purity, Cabot, Tuscola, IL) and potassium hydroxide pellets (Fisher Scientific, Hampton, NH) in deionized water until clear. Solutions were stored for a minimum of 24 hours prior to use to allow time for equilibration.

Polystyrene nanospheres were synthesized using the following chemicals: Styrene (Acros, Fairlawn, NJ) and divinylbenzene (DOW, Midland, MI) as monomer and cross-linker, Aerosol MA-80-I (Cytec, West Paterson, NJ) as a surfactant, sodium bicarbonate (Fisher Scientific, Hampton, NH) for use as a pH buffer, and ammonium persulfate (Fisher Scientific, Hampton, NH) to initiate the reaction.

Alkali Bonded Ceramic Synthesis

ABC samples were prepared by mechanically mixing natural or synthetic metakaolin with alkaline silicate solutions for a period of at least 15 minutes until a homogenous slurry had formed. In the case of natural metakaolin, molar ratios of SiO$_2$/Al$_2$O$_3$ = 4.3, K$_2$O/SiO$_2$ = 0.23, and H$_2$O/K$_2$O = 11 were used. In the case of synthetic metakaolin, molar ratios of SiO$_2$/Al$_2$O$_3$ = 4.0, K$_2$O/SiO$_2$ = 0.25, and H$_2$O/K$_2$O = 19 were used.

Membrane Design

Interconnected microporosity can be created in ABCs by using polymeric templates. PS nanospheres in a colloidal suspension are self-assembled to create closed packed layers of spheres. Once the solutions have evaporated, the polymeric template can be intruded with ABC slurry, to fill the space in between the nanoparticles. The ABC is allowed to cure and harden, and finally, the template is removed via dissolution. The resulting structure is expected to have a pore rating equal to the diameter of the necks between the contacted nanospheres. A schematic representation of the main steps of the procedure is shown in Figure 2. In this article, the first three steps will be discussed.

Characterization

Microstructural analysis was performed using a JEOL 6060LV Scanning Electron Microscope (SEM) (Jeol Inc., Peobody, MA). Polymeric templates were sputter coated (Emitech K575, Emitech, Ashford, England) with a gold/palladium alloy prior to SEM analysis. ABC samples were mounted in epoxy and sectioned using a slow speed diamond saw (Isomet, Beuhler, Lake Bluff, IL). The sections were then polished with progressively finer media prior to final polishing using one micron diamond paste. ABC sections were coated with carbon (Polaron SEM Coating System, Polaron, Watford, UK) for Energy Dispersive Spectroscopy (EDS) analysis in the SEM.

RESULTS AND DISCUSSION

Step 1: Synthesis of polystyrene nanospheres

Although a variety of particle shapes and materials could be used, PS nanospheres were chosen for these experiments. Monodisperse PS nanospheres with approximate diameters of 100 and 300 nanometers were synthesized using emulsion polymerization as described by Reese et al.[24] The concentration of spheres was determined by allowing the solvent of a known mass of suspension to evaporate, leaving behind a mass of spheres. The suspensions of 100 and 300 nanometer spheres had mass concentrations of 26% and 18%, respectively. As larger spheres synthesized with this method were less uniform in size and coagulated to a greater extent, 100 nanometer spheres were used in the self-assembly experiments reported in this article.

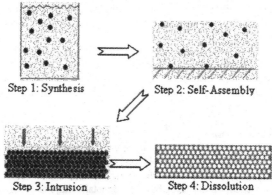

Fig. 2. Schematic showing different processing steps used to synthesize a microporous ABC sample for ultrafiltration applications. Step 1 involves synthesis of colloidal nanoparticles. These particles are then self-assembled into a polymeric template (Step 2). ABC slurry is intruded into the microporosity of the template in Step 3. Upon hardening of the ABC, the template is removed via dissolution (Step 4).

Step 2: Self-Assembly of Colloidal Templates

Diluted suspensions of nanospheres, sufficient to produce approximately ten monolayers, were prepared in deionized water (DI) and allowed to evaporate in air on a polystyrene substrate. Self-assembled layers formed using this suspension produced a non-uniform distribution of spheres. Additionally, most of the spheres were concentrated on a small percentage of the total area. This observation is believed to be a result of high surface tension between the substrate and the nanospheres. In order to produce a homogeneous distribution of PS nanospheres, small amounts of commercially available "Joy" detergent (Procter & Gamble, Cincinnati, OH) were added to the suspensions prior to evaporation. Remarkably uniform and ordered closed packed arrangements of PS nanospheres were observed in the resultant, self-assembled layers, as shown in Figure 3.

Directed assembly of PS nanospheres was also realized by vacuum filtration of the suspension to synthesize patterned templates. Dilute suspensions of nanospheres were vacuum filtered through standard nylon filter paper (Magna filter #1213762, General Electric, Fairfield, CT), with a pore rating of 100 nanometers, mounted on a plastic disc with holes of the desired pattern (see Figure 4). Upon drying, layers of spheres remained directly above the holes in the underlying plastic support for the filter paper. The micrograph in Figure 4 shows a plan view of the cylindrical deposits of spheres remaining after completion of this procedure. Negligible amounts of spheres were deposited in other regions.

Step 3: Intrusion of ABC

Samples of filter paper with patterned (cylindrical) deposits of spheres on the surface were covered with a layer of ABC slurry (natural metakaolin). These samples were then sealed and placed in a cold isostatic press. A pressure of 30,000 psi (200 MPa) was applied to the assembly for a period of one hour. The samples were then allowed to cure at 50°C for a period of 24 hours. Following curing, the samples were prepared for microstructural analysis. Since ABCs consist of aluminum, silicon, and potassium, EDS analysis of these elements was performed. Linescans showing distribution of aluminum, silicon, and potassium content respectively are overlaid on a cross sectional image of the filter paper in Figure 5a. Due to difficulties in imaging cross sections of deposited nanospheres, intrusion of ABC into the microporous filter paper itself was studied.

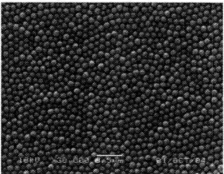

Fig. 3. SEM micrograph of self-assembled polystyrene nanospheres.

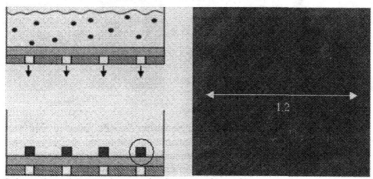

Fig 4: Schematic and SEM micrograph of patterned template created using vacuum filtration.

Three distinct regions can be seen in the micrograph in Figure 5a: the ABC layer, the filter paper, and the epoxy used to mount the sample. Vertical dashed lines have been added to highlight the interface between these regions. Both silicon and potassium were detected at significant levels within the pores of the filter paper. No intrusion occurred in the center of the filter paper due to the presence of a non-porous reinforcement mesh. Notable quantities of aluminum were not detected within the filter paper. This observation can be explained by a slow rate of dissolution of metakaolin particles in the ABC slurry. In order to confirm this, a linescan was performed on bulk ABC. The intensity of aluminum and silicon for each point taken was averaged and this number was subtracted from each recorded data point to obtain Figure 6. In general, the relative amounts of aluminum and silicon tended to fluctuate together. These variations can be directly related to local deviations in porosity. However, several regions within the scanned area exhibited a local excess of aluminum coupled with a local deficiency of silicon. This observation, we believe, is evidence for unreacted metakaolin within the hardened ABC. Metakaolin is the lone source of aluminum in ABCs but provides only one half of the total silicon content. The rest of the silicon originates in the alkali silicate solution.

A second set of ABC specimens was prepared using synthetic metakaolin. EDS analysis of the bulk material failed to identify any regions containing significant unreacted metakaolin, an expected result considering the very high surface area characteristic of this material. Samples of nylon filter paper with deposited spheres on the surface were intruded in the same way as the previous specimens. Analysis of the cross section of the intruded filter (not shown) indicated that aluminum intrusion was achieved, although to a lesser extent than necessary to form ABC. It was determined that an auxiliary source of aluminum was needed to supplement that provided by the metakaolin.

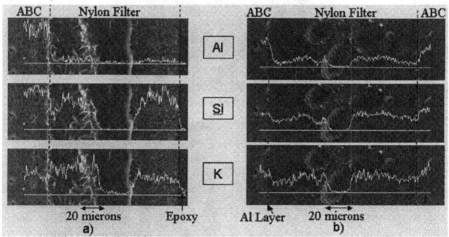

Fig. 5. EDS line scans of aluminum, silicon and potassium for ABC intruded filter paper. Image (a) corresponds to intrusion of ABC made with natural metakaolin and image (b) represents ABC made with synthetic metakaolin. In addition, an auxiliary aluminum source was used in sample (b). A significant increase in aluminum intrusion was thereby achieved.

In order to increase the amount of aluminum intruding into the filter paper and thus, available for ABC formation in the filter paper, a 50-100 nanometer layer of aluminum metal was first evaporated on to the surface of the filter paper prior to templating and intrusion. This layer was easily dissolved in the alkaline environment of the ABC prior to curing. Pressure was applied to the sample in the same manner, but in this case, ABC slurry was intruded from both sides of the filter paper. The resulting EDS analysis is shown in Figure 5(b). A significant increase in intruded aluminum was found. Although not directly verified, these results indicate the feasibility of forming ABC within a microporous polymeric template.

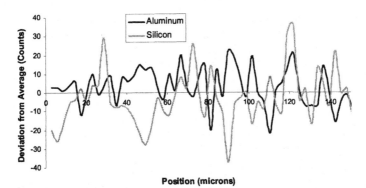

Position (microns)

Fig. 6. Deviation of aluminum and silicon from average EDS intensities in bulk ABC made with natural metakaolin. The intensities at each point taken across the length of the scan were averaged to determine the average intensity. Several regions in which there is locally an excess of aluminum coupled with a deficiency of silicon can be seen. This was attributed to unreacted metakaolin.

CONCLUSIONS

Monodisperse PS nanospheres of diameters 100 and 300 nanometers were synthesized and self-assembled to form microporous polymeric templates. Evaporation was used to create a template of constant thickness and vacuum filtration was used to create patterned templates. Intrusion of silicon and potassium into nylon filter paper was achieved using a cold isostatic press (CIP). The intrusion of aluminum into the filter pores (and possibly the template) was facilitated by using synthetically derived metakaolin as a raw material, or by using an auxiliary source of Al (evaporated). In the future, we plan to dissolve the PS templates to create the desired microporous ABC structure for membrane applications

ACKNOWLEDGMENTS
This work was supported by the AFOSR, under STTR Phase II under Grant number FA 9550-04-C-0063. Part of this research was carried out in the Center for Microanalysis of Materials, University of Illinois, which is partially supported by the U.S. Department of Energy under grant DEFG02-91-ER45439.

REFERENCES
[1] V. F. F. Barbosa, K. J. D. MacKenzie and C. Thaumaturgo, "Synthesis and Characterisation of Materials Based on Inorganic Polymers of Alumina and Silica: Sodium Polysialate Polymers," *Int. J. Inorg. Mater.*, **2**, 309-317 (2000).
[2] V. F. F. Barbosa and K. J. D. MacKenzie, "Synthesis and Thermal Behaviour of Potassium Sialate Geopolymers," *Mater. Lett.*, **57**, 1477-1482 (2003).
[3] P. Duxson et al, "Microstructural Characterization of Metakaolin-Based Geopolymers," Ceramic Transactions, vol. **165**. 71-85 (2005).
[4] M. L. Granizo and M. T. Blanco, "Alkaline Activation of Metakaolin - An Isothermal Conduction Calorimetry Study," *J. Therm. Anal. Calorim.*, **52**, 957-965 (1998).

[5] M. L. Granizo, M. T. Blanco-Varela and A. Palomo, "Influence of the Starting Kaolin on Alkali-Activated Materials Based on Metakaolin. Study of the Reaction Parameters by Isothermal Conduction Calorimetry," *J. Mater. Sci.*, **35**, 6309-6315 (2000).

[6] A. Palomo et al, "Chemical Stability of Cementitious Materials Based on Metakaolin *Cem. Concr. Res.*, **29**, 997-1004 (1999).

[7] H. Rahier, B. Wullaert and B. Van Mele, "Influence of the Degree of Dehydroxylation of Kaolinite on the Properties of Aluminosilicate Glasses," *J. Therm. Anal. Calorim.*, **62**, 417-427 (2000).

[8] M. Rowles and B. O'Connor, "Chemical Optimisation of the Compressive Strength of Aluminosilicate Geopolymers Synthesised by Sodium Silicate Activation of Metakaolinite," *J. Mater. Chem.*, **13**, 1161-1165 (2003).

[9] J. G. S. van Jaarsveld, J. S. J. van Deventer and G. C. Lukey, "The Characterisation of Source Materials in Fly Ash-Based Geopolymers," *Mater. Lett.*, **57**, 1272-1280 (2003).

[10] W. K. W. Lee and J. S. J. van Deventer, "Structural Reorganisation of Class F Fly Ash in Alkaline Silicate Solutions," *Colloids Surf., A*, **211**, 49-66 (2002).

[11] A. Palomo, M. W. Grutzeck and M. T. Blanco, "Alkali-activated Fly Ashes - A Cement for the Future," *Cem. Concr. Res.*, **29**, 1323-1329 (1999).

[12] J. W. Phair and J. S. J. van Deventer, "Effect of Silicate Activator pH on the Leaching and Material Characteristics of Waste-Based Inorganic Polymers," *Miner. Eng.*, **14**, 289-304 (2001).

[13] J. C. Swanepoel and C. A. Strydom, "Utilisation of Fly Ash in a Geopolymeric Material," *Appl. Geochem.*, **17**, 1143-1148 (2002).

[14] A. Allahverdi and F. Skvara, "Nitric Acid Attack on Hardened Paste of Geopolymeric Cements - Part 1," *Ceram. Silik.*, **45**, 143-149 (2001).

[15] T. W. Cheng and J. P. Chiu, "Fire-Resistant Geopolymer Produced by Granulated Blast-Furnace Slag," *Miner. Eng.*, **15**, 205-210 (2003).

[16] K. C. Goretta et al, "Solid-Particle Erosion of a Geopolymer Containing Fly Ash and Blast-Furnace Slag," *Wear*, **256**, 714-719 (2003).

[17] D. Roy, "Alkali-Activated Cements - Opportunities and Challenges," *Cem. Concr. Res.*, **29**, 249-254 (1999).

[18] J. Davidovits, "Geopolymers - Inorganic Polymeric New Materials," *J. Therm. Anal. Calorim.*, **37**, 1633-56 (1991).

[19] J. Davidovits, "Properties of Geopolymers Cements," *Proceedings First International Conference on Alkaline Cements and Concretes*, 131-149 (1994).

[20] Cheryan, M. "Utrafiltration and Microfiltration Handbook." Technomic Publishing Co., Lancaster, PA (1998).

[21] D.C. Comrie and W.M. Kriven, "Composite Cold Ceramic Geopolymer in a Refractory Application," *Ceramic Transactions*, **153**, 211-225 (2003).

[22] J. Blaakmeer, "Diabind: An Alkali-Activated Slag Fly Ash Binder for Acid-Resistant Concrete," *Adv. Cem. Based Mater.* **1**, 275-276 (1994).

[23] M. Gordon, J. Bell and W. M. Kriven, "Comparison of Naturally and Synthetically Derived, Potassium-Based Geopolymers," Advances in Ceramic Matrix Composites X, edited by J. P. Singh, N. P. Bansal and W. M. Kriven, *Ceramic Transactions,* vol.**165**, 95-106 (2005).

[24] C. E. Reese, C. D. Guerrero, J. M. Weissman, L. Kangtaek and S. A. Asher, "Synthesis of Highly Charged, Monodisperse Polystyrene Colloidal Particles for the Fabrication of Photonic Crystals," *J. Colloid Interface Sci.*, **232**, 76-80 (2000).

CONTROLLING MICROSTRUCTURAL ANISOTROPY DURING FORMING

Shawn M. Nycz, Richard A. Haber
Rutgers University
Ceramic and Materials Engineering
98 Brett Rd.
Piscataway, NJ 08854-8065

ABSTRACT

Microstructural anisotropy is a key feature in the control of specific properties of low thermal expansion ceramics. A study has been conducted that quantifies the particle orientation for anisometric particles when subjected to a shear in a fluid/paste. The preferential orientation which remains in a green body after processing can affect the tendency to crack during drying and firing, differential shrinkage and warpage, as well as a multitude of physical properties such as strength and density. An optical method was developed which reveals details about the preferred orientation that exists in thin samples. This technique was applied to aluminum oxide green bodies to determine orientation which was affected by experimentally varying rheology, solids loading, and shear rate. Computational fluid dynamics software was used to model the shear fields present in the experimental fluid during forming. Results show that microstructural anisotropy can be varied throughout a formed part by controlling cross-sectional dimensions and forming parameters.

INTRODUCTION

Studies have shown that grain orientation and particle orientation which exist in a ceramic body can affect many properties of the fired piece. For example, particle orientation has been demonstrated to affect firing shrinkage anisotropy, fracture toughness, thermal conductivity, and electrical conductivity.[1-6] Similarly, grain orientation or crystalline texture has been shown to affect properties in a body such as dielectric constant[7,8] and optical birefringence.[7-9]

While most of the above studies involve processes which intentionally develop high degrees of particle orientation and texture, other studies have shown that virtually any processing technique can form bodies containing oriented microstructures. Extrusion, injection molding, tape casting, uniaxial pressing and isostatic pressing have all been shown to induce at least small degrees of particle orientation to formed bodies comprised of anisometric particles.[5,10-13] The origin of the majority of the particle orientation found by these studies is the shear inherent in the forming process. The ability to control this shear-induced particle orientation can be of particular value as it could enable the tailoring of bulk properties to a particular application or the reduction of undesired anisotropy which can lead to residual stresses and cracking. Taking into account the range of affected properties and the degree to which they can be influenced, it is clearly important to consider particle orientation when developing or improving any form of ceramic processing.

The first step to understanding anisotropy is characterizing the green texture. Under typical processing conditions using nearly equiaxed particles, however, texture in the green state is often too weak to be detected by traditional X-ray diffraction techniques such as the Lotgering method or rocking curves. A more sensitive texture detection technique was developed by Uematsu which involves examination of a ceramic microstructure which has been impregnated with a matching refractive index fluid under cross-polarized light.[14] This technique has been applied in several studies to indicate the presence of textured microstructures resulting from the processing shear of several different forming methods.

Previous studies have reported that tape cast bodies are either more strongly[10,11,15] or less strongly[16] textured in the top (doctor blade) surface than in the bottom (substrate) surface. Most studies assume the flow in the blade gap during tape casting to be governed by Couette flow, which results in the conclusion that shear rate is constant throughout the blade gap. This assumption leads Böcker et al. to attribute the occurrence of greater texture strength in the top surface to surface energy effects tending to favor directional grain growth along the solid/air interface.[11] Investigation into the uniformity of the shear throughout the blade gap would be valuable to better understand a different possible source of the texture difference within tape cast bodies.

The present study aims to further understand the relationship between processing shear and the resulting texture and particle orientation. It examines the specific relationships between several processing parameters and the green texture induced by shear during the forming process. The system of study selected is an aqueous alumina tape casting composition. Texture in the green bodies is characterized by combining the previously mentioned polarized light technique with simple image analysis to obtain a quantitative value for the texture strength in a sample. The forming geometry is simulated using computational fluid dynamics modeling to predict the shear profile throughout the system. The green bodies are sampled from multiple locations to observe the relationship between localized shear rate and resulting shear-induced texture.

EXPERIMENTAL PROCEDURE

To prepare the tape casting slurry, a solvent/dispersant mixture was prepared from deionized water and citric acid then adjusted to a pH of 8.5 by addition of ammonium hydroxide. Alumina (A16SG, Alcoa Inc., Pittsburgh, PA) was mixed into the prepared solvent and then milled for 24 hours. Once milling was completed, the slurry was separated from the milling media and mixed with a binder (B-1000, Rohm and Haas Inc., Philadelphia, PA) for twenty minutes.

The composition was then tape cast at a rate of 7.3 cm/sec and allowed to dry under ambient conditions. The green tapes were heated to 1000 C at a rate of 300 C/hr and held for two hours to remove the organic components and begin interparticle neck formation to impart enough strength to allow the samples to be handled without breaking.

The texture is quantified using a modified version of the previously mentioned Uematsu technique. This technique involves thinning a green sample to approximately 100 μm then impregnating it with a matching refractive index fluid to reduce scattering from the surfaces of the constituent particles, resulting in a nearly transparent sample. Observing the sample under cross-polarized light reveals particles which are optically anisotropic in the plane perpendicular to the direction of light. Rotation of a sample in this plane reveals preferred directionality of this optical anisotropy through alternating directions of maximum and minimum extinction. The magnitude of the variation between maximum and minimum extinction is related to the strength of the preferred orientation of the crystals in the sample. Samples were prepared as described and observed with a standard petrographic research microscope (BH2, Olympus, Tokyo, Japan) and digital micrographs were recorded at minimum and maximum extinction. Care was taken to maintain similar microscope and camera settings when recording the micrographs to ensure representative analysis values between micrographs. To obtain a quantitative value representing the texture in the sample, the average grayscale brightness value was taken from the digital micrographs of the sample at maximum and minimum extinction. The difference between the average values was then taken, giving a single quantity which represents the magnitude of transmitted intensity change upon sample rotation, which is directly related to texture strength in the analyzed sample. The quantitative method is normalized such that a fully random sample will generate a value of zero while a sample fully textured with the crystals' c-axes perpendicular to the direction of light propagation will generate a value of one. Intermediate values are expressed as a percentage of maximum orientation, giving a relative texture value analogous to using relative peak heights as in the Lotgering method for X-ray diffraction.

Computational fluid dynamics (CFD) simulations were performed using POLYFLOW (Fluent USA, Lebanon, NH) which modeled the flow of slurry during tape casting. A mesh was generated representing the cross-sectional dimensions of the tape casting geometry used in this study. The rheology model used in the simulations was selected as power-law and its parameters were extrapolated from experimental results obtained from the compositions used in this study using a dynamic stress rheometer (AR-1000, TA Instruments, New Castle, DE). Since the characteristic length scales were several hundred times greater than the typical particle size, it was assumed that particle effects could be ignored when modeling the fluid. Boundary conditions appropriate for each of the surfaces were selected and the simulations were executed.

RESULTS AND DISCUSSION

CFD Shear Simulations

Figures 1 and 2 show the predicted shear intensity profiles for the cross-sectional tape casting geometry used in this study. From these figures it is seen that the CFD simulation disagrees with the previously assumed Couette model, which predicts that the shear throughout the blade gap should be constant. The simulations of both blade gap geometries show dramatically higher shear rates near the doctor blade than near the substrate surface, with a rapid decay of shear intensity starting approximately 20 μm below the tip of the doctor blade. To the best of the authors' knowledge, fluid modeling results such as these have not been previously reported. The local shear nonuniformity is predicted to be particularly pronounced for the 500 μm blade gap, with the shear rate at the blade tip being approximately 230 sec^{-1} compared to approximately 40 sec^{-1} near the carrier surface. In the 250 μm blade gap geometry, the shear is still seen to be dramatically nonuniform, but to a substantially lesser degree than the 500 um geometry, with shear rates of approximately 310 sec^{-1} immediately at the blade tip and 120 sec^{-1} near the carrier. By contrast, by Couette flow it is predicted that the shear rates between the doctor blade and substrate for the 500 and 250 μm blade gaps would be constant rates of 146 and 292 sec^{-1}, respectively. It is expected that if these simulations are representative of the flow conditions which exist experimentally, it should be shown that the shear-induced texture in the tape cast samples is more uniform in the bodies cast with a 250 μm blade gap than those cast with a 500 μm blade gap.

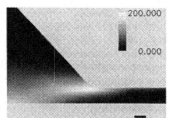

Figure 1: Shear rate profiles of tape casting geometries cast at 7.3 cm/sec with a blade gaps of 250 µm, black bar represents 100 µm

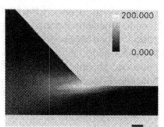

Figure 2: Shear rate profiles of tape casting geometries cast at 7.3 cm/sec with a blade gap of 500 µm, black bar represents 100 µm

Effect of Processing Parameters on Orientation

Figures 3-6 show the relationships between solids loading and orientation for the top and bottom surfaces of samples cast with different degrees of dispersion and different doctor blade gaps. These figures illustrate several orientation phenomena which result from the differing processing conditions.

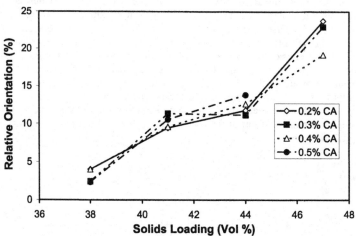

Figure 3: Effect of solids content and citric acid addition (% CA, based on total batch weight) on the top surface orientation for tapes cast with 250 µm blade gap

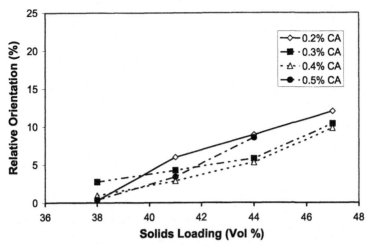

Figure 4: Effect of solids content and citric acid addition (% CA, based on total batch weight) on the bottom surface orientation for tapes cast with 250 μm blade gap

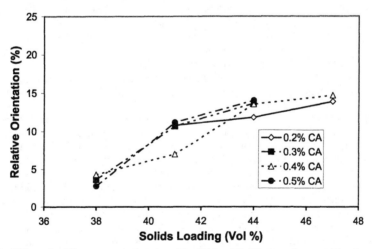

Figure 5: Effect of solids content and citric acid addition (% CA, based on total batch weight) on the top surface orientation for tapes cast with 500 μm blade gap

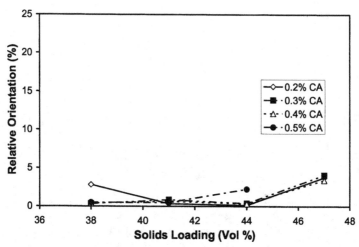

Figure 6: Effect of solids content and citric acid addition (% CA, based on total batch weight) on the bottom surface orientation for tapes cast with 500 μm blade gap

The first observation to note from these figures is that solids loading is the parameter most directly related to the grain orientation in a tape cast sample. This is in agreement with the findings of Goldsmith et al.[17] and Watanabe et al.[16] As solids loading is increased, it is seen that orientation rapidly and predictably increases. Goldsmith et al. explain that in a laminar flow, disk-shaped particles undergo angular rotation indefinitely until impinged upon by another particle. Presumably this explains the relationship between solids loading and orientation because the frequency of particle-particle interactions increases as the volume fraction of particles increases. Also in agreement with Watanabe et al. is the observation that there seems to be no significant relationship between slurry viscosity (varied in this study by changing the degree of slurry dispersion) and orientation.

Figures 3 and 4 show the orientation values for the top and bottom surfaces, respectively, of tapes cast with a 250 μm blade gap. Figures 5 and 6 show the orientation for top and bottom surfaces of tapes cast with a 500 μm blade gap. Samples cast with both blade gaps all show stronger orientation in the top surface than in the bottom surface. There are several possible explanations for this behavior, including effects of surface energy similar to the those described by Böcker et al. regarding grain growth during sintering,[3] and also the nonuniform shear distribution predicted by the flow simulations presented above. As mentioned before, the flow simulations predict that the difference in shear between the top and bottom surfaces decreases as the blade gap decreases. Comparing Figures 4 and 6 it is seen that the bottom surface of the tapes cast with a gap of 500 μm are virtually free of texture, while the bottom surface of tapes cast with a gap of 250 μm show definite orientation.

The agreement between the predicted shear profiles and the experimental texture is evidence that the predicted profiles indeed represent the actual state of shear in the tape cast system. If this is the case, then the greater shear rate near the doctor blade during tape casting is clearly a significant factor in the texture inhomogeneity reported in the literature.

While the system selected for this study is based around aqueous tape casting, the principles discussed can be applied more generally to a wide range of forming methods and geometries. The CFD modeling technique is broadly applicable to other forming methods involving flow, such as extrusion and injection molding. In many systems, as characteristic dimensions are reduced, the concept of bulk properties being more important than surface properties often loses validity as the entire system becomes effectively surface dominated. At the same time, as dimensions are reduced and bodies become more fragile, controlling anisotropy and residual stresses become more important considerations. It is expected that these techniques could find application particularly in environmental applications such as for catalyst substrates.

CONCLUSION

The influence of various processing parameters on microstructural anisotropy was examined. These parameters included solids loading, degree of dispersion, blade gap, and local shear rate. Computational fluid dynamics modeling was used to predict the shear inhomogeneity that exists in a tape casting geometry. A quantitative polarized light microscopy technique was used to determine texture strength in green samples. Solids loading, blade gap, and shear rate inhomogeneity were shown to influence the degree of shear-induced texture in green samples, while slurry viscosity and dispersion were shown to have little influence on green texture. It is predicted that anisotropy control will become more important as ceramic components are miniaturized and processes are optimized for cost and efficiency.

REFERENCES
1. T. Komeda, Y. Fukumoto, M. Yoshinaka, K. Hirota, O. Yamaguchi, "Hot pressing of Cr_2O_3 powder with thin hexagonal plate particles," *Mat. Res. Bull.*, **31** [8], 965-971 (1996).
2. T. Ohta, M. Harata, A. Imai, "Preferred orientation on beta-alumina ceramics," *Mat. Res. Bull.*, **11**, 1343-1350 (1976).
3. J. Patwardhan, "Anisotropic shrinkage characteristics of tape cast alumina," Ph.D. Thesis, Rutgers University, January 2004.
4. P. Raj, W. Cannon, "Anisotropic shrinkage in tape-cast alumina: Role of processing parameters and particle shape," *J. Am. Ceram. Soc.*, **82** [10], 2619-2625 (1999).
5. A. Shui, M. Saito, N. Uchida, K. Uematsu, "Development of anisotropic microstructure in uniaxially pressed alumina compacts," *J Euro. Ceram. Soc.*, **22**, 1217-1223 (2002).
6. T. Takenaka, K. Sakata, "Grain orientation and electrical properties of hot-forged $Bi_4Ti_3O_{12}$. *Jpn. J. Appl. Phys.*, **19**, 31-39 (1980).
7. G. Jiang, M. Gilbert, D. Hitt, G. Wilcox, K. Balasubramanian, "Preparation of nickel coated mica as a conductive filler," *Composites: Part A*, **33**, 745-751 (2002).
8. Y. Kan, P. Wang, Y. Li, Y. Cheng, D. Yan, "Fabrication of textured bismuth titanate by templated grain growth using aqueous tape casting," *J. Euro. Ceram. Soc*, **23**, 2163-2169 (2003).
9. Y. Goto, A. Tsuge, "Mechanical properties of unidirectionally oriented SiC-whisker-reinforced Si_3N_4 fabricated by extrusion and hot-pressing," *J. Am. Ceram. Soc.*, **76** [6], 1420-1424 (1993).
10. A. Böcker, H. Brockmeier, H. Bunge, "Determination of preferred orientation textures in Al_2O_3 Ceramics," *J. Euro. Ceram. Soc.*, **8**, 187-194 (1991).

11. A. Böcker, H. Bunge, J. Huber, W. Krahn, "Texture formation in Al$_2$O$_3$ substrates," *J. Euro. Ceram. Soc.*, **14**, 283-293 (1994).

12. A. Shui, Z. Kato, S. Tanaka, N. Uchida, K. Uematsu, "Sintering deformation caused by particle orientation in uniaxially pressed alumina compacts," *J. Euro. Ceram. Soc.*, **22**, 311-316 (2002).

13. K. Uematsu, S. Ohsaka, N. Shinohara, M. Okumiya, "Grain-oriented microstructure of alumina ceramics made through the injection molding process," *J. Am. Ceram. Soc.*, **80** [5], 1313-1315 (1997).

14. K. Uematsu, Y. Zhang, N. Ito, "Novel characterization method for the processing of ceramics by polarized light microscope with liquid immersion technique," *Ceram. Trans.*, **51**, 273-70 (1995).

15. F. Dimarcello, P. Key, J. Williams. "Preferred orientation in Al$_2$O$_3$ substrates," *J. Am. Ceram. Soc.*, **55** [10], 509-514 (1972).

16. H. Watanabe, T. Kimura, T. Yamaguchi. "Particle orientation during tape casting in the fabrication of grain-oriented bismuth titanate," *J. Am. Ceram. Soc.*, **72** [2], 289-293 (1989).

17. H. Goldsmith, S. Mason, "Particle motions in sheared suspensions: XIII The spin and rotation of disks," *J. Fluid Mech.*, **78**, 88-96 (1976).

CHARACTERISATION OF LZSA GLASS CERAMICS FILTERS OBTAINED BY THE REPLICATION METHOD

C. Silveira, E. Sousa, E. Moraes, A.P.N. Oliveira, D. Hotza[*]
Materials Science and Engineering Graduate Program (PGMAT)
Federal University of Santa Catarina (UFSC)
88040-900 Florianópolis, SC, Brazil

T. Fey, P. Greil
Institute of Glass and Ceramics
University of Erlangen-Nuernberg
Erlangen, Germany

ABSTRACT
    In this work the replication process to produce porous glass ceramics was used. Polymeric sponges of different sizes of pores (small, mixing of small and large, large) were impregnated, by immersion in a slurry of the precursor glass ceramic of composition $Li2O-ZrO2-SiO2-Al2O3$ and submitted to heat treatment. The characterization of porous glass ceramics structures was made by scanning electron microscopy (SEM), stereoscopy and image analysis, and permeability measurements.

INTRODUCTION

    Porous ceramics are high porosity fragile materials that presents cellular structure with closed, open and interconnected cells.[1] The increasing interest in porous ceramics is associated mainly with some specific properties, like high superficial area, high permeability, low density, low thermal conductivity along with the characteristics of the ceramic material, such as high refractoriness and chemical resistance. These properties make the porous ceramics good materials for various technological applications like filters, thermal insulation, membranes, gas sensors, catalyst supports, light structural material, implant material.[2,3] Composition, size, pore morphology and mechanical properties are important factors that influence the choice of appropriated technological application.
    Several process such as replication, bubbles generation in a slurry, controlled sintering, sol-gel, pyrolysis of organic additives, can be used for producing this class of materials. It is possible to obtain in each case a microstructure with different pore sizes in a range from nanometers to some millimeters.[1]
    The most known industrial process is the replication process. This technique also called polymeric sponge process was invented by Schwartzwalder and Somers.[4] It consists in the impregnation of polymeric or natural sponges with ceramic slurry following by thermal treatment that allow burning out the organic additives, and sintering the ceramic material, resulting in the

---

[*] Corresponding author (hotza@enq.ufsc.br).

---

replication of the original sponge. The optimization of the process steps: choice of the polymeric sponge, ceramic slurry preparation, impregnation, drying and thermal treatment allows the development of materials with desirable characteristics for specific applications.

Various kinds of ceramics such as cordierite, mullite, silicon carbide, alumina and some composite system (SiC-Al_2O_3, alumina-zirconia, alumina-mullite, mullite-zirconia) are produced by this process.[5,6] In this work, the replication method was applied for production of porous glass ceramic belong to Li_2O-ZrO_2-SiO_2-Al_2O_3 (LZSA) system. Glass ceramics are policrystalline materials containing some glass phase, which are obtained by melting, consolidation and crystallization of a parent glass composition. Glass ceramic materials present interesting properties, such as high hardness, high mechanical strength, excellent chemical resistance, large isolating power, high thermal shock strength,[7,8] finding applications in several industrial sectors.

Glass ceramic composition LZSA used in this work was chosen for presenting interesting propriety, such as flexural strength of 100 to 160 MPa, wearing strength of 40 to 80 mm^3, chemical resistance (0.5 wt.% loss in acid leaching and 1 wt.% in basis leaching. The coefficient of linear thermal expansion can vary approximately from 2 to 11 x 10^{-6}°C^{-1}. Other important aspect is the sintering and crystallization temperature, between 800 and 900°C.[8]

MATERIALS AND METHODS

Ceramic Raw Materials and Preparation of Slurry

A parent glass was used to obtain LZSA glass ceramics, supplied by CTCmat (Criciúma, SC, Brazil). Samples of glass ceramics were characterized by particle size measurement (laser diffraction,Cilas 1064L) and differential thermal analysis (DTA, Netzsch STA 409 EP) at 10 °C/min in synthetic air.

The slurry was prepared containing 60 vol.% parent glass (LZSA), 5 vol.% bentonite, as plasticizer, and 1 vol.% sodium silicate (Merck), as dispersant. The slurry was mixed for 2 min in a ball mill. The slurry viscosity was measured in a rheometer (Haake Polylab System/52p Rheomex).

Polymeric Raw Materials and Characterization

Three different commercial poly(urethane) sponges were used: small pores (smooth, Santa Maria), mixed porous (Fresh Skin) and large pores (rough, Santa Maria). Samples were cut with ~1 cm x 1 cm x 1 cm.

Polymeric sponges were submitted to thermogravimetric analysis (TG, Shimadzu TGA -50) at 10°C/min in nitrogen. Thermal decomposition analysis of PU sponges was accomplished in a tubular oven, connected to a spectrophotometer Perkin Elmer FT-IR 1600.

Impregnation and Thermal Treatment

The samples were impregnated by immersion in a parent glass slurry (LZSA). After impregnation, the slurry excess was removed. Samples were dried at room temperature for 24 h.

The thermal treatment was carried out in a laboratory electric oven (EDG 1800): from 25°C with a heating rate of 1°C/min till 450°C, holding for 30 min for organic material elimination; then, with a rate of 5°C/·min till 700°C, holding for 10 min; finally, with the same previous rate, till 900°C, maintaining for 10 min. Cooling followed without rate control.

Characterization of the Final Product

The final product was characterized by scanning electron microscopy (SEM, Philips XL 30) and optical microscopy (stereoscopy,Olympus SZ – CTV). In the latter case, samples were mounted under vacuum in a pigmented poly(ester) resin basis, and polished. Images obtained by stereoscopy were analyzed using the software Imago V 32 (gray scale).

Permeability in nitrogen was measured in a gas permeameter. The gas speed through the sample was varied by a flow controller. A multimeter (Agilent 34401A) connected to a transducer (Sensym SS X 500G) was used to measure pressure drop up to 2 atm.

RESULTS AND DISCUSSION

Characterization of Powder and Suspension

The powder particles were 100% smaller than 23 μm, presenting a mean particle size of 4.7 μm, allowing a good dispersion in an aqueous slurry. The rheological behavior is shown in Figure 1, shear stress as a function of shear rate for the LZSA slurry. It can be observed that the slurry presents a pseudoplastic behavior, that is, apparent viscosity decreased when shear rate is increased, due to agglomerates which are submitted to shear and then break, discharging water arrested inside.[10] Besides, this slurry presents a thixotropic behavior, characterized by hysteresis of the ascendant and descendant curve.

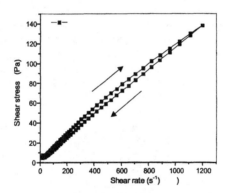

Figure 1: Shear stress behavior as a function of shear rate for LZSA slurry.

Decomposition of Sponges

Depending on the composition, poly(urethane) can show up 2 to 4 steps of decomposition in nitrogen, and about 13 wt.% decomposition residue.[11] Typical ATG and DATG curves (not shown) of PU sponges presented two weight loss plateaus, the first one of ~43 wt.% and the second one of ~54 wt.%, corresponding to the temperatures of 313°C and 397°C, respectively.

The products of sponge decomposition were determined using a tubular oven connected to a FTIR spectrophotometer. Typical spectrum (not shown) presented at 300°C peaks of CO^2 in 2360 and 670 cm⁻¹, traces of de CO in 2110 cm⁻¹ and NH_3 near to 930 cm⁻¹. A double peak was observed in 1560 and 1650 cm⁻¹ attributed, probably, to the isocyanate group.[11] Next to the range of 3400-3600 cm⁻¹ asymmetric and symmetric stretching of NH_3 group can be seen.

Increasing the temperature to 380°C, a decrease in the peaks of CO_2 in 2360 and 670 cm⁻¹ can be observed. In the region of 2700 and 3000 cm⁻¹, peaks due to the stretching of CH_2 can be identified; in 1750 cm⁻¹ a peak of of C=O, and in the regions of 3100 cm⁻¹ and 1150 cm⁻¹, a CH ring.

Elimination of CO_2 can be observed at 510°C. The increasing intensity of peaks in 3100 cm⁻¹ and 1150 cm⁻¹, referring to stretching of CH ring, indicate probable elimination of carbonyl groups. With higher temperatures the bounds break, generating polymeric fragments. The fragments with low molecular weight, corresponding to high vapor pressure, evaporate forming gas.[13]

Microstructure of Porous Ceramics

One method used for defining reproducibility of ceramic replicas is the so-called ω parameter.[14] This number expresses the quantity of slurry by volume of a sample of polymeric sponge. A high value of ω corresponds to a reticulate filled in excess occasioning increase of internal filament and of mechanical strength, consequently decreasing of permeability.

According to Innocentini,[14] samples with ω below 0.47 g/mL presented visible crack in the reticulated structure, with internal regions of coalescent pores. Samples with ω over 0.75 g/mL presented, on the other hand, a higher amount of filament densification.

In the measured samples, the parameter ω value were following: sample with small pores ω = 0.488 g/mL; with mixed pores ω = 0.587 g/mL; with large pores ω = 0.537 g/mL. The results obtained are coherent with the literature.[14] It lays inside the suitable range of 0,75 > ω > 0,45 g/mL.

Morphology of glass ceramics was obtained by scanning electron microscopy (SEM), Figure 2. Samples with small pores (a: ~1 mm,), with mixed pores (b: ~0,5 -1 mm) and large pores (c: ~2 mm) can be observed.

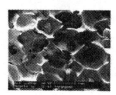 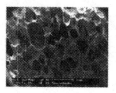 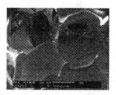

a) b) c)

Figure 2. SEM of samples with: a) small, b) mixed and c) large pores.

Permeability Measurements

Glass ceramic filters were analyzed in respect to the permeability. Constants of Darcian permeability, k_1, obtained experimentally were 2.15 x 10⁻¹¹ m² for the sample with small pores; 2.75 x 10⁻¹¹ m² with mixed pores; and 2.53 x 10⁻¹¹ m² with large pores.

The constants of non-Darcian, k_2, which represent inertial effect, were, respectively, 0.038 x 10^{-5} m, 0.046 x 10^{-5} m, and 0.042 x 10^{-5} m for the samples with small, mixed; and large pores. The permeability constant k_2, contribute, according to Innocentini,[14] up to 20% of total value of pressure drop. The results obtained for both constants are coherent with the literature, [14] in the range of 0.69 to 11.8 x 10^{-10} m^2 for k_1 and 0.61 to 10.78 x 10^{-5} for k_2.

Inertial or kinetic effects across pressure drop are caused by turbulence of the flowing fluid and/or for tortuosity of porous mean. In the first case, the turbulence is quantified by the Reynolds' number. When Re > 2100, the flow becomes turbulent or chaotic, which causes disturb of fluid layers and increased energy loss. As the kinetic energy of fluid increases, more turbulent will be then the flow and higher will be the pressure drop.[16]

The value for turbulence of fluid, quantify by Reynolds's number, is 1690 for the sample with small pores, 3076 for the sample with mixed pores and 4276 for the sample with large pores. Kinetic energy of flow was high for samples with mixed and large pores, corresponding to a turbulent fluid.

Stereoscopy and Image Analysis

After careful image binarization from Figure 3, the following results for porosity of reticulated ceramics were obtained: sample with small pores: 63.61%; sample with mixed pores: 62.25%; sample with large pores: 82.70%. It was expected that the sample with small pores shod lower porosity, followed by the sample with mixed pores and the sample with large pores. A possible explanation for these results is related to the ω parameter: a lower filament supplying means higher voids volume, causing higher porosity.

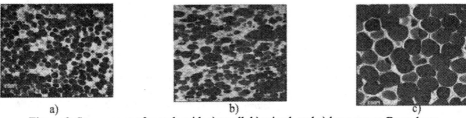

a)   b)   c)

Figure 3. Stereoscopy of sample with a) small, b) mixed, and c) large pores. Bar = 1mm.

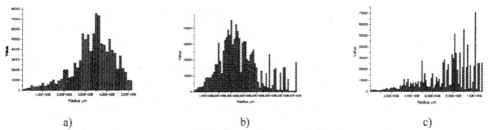

a)   b)   c)

Figure 4. Pore size distribution of samples with a) small, b) mixed, and c) large pores.

57

The mean pore diameters obtained by image analysis were: 525 μm (small pores), 825 μm (mixed pores) and 1087 μm (large pores). The size distribution of pores is plotted in Figure 4.

CONCLUSIONS

Porous glass ceramics can be obtained by the replication method using polymeric sponges as templates.

The slurry containing 60 wt.% parent glass, 5 wt.% bentonite, 1 wt.% sodium silicate showed rheological characteristics adequate for impregnation and production of porous glass ceramics.

By thermal analysis, plateaus of weight loss could be observed. Infrared spectroscopy analysis indicated CO_2, CO, NH_3 and isocyanate, as main products of thermal decomposition of poly(urethane) sponge.

SEM micrographs of glass ceramics samples identified open pores with a maximum size of 2 mm.

Results obtained for Darcian and non–Darcian constants of permealibility were according to values found in the literature. The flow was turbulent; the fluid lost energy due to differences of pore size. Samples with small pores showed higher porosity than the sponges with mixed pores, due to a lower ω parameter.

Such previous results characterize this glass ceramic porous material as potential use as filter for low temperatures applications.

ACKNOWLEDGEMENTS

The authors are grateful to CAPES and CNPq for financial support of this work.

REFERENCES

[1] J. Zeschky, F. Goetz-Neunhoeffer, J. Neubauer, S. H. Jason Lo, B. Kummer, M. Scheffler, and P. Greil, *Composites Science and Technology,* **63,** 2361-2370 (2003).

[2] P. Sepulveda, and G. P. Binner, *Journal of the European Ceramic Society,* **19,** 2059-2066 (1999).

[3] F. S. Ortega, A. E. M.Paiva, J. A. Rodrigues, and V. C. Pandolfelli, *Cerâmica,* **49,** 1-5(2003).

[4] K. Schwartzwalder, and A. V. Somers, *US Pat.* No 3 090 094, May 21 1963.

[5] X. Zhu, D. Jiang, and S. Tan, *Materials Science and Engineering A,* **323,** 232-238 (2002).

[6] V. R. Salvini, A. M. Pupim, M. D. M. Innocentini, and V. C. Pandolfelli, *Cerâmica,* **47,** 13-18 (2001).

[7] A. K. Varshneya. Fundamentals of Inorganic Glasses, Academic Press, New York, EUA (1994), p. 80, 143.

[8] A. P. Novaes de Oliveira, "Progettazione, Caratterizzazione ed Ottenimento di Vetri-Vetroceramici Appartenenti al Sistema LZS", Ph.D. Thesis, Modena, Italy, 1997.

[9] E.A. Moreira, M. D. M. Innocentini, and J. R. Coury, *Journal of the European Ceramic Society,* **24,** 3209-3218 (2004).

[10] F. S. Ortega, V. C. Pandolfelli, J. A. Rodrigues, and D. P. F. de Souza, *Cerâmica,* **43,** 5-10 (1997).

[11] M. Herrera, M. Wilhelm, G. Matuschek, and A. Kettrup, *Journal of Analytical and Applied Pyrolysis,* **58,** 173-188 (2001).

[12] M. L. Hobbs, K. L. Erickson, and T. Y. Chu, *Polymer Degradation and Stability,* **69,** 47-46 (2000).

[13] D. Becker, "Blendas PP/PU: Estudo do Efeito do Agente Compatibilizante e Reciclagem de Resíduos de PU," M.Sc. Thesis, UFSC, Florianópolis, 2002.

[14] M. D. M. Innocentini, "Filtração de Gases a Altas Temperaturas," Ph.D. Thesis, UFSCar, São Carlos,1997.

[15] V. R. Salvini, M. D. M. Innocentini, and V. C. Pandolfelli, *Cerâmica,* **48,** 22-28 (2002).

[16] M. D. M. Innocentini, P. Sepúlveda, V. R. Salvini, and V. C. Pandolfelli, *Journal of the American Ceramic Society.* **81,** 3349-52 (1998).

[17] C. R. Appoloni, C. P. Fernandes, M. D. M. Innocentini, and Á. Macedo, *Materials Research,* in press.

FRACTURE BEHAVIOR AND MICROSTRUCTURE OF THE POROUS ALUMINA TUBE

Chun-Hong Chen, Sawao Honda, and Hideo Awaji
Nagoya Institute of Technology
Gokiso-cho, showa-ku
Nagoya, Aichi, 466-8555

ABSTRACT

Porous alumina tubes with continuous and stepwise pore-gradient along the radial direction were successfully fabricated , where PMMA particles were used as pore forming agent. The tubes were expected to function as high temperature filters. The tube is designed in such a way that the inner part will act as the separation media while the outer part will function as mechanical support with the presence of the porosity gradient. Alumina and PMMA particles were mixed with water to form an aqueous slurry, compacted using the continuous and stepwise methods in centrifugal molding technique. The green body was dried in partial vacuum atmosphere, calcinated at 773 K to remove the organic component, sintered at 1623 K to obtain sintered porous α-alumina tubes. The microstructure observation on porous alumina tubes was carried out. The binary pores of 10μm and submicrometer in diameter were visible, which were formed by the burning-out of PMMA particles and partial sintering, respectively. The fracture behavior of the porous tube was observed by using the ring diametral compression testing. The experimental results show that fracture behavior varies with the amount of pore-forming agent, regardless of the pore-gradient along the radial direction. A reduction in fracture strength is observed with increasing the laminate number, and the minimum strength is found in continuous graded tubes.

INTRODUCTION

Porous ceramics are used as gas/liquid separators, catalyst supports and molecular sieves, which can be used at high temperature and under severe chemical conditions. There are several approaches for the fabrication of micro/macro porous ceramics, which can be classified into the following four groups [1]. (1) Controlling the particle size of ceramics to maintain constant pores among particles. (2) Reducing the forming pressure and/or sintering temperature. (3) Mixing the ceramic powders with a bubble former so that large and homogenous pores can be introduced. (4) Mixing ceramic powders with additional organic particles, which vaporize at relatively low temperatures and form small and homogenous pores. A sol-gel process with the presence of a surfactant has been developed to prepare mesoporous ceramics with a well-ordered arrangement of pores [2]. Recently porous ceramics have been developed with high-temperature stability, strength, catalytic activity, erosion resistance and corrosion resistance. In spite of these excellent properties, the potential of porous ceramics has not been fully realized because of their well-known problems, such as lack of pore size control, lack of continuous processing method, and absence of a model relating pore structure to mechanical properties.

In a Functionally Graded Material (FGM), the composition and structure gradually change over volume, resulting in corresponding changes in the material properties. It can incorporate incompatible functions such as the heat, wear, and oxidation resistance of ceramics and the high toughness, machinability and bonding capability of metals. In recent years, the FGM concept has been introduced for porous materials, where pores are the important structural ingredient for FGMs. Centrifugal molding is an economical and convenient method to fabricate a hollow cylinder. Hollow cylinders with a functional gradient were also fabricated using the centrifugal molding technique with the difference in the centrifugal force between powder mixtures to create a compositional gradient along the radial direction due to the difference in their densities and particle sizes [3,4]. The gradient of the tubes varying from an homogenous structure to a bilayered structure can be controlled by adjusting the slurry character [5].

Recently, we have proposed a model (Fig.1) to fabricate tubes with a porosity gradient to serve as high temperature filters [6]. Tubes are designed in such a way that the inner part will act as the separation media while the outer part will function as the mechanical support with the presence of the porosity gradient. These figures illustrate the material design model from forming to application. Firstly, the organic particles are mixed with the ceramic particles. Then, the mixture is poured into the mold and the gradient of organic particles is formed in the centrifugal field. The porosity gradient is formed with the vaporization of organic particles during the sintering procedure. As a filter, the unwanted particles are immobilized in the pores and cleaned gas is transported along the radial direction of the tube.

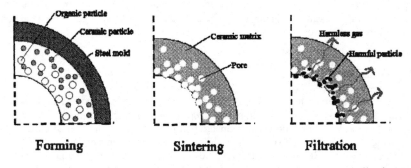

Forming   **Sintering**   **Filtration**

Fig. 1. Schematic of material design model for porous tube from forming to application

In this paper, porous ceramic tubes with different pore-gradients were fabricated by the continuous and stepwise methods in the centrifugal molding technique with the presence of an organic pore forming agent. The microstructure and fracture behavior with respect to the process were also characterized.

EXPERIMENTAL PROCEDURE

The α-alumina powders used in this study were AES-12 (Sumitomo Chemical Co. Ltd) with a BET surface of 6 m^2/g. The alumina powders have narrow particle size distributions with a mean particle size of 0.5 μm. PMMA (polymethylmethacrylate) powder with an average particle size of 10 μm was used as pore forming agent. PMMA particles are the desirable pore forming agent for macropores due to their homogenous particle sizes and low vaporization temperature [7]. For good dispersion, Seruna D-305 was added along with distilled water as solvent. 0.25 mass% of MgO was added as the sintering aid and grain growth inhibitor.

First, the alumina powders and PMMA particles were mixed with the dispersant by magnetic stirring for 1 h. Then, the mixed slurry was treated ultrasonically to break down particle agglomerates. Finally the slurry was degassed in a vacuum for 15 minutes and poured in the stainless steel mold of dimension in the inner diameter of 20 mm and the length of 70 mm. For easy removal of the sample after drying, the inner surface of the mold was pre-coated with vaseline as a lubricant. The cylindrical mold was centrifuged around the radial axis with a speed of 3000 rpm for 15 minutes. The residual liquid was poured out after centrifugation. In the stepwise method, the same procedure was repeated to form the next laminate. The finished green tubes were dried vertically inside the mold itself by keeping it in the vacuum chamber so that the shape was maintained. The temperature was allowed to rise to 373 K with a ramp of 283 K/h. The tubes shrunk and released easily from the mold during drying. The PMMA was dissociated completely by heating at 773 K for 1 h with a controlled heating rate of 303 K/h. The green tubes after calcination were sintered at 1623 K for 1 h in air atmosphere.

The density and porosity of sintered tubes were measured by Archimedes method with distilled water as media. Microstructures of the sintered specimen were examined by scanning electron microscope (SEM) on the cross-sectional surfaces in equal intervals along the radius.

For mechanical evaluation, the tubes were cut into 5 mm in length by a diamond cutter and subjected to ring diametral compression testing using a mechanical testing machine (Instron 5582, USA) with a crosshead speed of 0.5 mm/min.

The fracture strength of the tubes with the pore gradient can be estimated based on curved beam theory [8]. The hoop stress σ_θ of a graded ring specimen with a rectangular cross section subjected to diametral compression is expressed as follows,

$$\sigma_\theta = E(x)\frac{\varepsilon_0 \rho + y\omega}{\rho + y} \tag{1}$$

where ε_0 is a strain at a centroid, ρ is the radius of the center axis, y is the layer distance from the center line and ω is the angular strain, x is the non-dimensional distance along the radial direction and $E(x)$ is the radial gradient of Young's modulus. The analysis results show that the maximum stress occurs on the inner surface of the ring along the load axis and can be regarded as the fracture strength of the ring based on the maximum hoop stress.

RESULTS AND ANALYSIS
Microstructure Observation

The porous tubes that were prepared are depicted in Fig. 2. No any defects, such as delamination or cracks were observed after sintering.

Fig.2 The porous tube specimen.

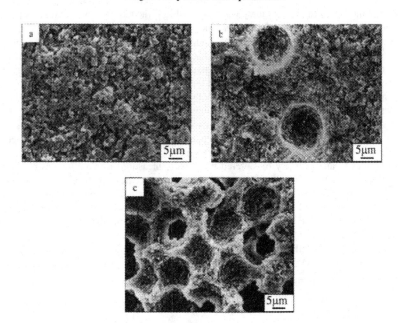

Fig. 3 SEM micrographs with the addition of
(a) 0 vol.% PMMA, (b) 10 vol.% PMMA, and (c) 40 vol.% PMMA.

    The experimental results show that in the absence of the dispersant, pore morphology was homogenous regardless of the variation of the solid load by volume. Figure 3 shows the bimodal pore structures of the samples with 0 vol.%, 10 vol.% and 40 vol.% PMMA additions in the absence of a dispersant. Grain growth of alumina particles is not observed and continuous but sub-micrometer pores exist among the alumina grains (Fig. 3a). The sub-micrometer pores mainly result from partial sintering. On the addition of the pore-forming agent PMMA, spherical pores with 10 μm in diameter could be clearly observed. The isolated spherical pores are connected with a sub-micrometer pore (10 vol.% PMMA). The spherical pores were observed to be uniformly distributed in the alumina matrix and to form a three-dimensional network, which rendered a body with open porosity regardless of the sub-micrometer pores (40 vol.% PMMA). It was also found that samples with more than 30 vol.% of PMMA were difficult to prepare by the continuous method.

Outer wall

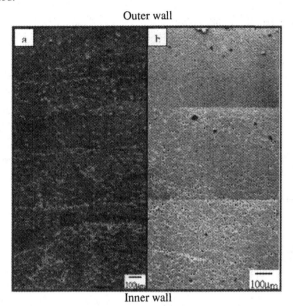

Inner wall

Fig. 4 Composed SEM photographs with 10vol.% PMMA for
(a) continuous graded and (b) stepwisely graded samples.

No difference in microstructures from the inner wall to outer wall of the tube is observed though the dispersant concentration varies in the absence of PMMA particles. Figure 4 shows the composed SEM microstructures of the continuous graded samples with 0.05 wt.% dispersant and stepwisely graded samples composed of the laminates with 5 vol.%, 10 vol.% and 15 vol.% PMMA additions along the radial direction. The spherical pores are graded along the radial direction owning to the difference in centrifugal force on the alumina and PMMA particles by the continuous method. Figure 5 shows that the number of pores decreases from the inner wall towards the outer wall of the tube by the continuous method. For the specimen of 20 vol.% PMMA, there is a continuous decrease along the radial direction. For the specimen with 30 vol.% PMMA, a minimum in pore density is observed. This phenomenon is due to the difference in the network formation process by particle settling in the slurry, where the formation mechanism is essentially attributed to collisions of dispersed particles and hence the difference in their settling velocities under centrifugal field [9,10]. The experimental results show that the pore density along the radial direction can be changed by selecting dispersant concentration and the solid load in the slurry.

Fracture Behavior

The results of ring diametral compression testing could be roughly classified into two types based on the shape of the load vs. load-point displacement curves as shown in Fig. 6. The failure appearance of type I is shown in Fig. 7, where (a) is after complete failure, (b) is immediately after the first maximum peak a. The results show that the peak a in Fig. 6a corresponds to the initiation of a crack at the

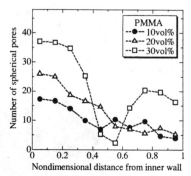

Fig. 5 Pore distributions for continuous graded samples.

inner surface A, which extends into the exterior along the load axis, resulting in two c-shaped pieces with linear crack growth. At peak b, cracks are developed on the outer surface B and propagate into the interior perpendicular to the load axis.

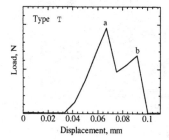
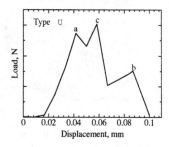

Fig. 6 Two types after the ring diametral compression testing.

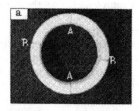
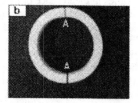
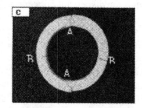

Fig. 7 Failure appearance of Type I and II, (a) after complete failure of Type I, (b) after the first maximum peak of Type I, (c) after complete failure of Type II

The failure appearance of type II is shown in Fig. 7c. It is found that peaks a and b indicate the initiation of crack on the inner and outer surface A and B. The crack deflection is seen clearly along the load axis, which corresponds with peak c. That is, the crack is arrested when a pore from PMMA agglomerate exists on the crack growth path, and, the crack was deflected along the periphery of the pore after the increase in load to peak c.

Therefore, the inner strength of O-ring specimens can be calculated by combining analyses formula with the first peak a in both types . The experiments show that with increase of PMMA additions, type I of fracture behavior becomes type II gradually, which shows that this phenomenon is independent of the number of the laminate in the tubes. The behavior deffers from that of laminated SiC tubes produced by electrophoretic deposition, where the graphite layers are set into SiC laminate [11].

Fracture Strength

The relation between the fracture strength and porosity are shown in Fig. 8. for continuous types (NL) and stepwise types, where specimens are divided into three groups (L4, L3, L2) by the laminate number. PMMA amount varies from 10%, 20%, and 30% by volume in each specimen. The hollow symbols indicate the laminated types containing an alumina single layer on the outer side, where the total PMMA amount is 10vol.%. It is observed that the porosity of specimens with the same PMMA additions varies with the number of laminate layers. With decrease in the number of laminate layers, the porosity is gradually increased resulting in greater air permeability. In the case of the same porosity, a reduction in fracture strength is observed with increase in the number of laminated layers, and the minimum strength is observed in continuous graded tubes. It is because the initial PMMA contents in the inner wall of tubes increase as shown in Table 6.1, and the radial gradient of Young's modulus introduced in Eq. (1) varies with the number of laminated layers. The experimental results show that the fracture strength and porosity are easy to be controllable by selecting the number of laminated layers and PMMA distributions in each layer. The error of fracture strength with high porosity is less than that with low porosity regardless of the laminated layers [12]. The reasons could be that fracture origin is limited in the spherical pore formed by PMMA particles.

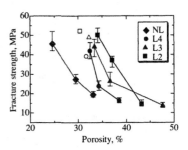

Fig. 8 Fracture strength vs. porosity

CONCLUSIONS

Porous alumina tubes were fabricated using commercial alumina powders with a narrow size distribution (AES-12) by centrifugal molding. The bimodal pore structure existed in the porous tubes, where sub-micrometer pores were introduced by the partial sintering, while spherical pores about 10 µm were formed with the evaporation of PMMA particles. The experiments also show that, type I of fracture behavior becomes type II gradually with increasing PMMA additions, which shows that this phenomenon is independent of the process. The fracture strength of O-rings increased with decreasing porosity regardless of the pore distribution of the tubes. In the case of the same porosity, a reduction in fracture strength is observed with increase in the number of laminated layers, and the minimum strength is observed in continuous graded tubes.

REFERENCES

[1]K. Ishizaki, S. Komarneni, and M. Nanko, Porous Materials–Process Technology and Applications, Kluwer Academic Publishers, Netherlands, 1998, Chap. 2.

[2]J. S. Beck, J. C. Vartuli, W. J. Loth, M. E. Leonowich, D. H. Lson, E. W. Sheppard, S. B. McCullen, J. B. Higgins, and J. L. Schlemckes, J. Am. Chem. Soc., 114, 10834-43 (1992).

[3]R. Sivakumar, T. Nishikawa, S. Honda, H. Awaji, and F. D. Gnanam, Processing of Mullite-Molybdenum Graded Hollow Cylinders by Centrifugal Molding Technique, J. Eur. Ceram. Soc., 23(5), 765-772 (2003).

[4]R. Sivakumar, T. Nishikawa, S. Honda, and H. Awaji, a Novel Technique Applied for Fabrication Alumina / Zirconia Continuous Graded Hollow Tubes, J. Ceram. Soc. Jpn., 110(5), 472-475 (2002).

[5]C. H. Chen, S. Honda, T. Nishikawa, and H. Awaji, a Hollow Cylinder of Functionally Gradient Materials Fabricated by Centrifugal Molding Technique, J. Ceram. Soc. Jpn., 111(7), 479-484 (2003).

[6]C. H. Chen, Fabrication and Characterization of Dense and Porous Ceramics-based Tubes of Functionally Gradient Materials, Doctoral Thesis, Nagoya Institute of Technology, (2005).

[7]T. Nishikawa, A. Nakashima, S. Honda, and H. Awaji, Effects of Porosity and Pore Morphology and Mechanical Properties of Porous Alumina, J. Soc. Mater. Sci. Jpn., 50(6), 625-629 (2001).

[8]C. H. Chen, K. Takita, S. Honda, and H. Awaji, Fracture Behavior of Cylindrical Porous Alumina with Pore Gradient, J. Eur. Ceram. Soc., 25, 385-391 (2005).

[9]J. Tsubaki, H. Mori, M. Kato, K. Okuda, Y. Yoshida, and T. Yokoyama, New Slurry Characterization Technique for Optimizing Wet-shaping Process (Part 2) – Dependency of Centrifugal Effect on Slurry Consolidation Process, J. Ceram. Soc. Jpn., 106(6), 616-620 (1998).

[10]T. Sugimoto, H. Mori, and J. Tsubaki, Network Formation Mechanism of Fine Particles in Suspension-Simulation on the Structure Formation in Binary Dispersions, J. Soc. Powder Technol. Jpn., 37,100-106 (2000).

[11]L. J. Vandeperre, and O. O. Van Der Biest, Graceful Failure of Laminated Ceramic Tubes Produced by Electrophoretic Deposition, J. Eur. Ceram. Soc., 18, 1915-1921 (1998).

[12]S. C. Nanjangud, R. Brezny, and D. J. Green, Strength and Young's Modulus Behavior of a Partially Sintered Porous Alumina, J. Am. Ceram. Soc., 78, 266 (1995).

TENSILE TESTING OF SiC-BASED HOT GAS FILTERS AT 600°C WATER VAPOUR

Pirjo Pastila, Antti-Pekka Nikkilä and Tapio Mäntylä
Tampere University of Technology
Institute of Materials Science
Korkeakoulunkatu 6
33720 Tampere, Finland

Edgar Lara-Curzio
Oak Ridge National Laboratory
High Temperature Materials Laboratory
1 Bethel Valley Road
Oak Ridge
TN 37831-6069, USA

ABSTRACT

Tensile testing at 600°C water vapour environment of SiC-based aluminous alkaline silicate bonded hot gas filters is reported. Tests were completed with as-received material and with material exposed 455 hours in water vapour at 850°C and thermal cycles. A deep notch was needed to guide the fracture to the mid section, and FE analysis of the notch geometry with ABAQUS CAE was also preformed. The fracture surfaces were investigated with optical and scanning electron microscopy and compared with fracture surfaces of room temperature tests. Sub-critical crack growth may have occurred in some of the water vapour exposed specimens, but the measured strength at the displacement rate of 0.002 mm/s was not significantly different from that at the rate of 0.02 mm/s.

INTRODUCTION

Hot gas filtration is needed in advanced coal and biomass based power generation in order to develop more efficient and less environmentally hazardous power plants. The filters may also be used in waste incineration process. Since both combustion and gasification process environments contain water vapour, knowledge of how water vapour affects the mechanical properties of filter materials is important for understanding the degradation process under the complex operation environments, for lifetime prediction, and for further materials development. In this work the importance of slow crack growth mechanisms for the strength of advanced SiC-based clay-bonded hot gas filter material at high temperature water vapour was addressed.

Sub-critical crack growth (SCG) stands for the stable propagation of cracks at stress intensity levels below the critical stress intensity for catastrophic failure[1,2]. There may be a threshold value for the applied load (stress intensity) below which no crack propagation occurs[1,2]. Three stages of propagation are typically found[1,2]. In the regions I and III the crack velocity v is described as a function of the stress intensity factor K_I by power law as[1-3]

$$v = A_I K_I^n \qquad (1)$$

where A_I and n are constants. In the region II the crack velocity is constant. Typical values of n for oxides are 10-20 in the region I and ~100 in the region III. At high temperatures the region III behaviour with n~10 is also observed[4]. It is possible to determine the sub-critical crack growth exponent n from strength experiments completed with different strain rates as[1]

$$\left(\frac{\sigma_{\dot{\varepsilon}i}}{\sigma_{\dot{\varepsilon}j}}\right)^{n+1} = \frac{\dot{\varepsilon}_i}{\dot{\varepsilon}_j} \tag{2}$$

where $\sigma_{\dot{\varepsilon}i}$ is the strength at strain rate of $\dot{\varepsilon}_i$.

The sub-critical crack growth of silica glass in water containing environments occurs via formation of hydroxyl groups when water molecules react with bridging oxygens[2,4]. At high temperatures thermally activated bond rupture, creep and viscous flow mechanisms aside the environmentally assisted stress corrosion mechanism may contribute to the sub-critical crack growth[5-7]. Since the binder of the SiC-based filters in this study was of amorphous aluminous alkaline silicate with some mullite and cristobalite[8], a clear tendency for sub-critical crack growth leading to lower strength at lower loading rates was expected.

MATERIALS AND METHODS

Tensile tests to study the sub-critical crack growth behaviour at high temperature water vapour were completed at 600°C, 1.2 bar water vapour pressure. The test set-up is schematically presented in Figure 1. The testing chamber was a glass tube that was slid over the specimen and the grips and was sealed with O-rings and vacuum grease. The chamber had an inlet for water vapour, outlets for vapour and water, and a place for the temperature controlling thermocouple. Water vapour was generated by pumping water with an electrical dosing pump into a separate furnace. The specimen was heated in one hour up to 600°C and temperature was allowed to settle for 5 minutes. Then water vapour was pumped into the chamber for 15 minutes at the water feed rate of 3 ml/min. During this period the pressure reached and stayed at 1.2±0.1 bar. One minute before the test started the pumping rate was increased to 6 ml/min. Constant crosshead displacement rates of 0.02 mm/s and 0.002 mm/s were used.

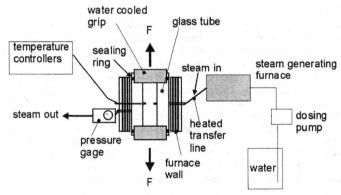

Figure 1.   Schematic presentation of the test set-up used for tensile tests in high temperature pressurized steam.

The material was commercial filter, type DSN 10-20 T of Schumacher Umwelt- und Trenntechnick GmbH, Germany. Three specimens of the material in as-received condition and after 455h in water vapour with 8 thermal cycles (<150°C - 850°C) were tested with each displacement rates. Specimens were 22.9 cm long tubes with a notch at the gauge section and grooves for the gripping, as shown in Figure 2. The in-house made grips had two water cooled half moon pieces of radius to fit the ground ends of the specimen and two aluminium half moon pieces to fit the shallow grooves machined to the specimen ends. The grooves were necessary to ensure the gripping at the high temperature tests and a notch was needed to ensure the failure at the gauge section. The notch in Figure 2 represents typical dimensions.

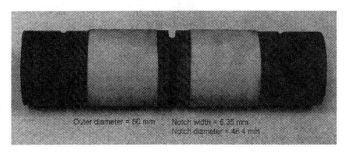

Figure 2.   The geometry of the specimens used in the tensile tests at 600°C water vapour.

The strength of the tubular and notched specimens was calculated from the maximum load and the dimensions of the specimen. The stress concentration was calculated for each specimen from the equation for a shaft with a groove[9] using the dimensions determined with the slide calliper. The notch caused a stress concentration factor of approximately 2.7. This was confirmed with a FE analysis with ABAQUS CAE (ABAQUS Inc., Rhode Island, USA). One half of the quarter section of the specimen was modelled with quadratic hexahedral elements assuming isotropic linear elastic material properties and evenly distributed pressure at the top of the specimen corresponding to the maximum tensile force. The analysis revealed that the stress decays very rapidly from the outer surface towards the inner surface of the specimen and the volume under the highest tensile stress is very small, Figure 3.

The force of 1150 N due to pressure inside the test chamber and the weight of the lower grip, 70 N, were added to the maximum load. The force was calculated from the cross sectional area of the test chamber and the target pressure level. The value corresponded well to the compression force detected after the specimen failed.

A SEM study of gold-coated fracture surfaces was completed for one sample of each testing condition. In addition to selected details on fracture surface it was imaged in every 20 degrees for classification of the fracture path. From the images classification of fracture path into categories of through SiC, through binder and rupture of binder from SiC revealing the SiC grain was made. These results were compared to an earlier classification of fracture path from room temperature impact loaded samples. Also, a visual comparison between SEM images of fracture surfaces of as-received material tested in tension at ambient air and in high temperature water vapour was made.

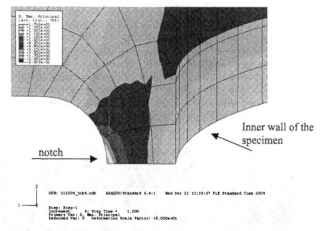

Inner wall of the specimen

notch

Figure 3. The distribution of the first principal stress at the notch. Note that the stress decays to 50% of its maximum value in approximately 1/3 of the wall thickness.

RESULTS AND DISCUSSION

The specimens were tested as pairs; each of the two specimens tested at the different rates were from the same filter element if possible since the variation of the initial strength from filter to filter could hide the effect of slow crack growth, as shown in Table I. On the other hand it was known that variation in strength within a single filter element is moderate (high Weibull modulus), Table I. Thus, each two specimens known to be from a single filter have to be compared. All tests were successful but unfortunately there were not enough as-received material to get three pairs from a single filter. The tensile strengths at 600°C water vapour are presented in Table II.

Table I   Modulus of rupture estimates of strength, Weibull modulus and the number of specimens for tensile tests at 600°C water vapour (HT) and Internal Hydraulic Pressure tests at room temperature (RT).

Test method and loading rate	As-received material			Exposed material		
	σ_0 [MPa]	m	# of specim.	σ_0 [MPa]	m	# of specim.
HT tensile 0.002 mm/s	21.3	14.2	3	17.9	9.6	3
HT tensile 0.02 mm/s	19.6	9.6	3	17.9	7	3
RT hoop* ~0.2 mm/s	17.5	9.6	100	14.4	8.3	30
RT hoop* ~0.2 mm/s	19.2[§]	18.5[§]	20[§]	-	-	-

* Hoop stress is $0.722\sigma_{IHP}$, the strength estimate determined with internal hydraulic pressure test.
[§] All specimens from single filter element. This set is included in the 100 specimen set presenting hoop strength of the as-received material.

Table II   Strength of specimen tested at 600°C 1.2 bar water vapour. The specimens were from the same filter element if not indicated otherwise.

Displacement rate [mm/s]	Pair 1 [MPa]	Pair 2 [MPa]	Pair 3 [MPa]
	As-received material		
0.02	20.3	19.5	17.3*
0.002	20.0	21.9	20.7*
	Exposed material		
0.02	18.1	14.8	18.7
0.002	18.7	16.2	17.4

* not from the same filter element

Most of the specimen pairs behaved "abnormally", that is the specimen loaded with 0.002 mm/s had a better strength than the one loaded with 0.02 mm/s. There is only one pair of the exposed material where a decrease in strength due to water vapour or sub-critical crack growth would be possible. In all other cases the difference in strength is not significant or the strength was better when tested with the slower displacement rate. The estimates of n calculated from pair 1 of the as-received material is 151 and pair 3 of the exposed material is 31. Both values are high if compared to typical value for oxides 10-20.

The mapping of fracture marks with stereo microscope and SEM suggests that several origins of fracture exist since marks for conflicting directions near each other could be found. The surface of all specimens studied had macroscopic "kinks" where the level of the surface was changed quite suddenly. Since the stress concentration affected a very limited volume the fracture planes remained relatively close to each other.

The fracture was clearly originated in individual SiC-grains on several locations, examples of such initiation is given in Figure 4. In all samples fractured SiC-grains with a lot of rib marks[10] were found as well as grains with essentially smooth surface. Usually the areas with smooth fractured grains did not contain grains with strong arrest lines[10], Figure 4a. The smoothly fractured SiC had sometimes a wavy geometry. It was not found clear markings from the fractured binder necks. Fractured corners of SiC as in Figure 4b were typical for the as-received material where as in the exposed material the fracture had passed completely through the grains. The rib marks in SiC were often arrest lines as the sharp edged curved lines in Figure 4c. Wallner lines are created when reflecting elastic wave interferes the crack propagation[10], the smoother frequently found wave like rib marks could be Wallner lines, Figure 4d. Very finely spaced rib marks were found from the samples tested with the lower displacement rate, Figure 4e.

Some oxidation of SiC occurred in all samples since the testing environment was high temperature water vapour. However, only from the exposed material clear signs of oxidation on fractured SiC were found. Strongly oxidised fracture surfaces were found on a limited area from the samples tested with both loading rates. The oxidation layers nearly 1 μm thick on the hackle marks in Figure 4f were found on the exposed material tested with the lower displacement rate. This suggests that the grains were fractured in early stage of the experiments, perhaps even during the settling period prior loading.

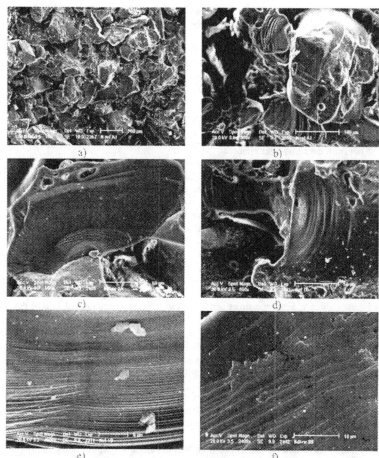

Figure 4.   Fracture markings on SiC in the selected samples studied with SEM. a) The as-received material tested with the displacement rate of 0.02 mm/s. An area of smoothly fractured SiC-grains. b) The as-received material tested with the displacement rate of 0.02 mm/s. The fractured corner of SiC-grain with arrest lines on the right. Possible Wallner lines and arrest lines on the other grains. c) Fracture origin with arrest lines in the exposed material tested with the displacement rate of 0.02 mm/s. d) Fracture origin with Wallner lines in the as-received material tested with the displacement rate of 0.002 mm/s. e) Very finely spaced rib marks in the as-received material tested with the displacement rate of 0.002 mm/s. f) Oxidation layers with wavy frontline typical for cristobalite on hackle marks in the exposed material tested with the displacement rate of 0.002 mm/s.

A classification of fracture paths to through SiC, through binder, and revealing not fractured SiC did not give clear changes in the fracture path. The major fracture path was through the binder in all samples and the fraction of through SiC-grains was high in all cases if compared with the fracture surfaces generated with impact load at room temperature. In the exposed material (especially that with the lower rate) significantly more fractures between SiC and the binder revealing cristobalite were found than in the as-received material; this fracture path was not classified since none were found from the as-received material. This suggests the binder is more strongly attached to the SiC in as-received condition than after the exposure to water vapour and thermal cycling.

Comparison of SEM images of fracture surfaces from as-received material tested in tension at ambient air (without notch) and at high temperature water vapour suggests that the high temperature fracture surface contains more fractured SiC than the ambient air surface. Further, the ambient air surface does not contain very strong rib marks as in Figure 4b but rather the fractured SiC shows hackle and cleavage fracture with only few rib marks.

CONCLUSION

In order to predict the lifetime of advanced SiC-based hot gas filter material knowledge about its failure mechanisms is needed. Tensile testing of at 600°C water vapour has been completed for three pairs of notched test bars with displacement rates of 0.02 mm/s and 0.002 mm/s to determine the significance of environmentally assisted sub-critical crack growth for this material. Specimens form as-received material and material exposed 455 hours at 850°C to water vapour and cooled 8 times below 150°C were tested. No significant effect of sub-critical crack growth was found, which is fairly surprising since the binder holding SiC-grains is mainly of amorphous aluminous alkaline silicate and SCG is known to occur in silicate glasses at water vapour. The exposed material was found more likely to suffer from SCG than the as-received material; the as-received material showed better strength when loaded with the lower displacement rate. Further, the estimates of SCG exponent n calculated from those specimen data pairs where both specimens were from the same filter element was 151 for the as-received material and 31 for the exposed material. These values are higher than expected for a silicate glass containing material. Because of the small number of specimens and very small volume of stressed material the results are only tentative.

ACKNOWLEDGEMENTS

The funding from Academy of Finland, Walter Ahlsrtömin säätiö, KAUTE and Jenny ja Antti Wihurin rahasto is gratefully acknowledged. Research at Oak Ridge National Laboratory was sponsored by the Office of Fossil Energy, U.S. Department of Energy under Contract DE-AC05-00OR22725 with UT-Battelle, LLC. Randy Parten is acknowledged for his numerous efforts in machining the specimens, completing the test set-up and assistance in testing. Ken Liu and Laura Riester are acknowledged for valuable discussions. Special thanks for the personnel of the Glass Shop at ORNL.

REFERENCES
[1] R. W. Davidge, Mechanical Behaviour of Ceramics, Cambridge University Press, Cambridge, UK. (1979).
[2] R. Gy, "Stress Corrosion of Silicate Glass: a Review", *J. Non-Cryst. Sol.*, **316**, 1-11 (2003).

[3] S. M. Wiederhorn," Influence of Water Vapor on Crack Propagation in Soda-Lime Glass", *J. Am. Ceram. Soc.*, **50**, 407-14 (1967).

[4] B. Lawn, Fracture of Brittle Solids, 2nd ed., Cambridge University Press, Cambridge, UK. (1993).

[5] K. Jakus, S. M. Wiederhorn and J. B. Hockey, "Nucleation and Growth of Cracks in Vitreous-Bonded Aluminium Oxide at Elevated Temperatures", *J. Am. Ceram. Soc.*, **69**, 725-31 (1986).

[6] K. D. McHenry and R. E. Tressler, "Fracture Toughness and High-Temperature Slow Crack Growth in SiC", *J. Am. Ceram. Soc.*, **63**, 152-56 (1980).

[7] J. C. Knight and T. F. Page, "Mechanical Properties of Highly Porous Ceramics Part II: Slow Crack Growth and Creep", *Br. Ceram. Trans. J.*, **85**, 66-75 (1986).

[8] P. Pastila, V. Helanti, A.-P. Nikkilä and T. Mäntylä, "Determining the microstructure of some SiC-based hot gas filter materials", to be published in Advances in Applied Ceramics: Structural, Functional and Bioceramics, **104** (2005).

[9] R. J. Roark and W. C. Young, Formulas for Stress and Strain, 5th ed., McGraw-Hill, New York, USA (1975).

[10] J. R. Varner, "Descriptive Fractorgraphy" in Engineered Materials Handbook vol. 4 Ceramic and Glasses, pp. 635-644, ASM International (1991).

QUASI-DUCTILE BEHAVIOR OF DIESEL PARTICULATE FILTER AXIAL STRENGTH TEST BARS WITH RIDGES

Gary M. Crosbie
Research and Advanced Engineering, Ford Motor Company
MD3083 SRL Bldg.; 2101 Village Road
Dearborn, Michigan 48124

Richard L. Allor
Research and Advanced Engineering, Ford Motor Company
MD3135 SRL Bldg.; 2101 Village Road
Dearborn, Michigan 48124

ABSTRACT

Since a high porosity is needed to maintain a low backpressure with a diesel particulate filter, mechanical strength becomes critical in DPF honeycomb structures. In our mechanical tests, we have observed a quasi-ductile stress-strain behavior of the cellular refractories used in exhaust gas aftertreatment. In particular, in 3X3-cell cross-section bend bars used for estimating the axial strength, we find a saw-tooth-type behavior. Associated with this behavior are monotonic decreases in bar stiffness with increasing load, as determined by repeatedly interrupting bar loading and measuring the small decreases in beam resonant frequency. Results are interpreted as the stopping of ridge-borne cracks when the greater crack-front length when the perpendicular wall is reached. This interpretation allows an improvement in a calibration factor for estimating overall honeycomb strength from bend testing of such small bars.

INTRODUCTION

To gain broader acceptance of the fuel-economy and CO_2-reduction advantages of compression ignition combustion with diesel and biodiesel fuels, filtration technologies[1,2] are needed to capture the particulate matter (soot) arising from use of those fuels. Unfortunately, the material-design features of diesel particulate filters that serve best in providing low backpressure of exhaust gas flow (high wall porosity and thin honeycomb walls) also tend to decrease the mechanical strength of the honeycomb. Based on known (typically transverse) filter cracking modes that arise during active regeneration (i.e., the periodic burning off of the soot), axial strength is seen as an important design parameter to use in comparing filter alternatives.

Although various practical constraints have led us to choose a small bend-bar test specimen design, one consequence of the small physical size of each sample is that (in the present work) only three complete honeycomb cell layers are present in the height of the bend bar. In particular, a non-continuum calculation basis for load-strength relationship is desired, but details (such as calibration factors) are needed. A second consequence is that, in the course of the experimental work, we have observed (in some samples) a stress-strain behavior that is more similar to the "graceful" failure of fiber-reinforced composites than to the tensile fracture of typical monoliths and honeycombs.

The focus of this paper is on quantitative estimates of 1) choices for the load-strength calculation of small bars and 2) the extent of the loss of stiffness before final fracture.

EXPERIMENTAL

Axial strength of monoliths can be measured in a bending test, to avoid gripping and alignment issues of direct tensile methods. The Swank, Caverly, and Allor test method[3] provides the operational convenience of fixed loading points, with measured frictional errors of less than 3% (Table II of Ref. 3). This design benefits from the error-minimization analyses[4,5] for solid bars, which represent a continuum approximation to honeycombs. Following Fig. 5 of Ref. 3, we used a four-point "B-type" fixture (as shown below in Fig. 1), where the inner span is 20 mm and the outer span is 40 mm. The contacts with the bar are cylindrical surfaces of radius 0.060 inch (1.524 mm), to minimize traction forces. Typical samples are approximately, 6 X 6 mm by 70 mm length, although the exact cross-section depends on the exact cell density.

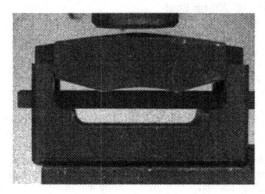

Figure 1. Four-point "B-type" fixture machined from hot-pressed SiC with an inner span of 20 mm and outer span of 40 mm. The contact surfaces are cylindrical to minimize traction forces. The corner posts keep the fixture halves in alignment.

To account for the finite number of cells in the small honeycomb samples, one needs to calculate of moment of inertia based on the individual cell walls and ridges, to estimate more accurately the fixture-sample calibration factor to convert load at failure to a honeycomb strength. In turn, the honeycomb strength is reduced from wall material strength by the fraction of material absent in the openings: honeycomb strength = $(1 - OFA)\cdot$"wall strength" , where OFA is the open frontal area (expressed as an area fraction).

For a solid beam, the load, P, at failure is related to the modulus of rupture (MOR, the maximum tensile stress in the bar at failure) by the moment of inertia (I) in the following equation:

$$MOR = M\,C\,/\,I, \tag{Eq. 1}$$

where the bending moment, $M = (P/2)\left[(L_{outer\ span} - L_{inner\ span})/2\right]$ and the centroid, $C = H/2$ (i.e., half the bar height in the diagram of Fig. 2).

The moment of inertia, I, is the geometric-contribution factor for stiffness. For a solid bar, moment of inertia, I, is the result of an integration from the neutral axis:

$$I = 2A \int_0^{H/2} z^2 f\, dz = H^3 A / 12,$$

(Eq. 2)

where A is the bar width (i.e., depth into in the image, Fig. 2) and the fraction of solids (in the infinitesimal layer in the integrand), $f = 1$ everywhere within the limits of integration, for a solid bar. In a honeycomb, the fraction, f, varies with position.

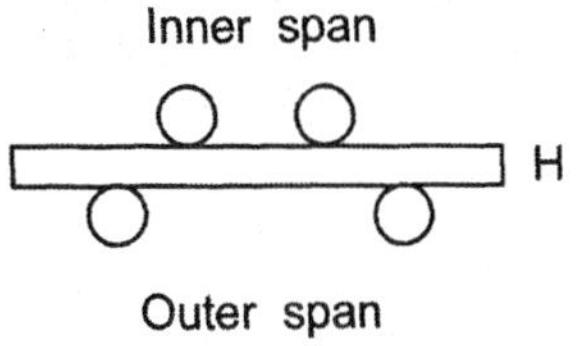

Figure 2. Schematic diagram used in defining terms for calculation of moment of inertia of a bar in four-point bending. Markings point out the bar height, H and the inner and outer spans.

A cross-section of a typical honeycomb bend sample is shown in Fig. 3. Note that small ridges help protect the as-sintered (and optionally coated), outermost honeycomb wall surfaces during machining and subsequent handling. In present tests, ridges were specified to be between 0.005 and 0.015 inch (0.13 – 0.38 mm). Geometrically-descriptive parameters for this sample design are:

w -- wall thickness
a_3 -- cell spacing [subscript 3 denotes transverse direction]
rhx -- ridge height in the width direction [x-direction]
rhz -- ridge height in the height direction [z-direction]

For a 3 X 3 cell honeycomb, the points in the integration direction where the fraction, f, of solid material change are noted by points marked "a" through "e" in Fig. 3.

Figure 3. Cross-section of a 3X3 honeycomb bend sample with ridges around the exterior left over from machining the bar. The designations "a" through "e" represent distances (away from the neutral axis) where the fraction, f, of solid material changes. The four ($\sim$0.25 mm) ridges shown at the top extend between positions "d" and "e" and are farthest from the neutral axis. [Scale: Repeat spacing of the 300 cpsi honeycomb is about 1.47 mm.]

We calculate the honeycomb moment of inertia, i, as a sum of the integrals from the neutral axis (or preceding point) to the identified points, using lower case "i" to distinguish the honeycomb moment of inertia, from that of a solid, "I", with $f = 1$ in Eq. 2. For example, in the region from "**c**" to "**d**", the integrated contribution is:

$$i_d = 2 \left[(3a_3 + w + 2\ rhx)/3 \right] \cdot (d^3 - c^3), \qquad \text{(Eq. 3)}$$

where the three parts of the summation quantity are for the solid wall across distances of, respectively, the three cell spacings, one wall thickness, and two x-direction ridges.

For integration from the neutral axis to "**d**", we find a honeycomb moment of inertia,

$$i = (2/3)w \left[36\ a_3^3 + 3\ a_3^2 (5\ rhx - 3w) + 6\ a_3 w^2 + (rhx - w)\ w^2 \right] \qquad \text{(Eq. 4)}$$

For integration to "**e**", we would need to add a term,

$$i_e = 2 \cdot (4w) \cdot (e^3 - d^3)/3, \qquad \text{(Eq. 5)}$$

wherein the $4w$ represents the solid across only the four ridges. The inclusion of this term would add about 10% to the (Eq. 4) moment of inertia, i, for a 3X3 cell bar of 300 cpsi/12 mil wall honeycomb [i.e., one with a cell spacing, $a_3 = 1.466$ mm and a wall thickness, $w = 0.305$ mm].

To determine which integration limit "**d**" or "**e**" should be used, mechanical tests [Instron Model No. 1122 using a 0.5 mm/min cross-head speed] were interrupted at load levels approaching (and above) average failure loads and tested for stiffness as inferred from resonant frequency.

As in the ASTM C1259 standard test method,[6] each bar (after each loading) was supported near the "free-free" vibration nodal points, here, on apices of polymer foam. An impulse was generated by striking with a miniature hammer (floating head mass 0.42 gram). A microphone held near the bar picked up the ringing as electrical input to an Ono Sokki CF-360 fast Fourier transform (FFT) analyzer. Due to the long ringing, "frequency zoom" software was able to be used at a setting of 8X to enhance resolution well beyond that of a 400 channel FFT.

RESULTS

A typical loading curve (for a silicon-bonded SiC -- NGK MSC11 at room temperature) is shown in Fig. 4. A saw-tooth shape is seen in the stress-strain curve as approximated by the raw loading [force] vs. cross-head displacement curve. The serrations are more prominent (and frequent) at the higher loading levels. The overall effect of the serrations tends to be that of graceful failure, as known for fiber composite materials. Likewise, the maximum load (ultimate load) in some samples was greater than the load at final failure.

The results of the interrupted loading stiffness tests (also at room temperature) are shown in Fig. 5, as a plot of 1 - (freq.)/(init. freq.). From other resonant beam methods, it is known that (frequency)^2 is proportional to the moment of inertia, i. The decrease of resonant frequency is larger for larger loadings. At 90% of max load, frequency drops by 0.25 to 0.45%. Therefore, we expect an 0.5 to 0.9% decrease in i at 90% maximum load.

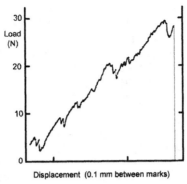

Displacement (0.1 mm between marks)

Figure 4. A load-fixture displacement curve for a 3X3 cell honeycomb bar of silicon-bonded SiC (NGK MSC11). Saw-tooth features (serrations) are seen in this loading [force] vs. cross-head displacement curve also point to ridges fracturing prior to final bar failure. Characteristic of quasi-ductile behavior, the final load (i.e., at the point of complete bar breakage)is lower than the maximum load, in this example. [Note: The location of the zero-displacement point on the x-axis is not determined.]

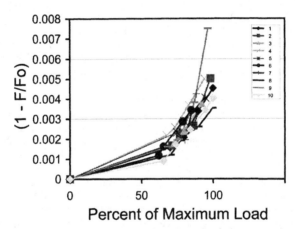

Figure 5. Viewed as an overlay plot (normalized by the unloaded resonant frequency of each bar and then subtracted from unity), decreases in resonant frequency appear as curves rising with increasing load (before each load interruption). For example, samples interrupted in loading at 90% of maximum load, the frequency drops by 0.25 to 0.45%. Since (frequency)^2 is expected to be proportional to the decrease in sample moment of inertia, i, decreases in stiffness can be calculated.

DISCUSSION

Honeycomb bend testing fixture design alternative
In the automotive literature of low-expansion, microcracked oxide substrates for catalysts and DPFs, Gulati and Helfinstine[7] have used larger bar samples for estimation of axial (and transverse) honeycomb strengths, over many years. In these tests of low strength materials, the bend bars are quite large: nominally 0.5 inch (13 mm) thick, 1.0 inch (25.4 mm) wide and 4 inch (101.6 mm) long. [To keep vestiges of walls after machining (ridges) small, the height and width of bars seemingly must vary from the nominal due to exact cell size.] In Ref. 7, the four-point test fixture has outer span of 3.5 inch (8.89 cm) and an inner span of 0.75 inch (1.905 cm) [which is uncommonly-narrow in proportion, giving more of the character of a three-point test]. A solid beam approximation appears to have been used to calculate strength. Because the bar height is that of several cells, rather than just three-cells here, the continuum approximation is satisfactory.

Various constraints led us to use a smaller bend-bar test specimen design. At early stages of development, only limited quantities of DPF pieces are available for destructive mechanical testing. The measurements are needed for each of a range of combinations of substrates, coatings, and treatments, in view of the intimate interposition of modern catalyzed DPF coatings throughout the interconnected pores of the filter walls. Still, quantities of ten bars were needed for statistics. Under these constraints, small tests bars were chosen.

The fixture we used (Fig. 1) is based on a fixed point loading design[3] that minimizes some sources of error: For example, the curvature of the contacts is chosen at radius of 0.060 inch (1.524 mm) to minimize the traction forces while the sample deflects under loading. As shown in Fig. 1, the outer span is 40 mm and the inner span is 20 mm. To preserve alignment of contact cylinder surfaces, posts are provided. Since the fixture is machined from hot-pressed SiC, it has been tolerant of hard materials and chemically-reactive coatings.

Choice of integration limit for finite (3X3 cell) bars
Working with small honeycomb bend samples has led to a need to take into account individual cell wall contributions to moment of inertia. To protect the (as-sintered and optionally coated) cell wall surfaces, a design retaining ridges (vestiges of walls after machining) was used. However, the inclusion of ridges ("**d**" to "**e**" in Fig. 3) adds potentially 10% to moment of inertia. Which integration limit ("**d**" or "**e**") should be used?

Load-crosshead displacement serrations (Fig. 4) suggest that something is intermittently fracturing prior to final failure. The decrease in stiffness (Fig. 5), likewise, points to a decrease in the (load-supporting) moment of inertia, as final fracture load is approached. Since witness marks from compression failure at the contacts are seen only with the low-expansion, microcracked materials, the lowering of stiffness from any ridge-damage at the four contact points is deemed to be at most a minor contribution for SiC-containing samples.

Qualitatively, a drop in geometrical stiffness as evidenced by the lower frequency (seen as an increase of $(1-F/Fo)$ in Fig. 5) is explained by multiple cracking of ridges on the tensile side of each bar. These cracks will tend to lower the i value as final fracture is reached. The cracks stop at position "d", where they encounter a larger cross-sectional area.

Therefore, integration limit of "d" is preferred for calculation from small samples. On this basis, for the 10% range between the two choices, we estimate an error of 3%.

Although it is the better approximation of the two, some ridge segments may remain between ridge cracks. To quantify this last detail, one would like to extrapolate curves of Fig. 5

out to the breaking point, but because the curves are concave upward, such extrapolation is itself subject to large errors. (It is also noted that the resonant beam loses its symmetry since ridge cracks are present only on the tensile side.) Larger samples (to minimize ridge residue effects) might be preferred to reach smaller errors in strength values, except that shear deformation arises in stubby bars.[8] Furthermore, in cases of distortions of honeycomb regularity, it may be impossible to machine regular shapes from larger samples. Given the spread of strength distributions and typically small sample counts, 3% error is seen as adequate for most cases.

Ridge-geometry-induced quasi-ductility

The tensile behavior of a honeycomb (or other cellular structure) of a brittle material is expected to be a single failure,[9] which is qualitatively the same as the fracture of the wall material in the form of a solid (i.e., non-honeycomb). In contrast, we find in our results for the bending of honeycomb samples with ridges that the loading curve is similar to that of the loading of fiber composites: 1) saw-tooth appearance, 2) a decrease in stiffness with loading (Fig. 5), and 3) a final fracture load that is (at least in some samples) below that of the maximum load (Fig. 4). It is of interest to understand how such a characteristic change in mechanical behavior can be obtained from a non-fiber-composite material in a particular (ridged) geometry.

The graceful failure of long-fiber-reinforced composites with poorly-adhering interfaces is well-known in this conference series, as exemplified by one of automotive interest.[10] Although individually both the higher-stiffness fiber and the matrix materials have brittle behavior, the same materials together in the composite material give rise to some characteristics of ductile behavior. Even though the individual microcracks in the composite are brittle failures, the composite can support tensile load after those cracks form. The term "quasi-ductile" (or graceful failure) serves to distinguish this property of the composite from the ductility found in metals. The loading curve (observed in Fig. 4) of the honeycomb samples with ridges appears to similarly justify the use of the term, "quasi-ductile."

An analogy of ridges to fibers is a starting point for resolution. Because the present test is in bending, tensile strain is not uniform across the bar cross-section: The uppermost portions ["e" in Fig. 3] of the ridges (residual wall vestiges left after machining) are farther away from the neutral axis of bending. As seen above in the integration of the moment of inertia, this uppermost material bears more tensile stress than elsewhere. In this regard, the ridges are like the fibers (of higher stiffness than the matrix) in a composite, and carry a disproportionately high share of the load.

However, unlike the composite in the analogy, the ridges (analogs of fibers) are made of the same material as the wall material (analog of the matrix), and therefore are likely to be the first places to crack. As noted above, the cracks can stop near the position "d" (Fig. 3), where the crack front must increase in length to traverse greater amount of material in the honeycomb wall [in the region "d" to "c" of Fig. 3].

This crack-stopping behavior is analogous to crack stopping in chevron-notched, long-crack fracture toughness specimens, where cracks can be induced to grow intermittently in both single-material and fiber-composite materials. The quasi-ductile behavior we find in bending is therefore taken to be a result of the ridge geometry. Because our samples have so few honeycomb cells, we see the ridge-breaking in the loading curves as quasi-ductile behavior.

CONCLUSIONS

Of two calculated limits, the cell wall position ["**d**" in Fig. 3] is chosen (instead of the ridge top,"e") as the better choice for calculation of wall strength, because evidence suggests that ridges are breaking before final fracture. Since portions of ridges may remain, this choice is still an approximation, with an estimated error of 3%.

Quasi-ductility has been observed as saw-tooth features in the load-fixture displacement curves and the breaking load (in some cases) is found to be less than the maximum load. Behavior can be rationalized in terms of cracks propagating through ridges but stopping at the enlarged crack front when attempting to pass from a ridge to going through a honeycomb wall.

Acknowledgments

The authors would like to express thanks to Steven Maertens (Ford Forschungszentrum Aachen) and Dale Hartsock (Ford R&A Dearborn) for discussion of moment of inertia calculation methods; to the DPF manufacturers; and, for support and encouragement in the work with honeycomb diesel particulate filters, to Brendan Carberry and Christoph Boerensen (FFA) and Bob Hammerle, Haren Gandhi, Ken Hass, Paul Killgoar, and Dick Baker (Ford R&A).

REFERENCES

[1] W. A. Cutler, "Overview of Ceramic Materials for Diesel Particulate Filters," *Ceram. Eng. Sci. Proc.*, **25** [3], 421-30 (2004).

[2] H. Sato, K. Ogyu, K. Yamayose, A. Kudo, and K. Ohno, "Soot Mass Limit Analysis of SiC DPF," *Ceram. Eng. Sci. Proc.*, **25** [3], 431-36 (2004).

[3] L. R. Swank, J. C. Caverly, and R. L. Allor, "Experimental Errors in Modulus of Rupture Test Fixtures," *Ceram. Eng. Sci. Proc.*, **11** [9-10], 1329-45 (1990). (See esp. Figs. 5 and 11 and Table II.)

[4] G. D. Quinn and R. Morrell, "Design Data For Engineering Ceramics: A Review of the Flexure Test," *J. Am. Ceram. Soc.*, **74** [9] 2037-66 (1991).

[5] F. I. Baratta, "Requirements for Flexure Testing of Brittle Materials," US Army Rept. No. AMMRC TR 82-20, U.S. Army Materials Technology Laboratory, Watertown, MA, April 1982.

[6] ASTM C1259, "Standard Test Method for Dynamic Young's Modulus, Shear Modulus, and Poisson's Ratio for Advanced Ceramics by Impulse Excitation of Vibration," in *Annual Book of ASTM Standards, Vol. 15.01*, pp. 296-310, Amer. Soc. for Testing and Materials, West Conshohocken, PA (2000).

[7] S. T. Gulati and J. D. Helfinstine, "High Temperature Fatigue in Ceramic Wall-Flow Diesel Filters," SAE Technical Paper No. 850010, February 1985, SAE International, Warrendale, PA, 7 pp.

[8] S.T. Gulati and K. P. Reddy, "Size Effect on the Strength of Ceramic Catalyst Supports," SAE Technical Paper No. 922333, Oct. 1992. (Also, pp. 129-34 in Automotive Emissions and Catalyst Technology, SAE SP-938 (1992).

[9] L. J. Gibson and M. F. Ashby, *Cellular Solids / Structure and Properties*, Cambridge Univ. Press, UK (1997). (See especially, p. 96.)

[10] G. M. Crosbie, J. M. Nicholson, and L. A. Deering, "Pseudo 3-D Reinforcement With Stretch-Broken Carbon Fibers in a Borosilicate Glass Matrix," *Ceram. Trans.*, **46** (Adv. in Ceramic-Matrix Composites II): 211-22 (1994).

Multifunctional Material Systems Based on Ceramics

MULTIFUNCTIONAL ELECTROCERAMIC COMPOSITE PROCESSING BY ELECTROPHORETIC DEPOSITON

Guido Falk, Michael Bender, Rolf Clasen
Saarland University, Department of Powder Technology
Geb. 43, D-66123 Saarbruecken, Germany

ABSTRACT

Aqueous colloidal processing of nanosized and submicron ceramic particles by means of electrophoretic deposition is considered a new process for manufacturing multifunctional electroceramics. Electrophoretic shaping process emphasizes the integration of colloidal particles into multifunctional structures and the production of novel architectures at finer scales in monolithic, composite and thick film form. Designing of advanced monolithic ceramics, layers and composites by electrophoretic deposition includes research of interaction between ceramic particles, organic and/or inorganic dopants and substrate surfaces influenced by external electrical fields.

This work represents a brief summary of our continuous research on electrophoretic processing of multifunctional electroceramic materials. It focuses on tailoring electrochemical and dielectric properties of functionally graded mixed ionic-electronic conductors and ceramic piezoelectrics for use as sensors and actuators. Based on impedance measurements and transmission electron microscopy analysis qualitative and quantitative microstructure-property relations for both application fields are given. Prospective applications of EPD processing methods are derived, by comparing these results with mixed ionic-electronic conductors and $xPbZrO_3$-$(1-x)PbTiO_3$ ($0<x<1$) binary system characteristics given in the state of the art.

INTRODUCTION

Doped ceria is considered a realistic promising candidate for fuel cell applications because of its favorable ion transport properties [1].

More specifically n-type conductivity of Gadolinium doped ceria (CGO) can be suppressed in the temperature range below 700 °C that is suitable for the operation of intermediate temperature oxide fuel cells (IT-SOFC) [2]. In order to reduce the overpotential loss significantly development of electrodes compatible with ceria is important. Comparison of anodic overpotential of Ni/yttria stabilized zirconia (YSZ) and Ni/CGO with hydrogen fuels is given in [3]. With Ni-CGO anodes, favorable performance of 0.4 Wcm^{-2} at 650 °C on dry methanol fuel, area-specific resistance (ASR) of 0.2 Ωcm^{-2} at 500 °C on syngas according to [2] were achieved. Furthermore effective suppressing of carbon deposition as well as high activity towards methane steam reforming with the onset of methane activation occurring at temperatures as low as 209 °C can be attained [3]. In order to overcome a shrinkage mismatch of functionally graded Ni-CGO anodes and the CGO electrolyte laminates, the sequence of well defined processing operations is of inherent importance in order to achieve superior Ni-CGO-based anode materials [2, 3]. In the field of SOFC processing, the electrophoretic deposition (EPD) is expected to fulfill the following requirements:

- coatings can be made in planar or tubular shape

- porous coatings as electrode as well as dense coatings as electrolyte can be processed by controlling deposition conditions

- well defined gradients of electrophoretic deposited CGO particles within a porous matrix can be achieved. Thereby, the motivation is to attain higher triple phase boundary (TPB) contributions and reduce the risk of overpotential loss. Therefore significant interest has recently been initiated in application of this method for solid oxide cells [4, 5].

In this paper, some approaches to minimize the shrinkage mismatch of electrophoretically deposited doped CGO are presented. The processing of CGO electrolyte for IT-SOFC applications is realized by reactive electrophoretic deposition (REPD).

A second group of multifunctional ceramics considered in the paper are piezoelectric/ electrostrictive materials since the miniaturization of microelectromechanical systems (MEMS) has resulted in extensive use of these ceramics. Examples of miniaturized lead-zirconate-titante (PZT) ceramics include components for micromotors, acoustic imaging devices, atomic force microscopes (AFM), and micropumps.

The mechanical and electrical properties of polymer-PZT composite depend strongly on characteristics of each phase and the manner in which they are connected. Integration of a piezoelectric ceramic with a polymer allows tailoring the piezoelectric properties of composites. The most extensively studied composites used in transducer applications are those having 1-3 connectivities. In ultrasonic transducer applications, it was shown that ultrasonic imaging is significantly improved by miniaturizing the rod size of polymer filled PZT arrays with high aspect ratio. The smallest rods have dimensions of 20 – 50 μm and techniques like rod placing, dice and fill, lost model processes, injection molding, laser cutting, solvent-air jet machining of green tapes and ceramic fibre processing were applied.

In this paper, PZT ceramics were formed by EPD of microscale PZT powder on dialysis membrane.

THEORETICAL CONSIDERATIONS

The improvement of processing, control and characterization of ceramic microstructures made of nanosized and submicron particulate materials is a multi-disciplinary challenge. One of the main objectives comprises the formation and control of structural and functional units in microscale dimensions built up by nano scaled architectures.

In wet nanopowder processing specific variation of parameters as solids loading, pH value, salinity, vacuum pressure, shear force, ultrasonic force and electric fields contributes to better microstructural homogeneity of the ceramic components. Investigation of electrophoresis, which involve especially double layer charging mechanisms, electro kinetics, ionic and macromolecular adsorption at interfaces and colloid stability, is one approach to design multifunctional ceramic monoliths, composites, laminates and functionally graded materials with tailored properties.

It is believed that EPD gives potential advances in a number of novel applications including piezoelectric and biomedical ultrasound devices, chemical sensors and solid oxide fuel cells.

EPD comprises two basic steps: the well understood migration of charged particles in the applied electric field (electrophoresis) and the less understood settling of the particles to form an adherent layer on the substrate (deposition).

Electrode reaction of water results in gas bubble formation near the electrodes and in heterogeneous deposits. An alternative solution to the above problem is the membrane-method. In this case the electrophoretic chamber is subdivided by a porous, ion permeable, membrane into two chambers that contain the suspension and another fluid respectively. Deposition occurs at the membrane whereas the ions pass the membrane so that the recombination of the ions and the generation of gas bubbles occurs at the electrodes. No gas bubbles are incorporated into the deposit.

EPD membrane method

Concerning the deposition mechanism, it is suggested that deposition onto porous membranes is a kind of sedimentation, activated by pressure onto particles approaching the membrane surface [6]. This pressure is initiated by electrophoretically activated incoming particles and particles in the outer layer. It is believed that the zeta potential of the particles shifts towards the isoelectric point when the solid loading is increased [7]. Although EPD on membrane substrates has not been applied yet on industrial scale silica [8, 9], alumina [10], clay [11, 12] and zirconia [13, 14] were successfully electrodeposited onto membranes in laboratory scale. Functional ceramics processed by EPD onto membrane substrates are presented in the literature as alumina and zirconia optoceramics [15, 16] and all-ceramic dental implants.

EPD electrode method

In spite of extensive research efforts, a general comprehensive mechanism explaining the coagulation of electrophoretically deposited particles on electrode substrates has not been proposed yet. Brown and Salt [17] explained the electrophoretic formation of initial monolayers onto electrodes by charge neutralisation as particles are deposited. According to Koelmans [18], an increased electrolyte concentration around the particles near the electrode causes a reduction of the zeta potential and induces flocculation. Other theories include particle coagulation caused by distortion of the double layer [19] and electrochemical coagulation by depletion of ions near the electrode leading to pH change towards the isoelectric point [20]. Zhitomirsky summarized these mechanisms and theories in respect of cathodic electrodeposition of ceramic and organoceramic materials [21].

Investigations of electrophoretic deposition of ceramic particles comprise processing of ceramic layers, monolithic ceramics and ceramic composites [22]. With respect to functional layered ceramics, thermal insulations [23], dielectric thick films [24], superconducting layers [25], photovoltaic layers [26], fuel cell layers [27], display layers [28] and biomedical layers [29] were manufactured.

Examples of electrophoretically deposited monolithic ceramics include sanitaryware, tubes and plates [30], ferrules [31], dental ceramics, piezoelectric actuators [32], piezoelectric microstructures [33], alumina microstructures [34], membranes [35], monolithics nanostructured ceramics [13, 36-38], optical and/or electronic micropatterns [39].

Manufacturing of ceramics composites by electrophoretic deposition includes mainly alumina/zirconia composites [40], functionally graded materials [41], electrophoretically impregnated porous ceramic substrates [42] and fibre reinforced ceramics [43].

Within the scope of fundamental research there is still a need to investigate electrochemical as well as colloidal chemical processes at so called amphifunctional substrates with combined polarizable and reversible double layer properties. Partially oxidized metals and semiconducting oxides are illustrative examples.

On a more practical level, such double layer knowledge opens the perspective of optimising industrial processing of multifunctional ceramics. A comprehensive understanding of the colloidal and electrochemical characteristics of amphifunctional interface systems is of inherent importance. Electrophoretic deposition of colloidal particles on non-conducting (membranes) and conducting substrates are presented as illustrative examples of such processes.

EXPERIMENTAL AND RESULTS

CGO electrolyte laminates by EPD-method

Rheology of Gadolinium doped Ceria suspensions. Commercially available Gadolinium doped Ceria CGO-10 (Rhodia, Frankfurt a. M., Germany) was used with a mean particle size of 270 nm, a specific surface of 31 m^2/g and SiO$_2$ impurities in the range of 50 ppm. In order to attain colloidal stabilized CGO slurries with solids contents of 50 wt.-%, the rheological properties were analyzed, such as shear stress and viscosity of the Polyacrylate (PAc)-CGO slurries as a function of pH.

According to Figure 1, the position of the isoelectrical point (IEP) shifts significantly towards lower pH values. Figure 1 also demonstrates that the ζ-potential has a local maximum in the basic branch at pH 11 that indicates adequate suspension stability. Rheological analysis illustrates that the shear stress increases monotonically with the increase of the shear rate for pH far from IEP.

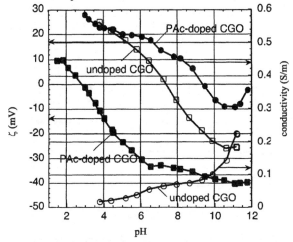

Figure 1: Variation of ζ-potential as a function of pH (a) and variation of shear stress as a function of shear rate at various pH (b) with $c_{PAc} = 1$ mg/m^2 and a solids loading of 50 wt.-% CGO

At higher shear rates, the slurry shows shear thickening behavior. As it is shown in Figure 2 for a weight concentration of 50 %, the suspension viscosity has a minimum between pH 9 and 10. However, if the pH gets closer to the IEP value at lower PAc concentrations, the viscosity increases and the dispersion acts like a non-Newtonian fluid. Based on the results shown in Figure 2, a PAc concentration of 1 mg/m^2 at pH values between 9 and 10 is a favorable condition to achieve homogeneous, high viscosity CGO slurries suitable for EPD process.

Electrophoretic deposition of Gadolinium doped Ceria. In order to achieve good homogeneity of deposited films, CGO suspensions were doped with polyvinylpyrrolidone (PVP). PVP was found to improve crack-free thin film formation. As stated before, PVP has no negative effect on rheological characteristics of PAc stabilized CGO suspensions.

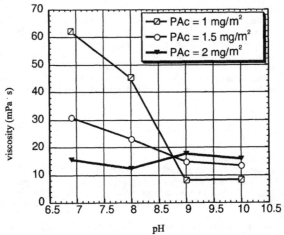

Figure 2: Variation of viscosity as a function of pH and PAc concentration related to specific CGO surface at constant shear rate of 45 s^{-1} and 50 % solids loading of CGO

The depostion rate of CGO powder used was determined first, because only an exact control of this parameter can guarantee reproducible manufacturing of CGO layers. The deposition rate for an aqueous suspension containing 50 wt.-% of CGO is shown in Figure 3.

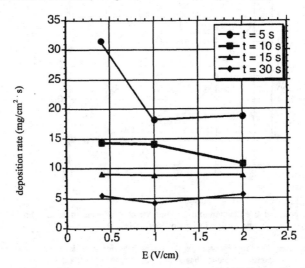

Figure 3: Deposition rate as a function of the externally applied electrical field strength and deposition time

Figure 3 illustrates that the deposition rate at reduced external electrical field strengths depends on deposition time. This result indicates that colloidal CGO particles are spontaneously deposited at high deposition rates in the beginning of the EPD process. When deposition times were increased, deposition rates were reduced continuously. The density of the green body of electrophoretic deposited CGO powder depends on solid content of CGO suspensions. With solids loading of 50 wt.-% an average green density of 36,7 % of theoretical density (TD) is attained.

Sintering characteristics of Co- Cu- and Mn doped CGO. The reduction of the co-sintering temperature decreases significantly the risk of desintegration between Ni-anode materials and CGO laminates. Therefore the CGO suspensions described above were doped with Co-, Mn- and Cu-salts and co-deposited electrophoretically according to the electrode EPD concept.

As it was expected according to [44], densified CGO laminates with densities up to 90 % TD were sintered at maximum temperature of 1300 °C. The results indicate that the co-deposition of particulate ceramics and reactive salts can lead to a significant improvement of the sintering characteristics (see Figure 4).

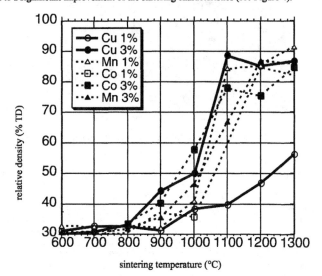

Figure 4: Relative density of doped CGO related to temperature

PZT ceramics by EPD membrane-method
The flow chart of processing piezoelectric ceramics by electrophoretic deposition is given in Figure 5.
In order to achieve suitable EPD processing conditions, an adequate electrical conductivity of the purified particulate raw materials is important.

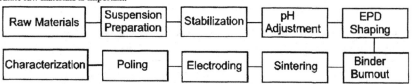

Figure 5: Flow chart for EPD processing of piezoelectric ceramics

Commercially available PZT powder PZT 856, EPC Ltd. with the stoichiometry $PbZr_{0,58}Ti_{0,42}O_3$ was purified and suspended with a dissolver (Dispermat N1-SP, VMA Getzmann GmbH, Germany) in bidestilled water which contained an steric stabilizing agent Dispex A (Zschimmer & Schwarz GmbH, Germany). The electrical conductivity of the purified powder was 150 µS/cm compared to 450 µS/cm of the unpurified powder. A solids loading of 80 wt.-% was realized by pH 12 adjustment with tetramethylammoniumhydroxide (TMAH).

To minimize the problem of PbO volatility at about 800 °C, the PZT samples are sintered in the presence of lead source, a powder mixture of 3 % Pb and 97 wt.-% PZT 856, and placed in closed crucibles. Saturation of the sintering atmosphere with PbO minimizes lead loss from the PZT bodies. The PZT samples were sintered at heating rate of 1 K/min up to 600 °C and up to 1270 °C at heating rate of 1.8 K/min at a dwell time 150 minutes. The cooling rate to room temperature was 0.7 K/min. A relative density of 96 % was achieved.

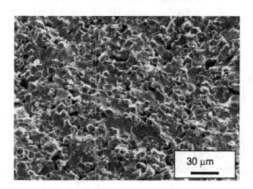

Figure 6: SEM microstructure of EPD processed and sintered monolithic PZT ceramic

After cutting and machining into desired shapes, Ni electrodes were applied by magnetron sputtering. Finally a strong dc field of 2-3 kV/mm was used at room temperature to orient the domains in the polycrystalline ceramic

In order to compare characteristics of EPD processed and dry pressed piezoelectric ceramics of the quality PZT 856 discs with a diameter of 11 mm and a thickness of 2 mm were processed with both fabrication methods.

EPD processed piezoelectric ceramics according to the methods described above were characterized in respect of d_{33} constant with a Berlincourt meter. d_{33} values up to $783 \cdot 10^{-12}$ C/N were achieved. In order to provide the means to evaluate the piezoelectric and elastic properties of the samples impedance measurements, mechanical Q-coefficients and dielectric permittivity coefficients, were conducted with the impedance measurements setup HP 4194. As it was expected, the resonant frequency where the impedance of the body is at a maximum and the oscillatory amplitude is at a minimum was in the range of 1 MHz (see Figure 7). Related to the sintering atmosphere and consequently to stoichiometry as well as impurities in the sintering atmospheres the characteristics of EPD processed PZT can vary in specific ranges according to Table I.

Table I: Typical properties of common dry pressed piezoelectric material (reference) and EPD processed PZT 856

	PZT 856 dry pressed	PZT 856 EPD shaping
dielectric constant ε ($\cdot 10^3$)	4.19	3.00 – 3.27
piezoelectric constant d_{33} (10^{-12} C/N)	550	420 – 783
mechanical Q	67.02	59.08 – 86.67
maximum impedance (Ohm)	3287	1800 - 3930

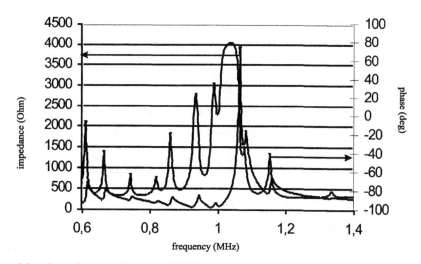

Figure 7: Impedance of piezoelectric ceramic PZT 856 shaped by EPD at resonance

COCLUSIONS AND OUTLOOK

Colloidal processing is demonstrated by reactive electrophoretic deposition (REPD) of metal ions into the matrix of anodic electrophoretically deposited CGO green bodies resulting in significant reduction of sintering temperature of SOFC electrolyte layers. Within the scope of fundamental research there is still a need to investigate electrochemical as well as colloidal chemical processes at so called amphifunctional substrates with combined polarizable and reversible double layer characteristics. Partially oxidized metals and semiconducting oxides used as EPD electrodes and/or substrates are illustrative examples.

In the field of PZT ceramics it is demonstrated that EPD processing is feasible to adjust the manufacturing process of PZT ceramics to produce tailored materials. In order to exclude uncertainties in the PZT characteristics there is a great need for the production of reliable PZT ceramics whose properties and microstructures are optimal. In the given example it is shown that PZT with significant high d-values are manufactured by EPD. Since the d-coefficient is the proportionality constant between electric displacement and stress, or strain and electric field high d coefficients are desirable in materials used as actuators, such as in motional and vibrational applications. By comparing PZT characteristics of EPD processed and dry pressed piezoelectric material it is shown that there is a variation of PZT parameters in certain ranges. It is to be hoped that microshaping technique of nanosized particles based on electrophoretic deposition will be a great deal of developing homogeneous PZT microstructures for transducer and advanced MEMs applications in near future.

A useful future strategy of processing multifunctional submicron and nano-particulate ceramics by electrophoretic deposition (EPD) is based on the formation of patterns of deposited colloidal particle layers on solid/liquid interfaces by variation of external electric variables. Combining electrochemical experiments with structure sensitive techniques, such as scanning probe microscopy, cryogenic SEM, surface X-ray scattering and vibrational spectroscopy it is aimed to develop an understanding of how layered manufacturing and solid freeform fabrication of multifunctional ceramics evolve from the macroscopic through the microscopic to an atomic and molecular level.

The first goal of research comprises the direct characterization of 2D and 3D assembly of macroscopic structures of particulate materials and ions on potentially controlled surfaces, electrodes, and membranes with atomic resolution employing STM and spectroscopy. The second goal is focused on the local addressing and tailoring of single and multiphase microstuctures by applying modified green body processing (i.e. electrophoretic impregnation, EPI), innovative sintering methods (i.e. laser sintering) as well as specific manipulation of sintering and densification mechanisms.

REFERENCES

[1]   S. M. Haile, "Fuel cell materials and components," *Acta Mater.*, **51** 5981-6000 (2003).
[2]   B. C. H. Steele, "Appraisal of $Ce_{1-y}Gd_yO_{2-y/2}$ electrolytes for IT-SOFC operation at 500°C," *Solid State Ionics*, **129** 95-110 (2000).
[3]   W. Z. Zhu and S. C. Deevi, "A review on the status of anode materials for solid oxide fuel cells," *Mater. Sci. Eng. A*, **362** 228-239 (2003).
[4]   J. Will, M. M. Hruschka, L. Gubler and L. J. Gauckler, "Electrophoretic Deposition of Zirconia on Porous Anodic Substrates," *J. Am. Ceram. Soc.*, **84** [2] 328-332 (2001).
[5]   I. Zhitomirsky and A. Petric, "Electrochemical deposition of ceria and doped ceria films," *Ceram. Int.*, **27** 149-155 (2001).
[6]   P. S. Nicholson, Y. Fukada, N. Nagarajan, W. Mekky, Y. Bao and H.-S. Kim, "Electrophoretic deposition-mechanism, myths and materials," *J. Mater. Sci.*, **39** 787-801 (2004).
[7]   O. O. V. d. Biest and L. J. Vandeperre, "Electrophoretic Deposition of Materials," *Annu. Rev. Mater. Sci.*, **29** 327-352 (1999).
[8]   R. Clasen, "Forming of compacts of submicron silica particles by electrophoretic deposition"; pp. 633-640 in *2nd Int. Conf. on Powder Processing Science*, ed. Edited by H. Hausner, G. L. Messing and S. Hirano. Deutsche Keramische Gesellschaft, Köln, Berchtesgaden, 12.-14. 10. 1988, 1988.
[9]   R. Clasen, "Verfahren zur Herstellung von Glaskörpern," EP Pat. 197 585 B1, 21.03.1986, 1986.
[10]   A. P. Gerk, "Formation of alumina products from a liquid dispersion through the use of membrane electrophoretic deposition," EP Pat. 0107 352 A1, 23.09.1983, 1984.
[11]   I. Hector and R. Clasen, "Electrophoretic deposition of compacts from clay suspensions," *Ceram. Eng. Sci. Proc.*, **18** [2] 173-186 (1997).
[12]   R. Clasen, "Keramische Formgebung durch elektrophoretische Abscheidung durch wässrige Suspensionen: Grundlagen und Anwendungen," *Fortschrittsberichte DKG*, **12** [2] 139-155 (1997).

[13]   R. Clasen, S. Janes, C. Oswald and D. Ranker, "Electrophoretic deposition of nanosized ceramic powders"; pp. 481-486 in *Ceram. Transactions*, ed. Edited by H. Hausner, G. L. Messing and S. Hirano. Am. Ceram. Soc., Westerville (USA), 1995.

[14]   K. Moritz, R. Thauer and E. Müller, "Electrophoretic deposition of nano-scaled zirconia powders prepared by laser evaporation," *cfi/Ber. DKG*, 77 [8] E8-E14 (2000).

[15]   G. Falk, A. Braun, M. Wolff and R. Clasen, "Sintering of alumina rich composites produced by electrophoretic deposition and advanced sintering techniques"; pp. in *3rd International Conference on the Science, Technology and Applications of Sintering*, ed. Edited by PennState University, USA, 2003.

[16]   M. Wolff, A. Guise, G. Falk and R. Clasen, "Sintering conditions of translucent zirconia ceramics made of nanosized powders"; pp. in *3rd International Conference on the Science, Technology and Applications of Sintering*, ed. Edited by PennState University, USA, 2003.

[17]   D. R. Brown and F. W. Salt, "The mechanism of electrophoretic deposition," *J. Appl. Chem.*, 15 [40] 40-48 (1965).

[18]   H. Koelmans, "Suspensions in non-aqueous media," *Philips Res. Reports*, 10 161-193 (1955).

[19]   P. Sarkar and P. S. Nicholson, "Electrophoretic Deposition (EPD): Mechanisms, Kinetics, and Application to Ceramics," *J. Am. Ceram. Soc.*, 79 [8] 1987-2001 (1996).

[20]   D. De and P. S. Nicholson, "Role of Ionic Depletion in Deposition during Electrophoretic Deposition," *J. Am. Ceram. Soc.*, 82 [11] 3031-3036 (1999).

[21]   I. Zhitomirsky, "Cathodic electrodeposition of ceramic and organoceramic materials. Fundamental aspects," *Adv. Colloid Interface Sci.*, 97 279-317 (2002).

[22]   A. R. Boccaccini and I. Zhitomirsky, "Application of electrophoretic and electrolytic deposition techniques in ceramic processing," *Solid State Mater. Sci.*, 6 251-260 (2002).

[23]   E. R. Kreidler and V. P. Bhallamudi, "Application of ceramic insulation on high temperature instrumentation wire for turbin engines," *J. Ceram. Proc. Res.*, 2 [3] 93-103 (2001).

[24]   J. Zhang and B. I. Lee, "Electrophoretic deposition and characterization of micrometer-scale BaTiO₃ based X7R dielectric thick films," *J. Am. Ceram. Soc.*, 83 [10] 2417-2422 (2000).

[25]   H. Negishi, N. Koura and Y. Idemoto, "Electrophoretic deposition and the deposition mechanism of Tl-2223 superconducting powder," *J. Ceram. Soc. Jpn.*, 105 379-383 (1997).

[26]   D. Matthews, A. Kay and M. Grätzel, "Electrophoretically deposited titanium dioxide thin films for photovoltaic cells," *Aust. J. Chem.*, 47 1869-1877 (1994).

[27]   M. Matsuda, O. Ohara, K. Murata, S. Ohara, T. Fukui and M. Miyake, "Electrophoretic fabrication and cell performance of dense Sr- and Mg-doped LaGaO₃-based electrolyte films," *Electrochem. Solid State Lett.*, 6 [7] A140-A143 (2003).

[28]   G. J. Verhoeckx and N. J. M. v. Leth, "Hamaker's law and throwpower in deposition of nano-sized alumina particles"; pp. 118-127 in *Electrophoretic Deposition: Fundamentals and Applications*, ed. Edited by A. R. Boccaccini, O. v. d. Biest and J. B. Talbot. The Electrochemical Society, Inc., Pennigton, N. J. (USA), 2002.

[29]   M. Wei, A. J. Ruys, B. K. Milthorpe, C. C. Sorrell and J. H. Evans, "Electrophoretic Deposition of Hydroxyapatite Coatings on Metal Substrates:
A Nanoparticulate Dual-Coating Approach," *J. Sol-Gel Sci. Technol.*, 21 39-48 (2001).

[30]   F. Harbach and H. Nienburg, "Homogeneous Functional Ceramic Components through Electrophoretic Deposition from Stable Colloidal Suspensions - II Beta-Alumina and Concepts for Industrial Production," *J. Eur. Ceram. Soc*, 18 685-692 (1998).

[31]   J. M. Andrews, A. H. Collins, D. C. Cormish and J. Dracass, "The forming of ceramic bodies by electrophoretic deposition," *Proc. Brit. Ceram. Soc.*, 12 211-229 (1969).

[32]   Y. H. Chen, T. Li and J. Ma, "Investigation on the electrophoretic deposition of a FGM piezoelectric monomorph actuator," *J. Mater. Sci.*, 38 2803-2807 (2003).

[33]   J. Laubersheimer, H.-J. Ritzhaupt-Kleissl, J. Haußelt and G. Emig, "Electrophoretic Deposition of Sol-Gel Ceramic Microcomponents using UV-curable Alkoxide Precursors," *J. Eur. Ceram. Soc*, 18 255-260 (1998).

[34]   H. v. Both and J. Haußelt, "Ceramic Microstructures by Electrophoretic Deposition of Colloidal Suspensions"; pp. in *International Conference On Electrophoretic deposition: Fundamentals and Applications*, ed. Edited by Electrochemical Society, Banff, Canada, 2002.

[35]   T. Seike, M. Matsuda and M. Miyake, "Preparation of FAU type zeolite membranes by electrophoretic deposition and their separation properties," *J. Mater. Chem.*, 12 366-368 (2002).

[36]   R. Clasen, *Electrophoretic deposition of compacts of nanosized particles*, ed. Am. Ceram. Soc., Westerville (USA), 1995.

[37]   K. Moritz, J. Tabellion and R. Clasen, "Elektrophoretische Abscheidung keramischer Feinst- und Nanopulver - Anwendungen als Formgebungsverfahren und Infiltrationsmethode," *Fortschrittsber. Dtsch. Keram. Ges.,* **17** [1] 48-55 (2002).

[38]   C. Oetzel and R. Clasen, "Manufacturing of zirconia components by electrophoretic deposition of nanosized powders," *Ceram. Eng. Sci. Proc.,* **24** [3] 69-74 (2003).

[39]   K. Ishikawa, H. Harada and T. Okubo, "Formation Process of Three-Dimensional Arrays from Silica Speres," *AIChE J.,* **49** [5] 1293-1299 (2003).

[40]   B. Ferrari, A. J. Sánchez-Herencia and R. Moreno, "Electrophoretic Forming of Al$_2$O$_3$/Y-TZP Layered Ceramics from Aqueous Suspensions," *Mater. Res. Bull.,* **33** [3] 487-499 (1998).

[41]   C. Kaya, "Al$_2$O$_3$-Y-TZP/Al$_2$O$_3$ functionally graded composites of tubular shape from nano-sols using double-step electrophoretic deposition," *J. Eur. Ceram. Soc.,* **23** 1655-1660 (2003).

[42]   P. K. Das, S. Bhattacharjee and W. Moussa, "Electrostatic Double Layer Force between Two Spherical Particles in a Straight Cylindrical Capillary: Finite Element Analysis," *Langmuir,* **19** 4162-4172 (2003).

[43]   C. Georgi, U. Schindler, H.-G.Krüger and H. Kern, "Nanokomposite für die Herstellung oxidischer Matrices keramischer Faserverbundwerkstoffe," *Mat.-wiss. u. Werkstofftech.,* **34** 623-626 (2003).

[44]   C. Huwiler, E. Jud, L. P. Meier and L. J. Gauckler, "Constant heating rate sintering of ceria-nanopowders"; pp. 61-73 in *Grain Boundary Engineering in Electronic Materials,* ed. Edited by R. Freer, J. V. Herle, J.Petzelt and C. Leach. Institute of Materials, Minerals and Mining, 2003.

TRANSPARENT ALUMINA CERAMICS WITH SUB-MICROSTRUCTURE BY MEANS OF ELECTROPHORETIC DEPOSITION

Adelina Braun
Saarland University
Building 43
Saarbruecken, 66041, Germany

Matthias Wolff, Guido Falk, Rolf Clasen
Saarland University
Building 43
Saarbruecken, 66041, Germany

ABSTRACT

Alumina is a multifunctional ceramic material combining hardness, corrosion and temperature resistance. Furthermore, the optical properties of aluminium oxide ceramic are of great interest. Therefore processing has to be optimized.

Electrophoretic deposition (EPD) is a colloidal process, where ceramic bodies are directly shaped from a stable suspension by an applied electric field. Forming bodies via electrophoretic deposition allows obtaining homogeneous uniform green microstructures with high density. It is a facile and precise technique to synthesize not only monoliths, but also composites with complex geometries [1].

Alumina green bodies were deposited from stabilized aqueous suspensions with and without doping. Green compacts of alumina were evaluated through their pore size distribution and density. Densification behavior was characterized by dilatometric studies, which were conducted at constant heating rate. Doping has a negative influence on suspension preparation, but contributes to the final densification of alumina. Samples were sintered at different temperatures with subsequent post densification via hot isostatic pressing. Transparency was evaluated by spectroscopic measurements. The measured in-line transmission of the samples at 645 nm was more than 50% and this is 58% from the value of sapphire. Mechanical properties of the samples were tested. The influence of dopings on transparency was investigated.

INTRODUCTION

Polycrystalline alumina ceramic (PCA) with its unique combination of low microplastic deformability, high subcritical crack growth resistance, corrosion resistance, thermal stability and optical properties makes it widely used in industry. One of the most important applications is as lamp envelopes in lighting equipment and windows.

Conventional alumina ceramics for lamp envelopes and windows are sintered at very high temperatures (>1700 °C) with resulting coarse grains (>15 μm). Cubic materials, such as spinel, can be transparent with such grain sizes, but the in-line transmission of sintered coarse-grained corundum is less than 10 % and the grain size has to grow to 200 μm to increase the in-line transmission to still low 15-20 % values [2]. Mechanical properties are important for windows and for envelopes in energy saving lamps with a high internal pressure of the discharged plasma. Unfortunately, the coarse grained translucent alumina, the coarse spinel and ALON materials possess lower hardness and limited strength. Improving the mechanical strength and the transparency of Al_2O_3 can be achieved by decreasing the grain sizes in the sintered alumina ceramic [3]. An analytical light-scattering model based on Rayleigh-Gans-Debey light theory gives a quantitative explanation for the increase of the transparency at decreasing grain size [4]. As a second most important condition for high in-line transmission is pointed out the amount of the residual porosity and the size of the pores. This was demonstrated by Mie calculations, which reveal that only 0.1 % residual porosity and pore sizes in the region of the visible wavelength of light can completely deteriorate the transparency.

MgO is the most used grain-growth inhibitor in alumina ceramic for obtaining small grain-sizes, which is an important prerequisite in producing transparent ceramic. On the other side by the high operating temperatures at many hours of the lamp envelopes it hinders the corrosion resistance causing darkening and reduced light output [2] and deteriorating the mechanical properties [5]. Therefore alternative dopants like La_2O_3, ZrO_2, Y_2O_3 etc. are of great interest to substitute MgO in order to improve the light transmittance, mechanical strength and to retain small grain sizes for a long period of time in transparent sintered alumina.

The present study demonstrates that highly dense transparent aluminium oxide ceramic can be produced from electrophoretically deposited green bodies. EPD combines the advantages of the colloidal processing and the electric field assisted processing of ceramics [6]. Stabilized and charged particles, suspended in a liquid are forced to move toward an oppositely charged electrode by applying an electric field to the suspension and form a stable deposited green body [7].

EXPERIMENTAL

The powder used is commercially available TM-DAR α- Al_2O_3 powder (Boehringer Ingelheim Chemicals, Tokyo, Japan). It has alpha-crystalline structure, purity > 99,99 %, $d_{50} = 165$ nm and a measured BET surface of 14,69 m^2/g. The powder used for doping was nanopowder with the average particle size of the dopant, preferably chosen smaller than the alumina particle size: VPPH ZrO_2 (Degussa, Germany) with a monocline crystal phase, $d_{50} = 30$ nm and Nanomyte^{TM} MgO (NEI Corporation, USA) with a cubic crystal phase, particle size 25-35 nm and purity >99,95 %.

The dispersant used is ammonium polymethacrylate, with a commercial name Darvan C (R.T.Vanderbilt Company, Inc, USA); provided as aqueous solution, from which 25 % is active; pH 7.5-9. The quantity of the dispersant used was 0,5 mg/ m^2 and quoted in mg of active dispersant content used per m^2 of alumina powder. Suspensions with different solids loadings were prepared. First step was mixing the calculated dispersant quantity and bidistilled water until the dispersant was uniformly distributed. Then the oxide used for doping was added. After that the alumina powder was gradually added and dispersed at low temperature conditions (10 °C) by means of a dissolver (Dispermat N1/SIP, VMA Getzmann GmbH, Germany). All the suspensions were ultrasonicated with a Branson Sonifier W450 (Branson Scientific, Inc., Plainview, NY) to destroy the agglomerates in the suspension and finally degassed at 200 mbar (20 kPa).

Suspension rheology was controlled using a viscosimeter (RheoStress 1, Thermo Haake, Germany with concentric cylinder) measuring shear stress and viscosity as a function of shear rate.

Shaping of the green compacts was performed by EPD according to the membrane method from aqueous suspensions [8]. The electrophoresis cell was subdivided into two chambers by a cellulose membrane that was permeable for ions but not for the Al_2O_3-particles which deposit onto the membrane. The green bodies were annealed at 800 °C in air to remove the organic additives and after that sintered for 2 hours in air at 1200-1300 °C. Finally the samples were subjected to hot isostatic pressing (HIPing) at 1200-1250 °C and 160 MPa for 2 hours in argon. Green and sintered densities were measured by the Archimedes method.

The hardness of the samples is measured on ground and polished samples with a Fischerscope H100 (Helmut Fischer GmbH+Co, Germany). This is the universal hardness of the material expressed in GPa. Spectroscopic measurements were performed with a Lambda 35 UV/VIS spectrometer (PerkinElmer USA) at a sample thickness 0.8 mm. The average grain-size was determined from the average linear intercept length multiplied by 1.56 from HR-SEM images of the samples.

RESULTS AND DISCUSSION

Electrophoretic deposition is a colloidal process, which is easy to control, fast and cost effective. To obtain high green densities and homogeneous microstructure it requires stable suspensions. The viscosity of 45 vol.% suspensions prepared with and without doping oxides is shown in table 1. For comparison are taken the values for 250 ppm quantity of the doping oxides. The viscosity of the suspension with MgO is higher compared with the alumina and ZrO_2 doped suspensions. This comes from the difficulty to stabilize the MgO nanopowder, because of its strong basic nature displayed by a very high isoelectric point pH~10.5-12 [9]. MgO has always a lower surface charge density than alumina at given pH and also dissolves easily in the suspension causing flocculation of it. Therefore MgO has to be added before the alumina powder into the solution of dispersant and water. The decrease in suspension stability by MgO addition is further confirmed by the green density of the EPD deposited samples. Pure alumina samples have the highest value of relative green density - 65.30 % TD, followed by the 250 ppm ZrO_2 doped samples- 65 % TD and, as lowest, the 250 ppm MgO doped samples- 63 % TD.

Parameter	Al_2O_3	250 ppm ZrO_2	250 ppm MgO
Viscosity 15 1/s (mPa·s)	793	798	1253
ρ_{green} (% Theoretical)	65.30	65.00	63.00

Table 1: Parameters of pure alumina powder suspensions and doped suspensions with ZrO_2 and MgO nanopowder, and respective green density of deposited samples.

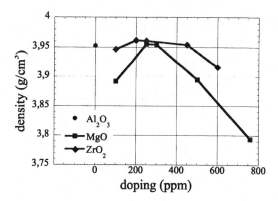

Figure 1: Density after sintering at 1600 °C of alumina compacts and doped alumina compacts versus different doping levels.

The density of the doped and nondoped compacts after sintering at 1600 °C (5 K/ min) with dwelling time 1 hour are compared in fig. 1. The highest density values for samples doped with MgO are obtained for the 250 and 300 ppm MgO and are 99.21 % TD for both, similar to the non-doped samples 99.15 % TD. ZrO_2 doped samples have the highest density for doping ZrO_2 levels from 200 till 450 ppm with a 99.40 % TD for the 200 ppm ZrO_2, which is the highest measured value. Both dopants enhance the densification of PCA in spite of the lower starting green density in comparison with alumina compacts without doping.

Dilatometric measurements were performed to find out the exact amount of the MgO in aluminium oxide leading to the best densification. Dilatometry-derived densification curves showing density as a function of time for annealing at 1600 °C at a constant heating rate (10 °C/min) are shown in fig. 2. The highest density is achieved for the 250 ppm MgO doped alumina compact. MgO does not influence the temperature of the most active sintering, which for pure aluminium oxide samples takes place at 1200 °C.

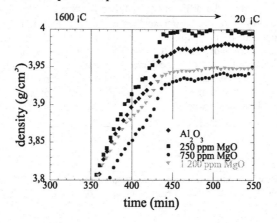

Figure 2:   Effect of magnesium oxide content on densification.

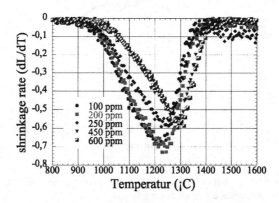

Figure 3:   Shrinkage rate versus temperature for different doping levels of ZrO_2 in alumina.

ZrO$_2$ influences the temperature of most active sintering. It can be seen from fig. 3 that by dopings 100, 200 and 250 ppm ZrO$_2$ the temperature for sintering is 1250 °C, only 50 °C higher than that of alumina oxide samples without doping. It increases with 100 °C by 450 and 600 ppm ZrO$_2$

The sinter density and respectively the grain sizes of the samples doped with ZrO$_2$ by different ppm levels are given in fig. 4. All samples were sintered at 1300 °C with 5 K/min and 2 hours dwelling time. The samples with 450 ppm ZrO$_2$ in PCA showed the highest density of 99,13 % TD and respectively the lowest grain size of. 455 nm.

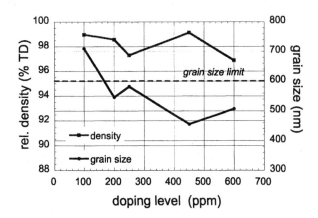

Figure 4:   Relative density and grain size of ZrO$_2$ doped samples versus different doping levels.

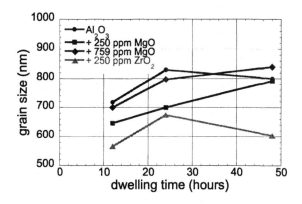

Figure 5:   Grain size of doped and undoped PCA samples versus different dwelling times.

Doped and undoped samples were sintered at 1200 °C with 5 K/min and different dwelling times (12, 24 and 48 hours) to investigate the influence of dopants on the grain size development with the time. From fig. 5 can be seen that the smallest grains were retained in the sample with ZrO₂ doping. After 48 hours sintering at 1200 °C the sample with ZrO₂ doping had 600 nm an average grain size of 600 nm in comparison with 791 nm for the 250 ppm MgO grains and 800 nm for the non-doped Al₂O₃.

The residual porosity and the grain sizes in the samples influence strongly the mechanical properties of the material. It can be seen from fig. 6 that with increasing of the porosity the hardness is decreasing. Samples with equal grain sizes, for instance 528 nm, have a different hardness influenced by the porosity content in the sample. It is obvious from the figure that smaller grain sizes have a higher hardness value in spite of the slightly higher porosity values.

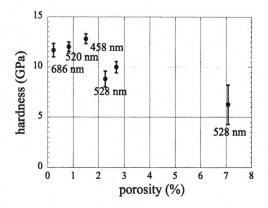

Figure 6:   Hardness of only alumina samples versus the porosity content in the sample

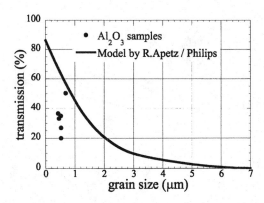

Figure 7:   Influence of grain size on the in-line transmission of sintered Al₂O₃ samples.

The grain-size dependence of the measured in-line transmission values of PCA samples without doping is compared with the analytical light scattering model of alumina for grain-boundary scattering at zero porosity based on the Rayleigh-Gans-Debey theory (fig.7). In accordance with the work of Apetz and van Bruggen, the in-line transmittance depends on the grain size and on the amount and size distribution of the few residual pores. The point, which is nearest to the model, possesses only 0.04% porosity. All other samples have a higher percentage of porosity and the higher it is the further are they positioned from the theoretical curve on the graphic. To characterize experimentally the size-distribution of these pores is unfortunately not possible. Microscope images were made to obtain information about the size of the pores. They were in the range of 30- 100 μm and not in the critical range of pore-sizes equal to the wavelength of the visible light. According to the model for such big pores the back-scattered light is only 5 % [4].

Figure 8:   Sintered and HIPed nondoped Al₂O₃ and high resolution SEM image of it.

In fig. 8 is shown the PCA sample without doping (left) and the HR-SEM image (on the right). The microstructure reveals to be fully dense, pore free (only 0.04 % residual porosity) and with an average grain size of 458 nm.

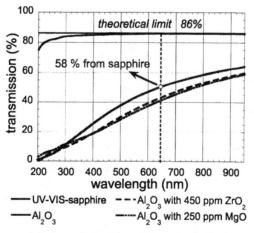

Figure 9:   In-line transmission of sapphire, fine-grained alumina ceramic, and alumina doped with 450 ppm ZrO₂ and 250 ppm MgO as a function of the wavelength.

The highest transmission value is obtained for the nondoped PCA, which is 58 % of that of sapphire and shown in fig. 9 for sample thickness 0.8 mm at 645 nm. The sample with 450 ppm ZrO_2 shows a higher value of 43 % from sapphire, compared with MgO, which has 41 %.

CONCLUSION

Alumina samples were prepared through electrophoretic deposition with and without doping. MgO doping influences negatively the stability of the alumina suspension more strongly in comparison with the ZrO_2 doping.

The influence of MgO and ZrO_2 dopings on the densification and microstructure development has been studied at different dopant levels. 250 ppm is the optimal doping level for MgO and 450 ppm for ZrO_2 both contributing to the highest density in the final sintered alumina samples. Abnormal grain growth in alumina, decreasing the long-time stability, is possible to be avoided with ZrO_2 doping, which was investigated as alternative dopant to substitute MgO. ZrO_2 doped PCA samples need higher temperatures of sintering and HIPing. It makes ZrO_2 an interesting dopant because the submicrometer microstructure and the optical properties can be retained at 1200 °C while shifting the sintering temperature to higher values and preventing in the meantime the additional grain-growth.

Transparent alumina with fine and homogeneous microstructure is produced from extremely pure submicropowders via sintering at lower temperatures. The microstructure of such samples reveals uniformity, an average grain size of 480 nm and only 0,04 % residual porosity. The residual porosity has a substantial influence on the mechanical properties and the transmission. The measured transmission of the transparent alumina sample without doping is 58 % of the value of sapphire. 450 ppm ZrO_2 doped PCA has a higher transmission value, which is 43 % , in comparison to 41 % for the MgO doped sample.

REFERENCES

[1] J. Tabellion and R. Clasen, "Electrophoretic deposition from aqueous suspensions for near-shape manufacturing of advanced ceramics and glasses - applications," *J. Mater. Sci.*, **39** 803-811 (2004).

[2] G. C. Wei, A. Hecker and D. A. Goodman, "Translucent polycrystalline alumina with improved resistance to sodium attack," *J. Am. Ceram. Soc.*, **84** [12] 2853-2862 (2001).

[3] A. Krell, P. Blank, H. W. Ma, T. Hutzler, M. P. B. v. Bruggen and R. Apetz, "Transparent sintered corundum with high hardness and strength," *J. Am. Ceram. Soc.*, **86** [1] 12-18 (2003).

[4] R. Apetz and M. P. B. v. Bruggen, "Transparent Alumina : a light scattering model," *J. Am. Ceram. Soc.*, **86** [3] 480-86 (2003).

[5] O. Asano, "Ceramic envelope for high intensity discharge lamp and method for producing a polycrystalline transparent sintered alumina body," EP Pat. 03251264.2, 10.09.2003, 2003.

[6] R. Clasen and J. Tabellion, "Electric-Field-Assisted Processing of Ceramics, Part I: Perspectives and Applications," *cfi/Ber. DKG*, **80** [10] E40-E45 (2003).

[7] P. Sarkar and P. S. Nicholson, "Electrophoretic Deposition (EPD): Mechanisms, Kinetics, and Application to Ceramics," *J. Am. Ceram. Soc.*, **79** [8] 1987-2001 (1996).

[8] S. Musikant, *Optical materials*, ed. Marcel Dekker, Inc., New York, Basel, 1985.

[9] G. Tari, J. M. F. Ferreira and O. Lyckfeldt, "Influence of magnesia on colloidal processing of alumnina," *J. Eur. Ceram. Soc.*, **17** [11] 1341-1350 (1997).

FUNCTIONAL NANOCERAMIC COATINGS ON MICTROSTRUCTURED SURFACES VIA ELECTROPHORETIC DEPOSITION

H. von Both
Wieland Dental + Technik GmbH & Co. KG
PO Box 2040
D-75179 Pforzheim

A. Pfrengle and J. Haußelt
Forschungszentrum Karlsruhe, Institute for Materials Research III
PO Box 3640
D-76021 Karlsruhe

ABSTRACT

In micro process and micro chemical engineering fluids or gases are to be transported through microstructured channels, e.g. in micromixers, microreactors as well as in micro heat exchangers. Usually they are manufactured from metallic substrates by mechanical milling or by chemical etching. Especially in applications with aggressive media the surfaces have to be protected against oxidation or corrosion. Because of their chemical inertness, ceramics are considered suitable as protective layers. An electrically conductive surface like steel can be coated by electrophoretically depositing charged particles from a ceramic suspension applying an electric field. Fully dense ceramic layers with a thickness of a few microns or less and sintered at lower temperature require fine to nanosized ceramic powders.

The present paper focuses on electrophoretic deposition (EPD) of submicron and nanosized zirconia particles onto micro heat exchanger substrates. The layers on microstructured steel substrates are dried and sintered at 1000°C. For this, aqueous electrostatically stabilized suspensions containing nanosized zirconia particles are developed and characterized, parameters of the EPD and of a drying and sintering process are examined. Furthermore the electrophoretic mobility and the zeta-potential of both nanosized and coarser particle fractions are analyzed. It can be shown that the combination of optimized ceramic suspensions with defined deposition kinetics allows the production of ceramic layers on microstructured steel substrates. The characterization and analysis of the surface and structure dimensions by SEM and white-light interferometry demonstrates that sintered ceramic coatings a few microns thick can be achieved.

INTRODUCTION

Thin ceramic layers are used in different applications, for example as protective layers in micro process engineering, as piezoelectric layers or as porous catalytic layers for micro reaction applications.

Amongst several processes like sol-gel processes, physical or chemical vapor deposition (PVD or CVD), the electrophoretic deposition (EPD) is a widely used method to deposit bodies or layers out of a ceramic suspension[1]. The suspension has to contain ceramic particles with a surface charge, due to which the particles move in an externally applied electrical field and deposit on one of the electrodes. Advantages of EPD are the processibility of submicron to nanosized powders and the use of suspensions with low solids content for the production of compacts with high green densities. This is especially convenient as suspensions containing a high volume fraction of nanosized particles are difficult to realize.

The goal of this work is to deposit thin zirconia layers on microstructured steel substrates, which are used as heat exchangers in the field of micro process engineering. Because of the requirement of producing layers as thin as possible (because of the low thermal conductance of ceramic compared to steel), it is necessary to use very fine powders. Another advantage of using very fine particles is their high sintering activity[2] due to their high specific surface of. Therefore, a relatively low sintering temperature of 1000°C can be applied. This is necessary because high temperatures thermally affect the metallic substrate and lead to annealing or recrystallization. The electrophoretic deposition of a ceramic layer on a metallic substrate with on-substrate-sintering is not new[3]. The problem that temperatures necessary for sintering may have a negative impact on the substrate can, for example, be solved by applying laser processing after sintering[4].

At first, to electrophoretically deposit a ceramic layer, a ceramic suspension has to be produced. This suspension is characterized with respect to electrical suspension conductivity, pH-value, particle size distribution (PSD) and electrophoretic mobility (EM)[5-7]. Afterwards, the movement and the deposition of the particles in the suspension is carried out by applying an electrical field across the suspension, whereby the cathode is the substrate on which the deposition takes place.

EXPERIMENTAL PROCEDURE

A suspension is produced by dispersing zirconia powder (stabilized with 5 mol% yttria, Unitec PYT05.0-001H, UCM Group PLC, Stafford, Great Britain) together with deionized water into a beaker. The powder has a theoretical density of 6.05 g/cm³ and a specific surface of 12.6 m²/g. The refractive index, which is important for the particle size distribution analysis, is 2.2 real and 0.01 imaginary.

For stabilization, 0.1 molar hydrochloric acid (HCl) is added to shift the pH-value to approximately 5 to examine its effects on deposition. Influences of solids content are also analyzed and changed between 4 and 8 vol%.

To break existing particle-aggregates, the suspension is exposed to ultrasonic treatment (Branson Sonifier W 450). The amplitude is set to 60 % for a total duration of 5 to 20 minutes with 50 % duty cycle. After treatment, the pH-value and electrical suspension conductivity are measured (pH-value measured with SenTix HW pH-probe and inoLab pH Level 2 gauging station, WTW, Weilheim, Germany; conductivity measured with conducting meter 703, Knick, Berlin, Germany)

The particle size distribution is characterized with LS230 (Beckmann Coulter, Krefeld, Germany), which is a laser light scattering method. The suspension has to be diluted for measurement. Refractive index and density of the particles are set to the values of zirconia in water. The electrophoretic mobility is measured with the ZetaSizer 3000 HSa (Malvern Instruments, Herrenberg, Germany)[8]. As for particle size measurement, it is necessary to dilute the suspension prior to measurement. A solids content of ca. 10^{-5} volume fraction is appropriate. The Zeta-Potential is calculated according to the Helmholtz-Smoluchowski-theory from the measured EM.

EPD is performed with a 3-electrode system with a Jaissle Potentiostat/Galvanostat 1002 PC.T (Waiblingen, Germany). The reference electrode is located beside the working and counter electrode. The latter two are opposite to each other with a distance of 15 mm. The electrode area of working and counter electrode is 14 x 14 mm², but only the lower half of the electrode is put into the suspension. The counter electrode is the anode (plain 1.4301 steel foil), the working electrode the cathode. The structure of the cathode, where the layer is deposited, is shown in figure 1. It consists of rounded microtrenches that are 160 microns wide with 65 microns base width and 40 microns depth.

 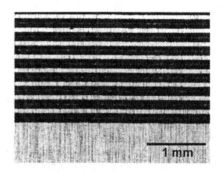

Fig. 1. Microstructured steel substrate used for EPD (right: LIM picture).

The potential of the working electrode at which EPD is carried out is kept below –1.5 Volts to prevent electrolysis of water. The duration of the deposition is varied between several seconds and one minute to get layers of variable thickness.

The deposited layers are taken out of the suspension and put in horizontal position into a vacuum chamber for drying. For sintering, a sintering furnace CWF 1300 (Carbolite, Ubstadt-Weier, Germany) is used. Sintering is performed at 1000°C for 15 minutes. Afterwards, the sintered layers are investigated by light microscopy (LIM) and scanning electron microscopy (SEM).

RESULTS AND DISCUSSION
    The characterization of the particle size distribution (PSD) is performed on a stabilized suspension at a pH-value 6 (figure 2). The PSD tends to get finer with increase of ultrasonic time. While at 10 minutes of ultrasonic treatment most of the particles are still aggregated, a treatment of 20 minutes results in a finer distribution, where 70 % of the particles are below a diameter of 80 nm and 90 % are below 400 nm.

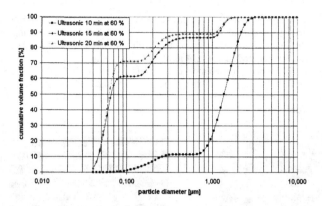

Fig. 2. Particle size distribution of the suspension after ultrasonic treatment.

The Zeta-Potential of the particles, calculated after the Helmholtz-Smoluchowski-theory, is shown in figure 3 as a function of pH-value. The isoelectric point (IEP), where the zeta-potential is zero, is at a pH-value of 6.5. In acidic milieu, the zeta-potential is positive and converges to about 50 mV at a pH-value of 2.5, whereas in basic milieu it converges to 50 mV for a pH-value 11.

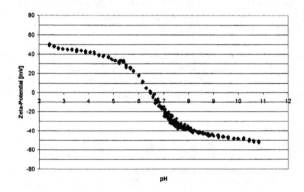

Fig. 3. Zeta-Potential depending on pH-value.

If only the zeta-potential, which is responsible for stabilization of the suspension, is taken into account, either a very high or a very low pH-value would be appreciated. But as shifting the pH-value means adding ions to the suspension, this increases the electrical suspension conductivity, which in turn shows negative effects for EPD. Experiments are performed in acid milieu by adding HCl. The experiments show that for pH-values larger than ca. 6, the suspension is not stable. For pH-values below 5, no deposit adheres to the substrate. The electrical suspension conductivities that are suitable for deposition are below a maximum of 300 µS/cm (better 250 µS/cm) for a 6 vol% suspension.

Another phenomenon is related to the solids content of the suspension. Experiments show, that depositing from suspensions with solids contents below 4 vol% does not lead to a homogeneous layer. Deposits show local defects, where no particles adhere to the substrate. On the other hand, suspensions with solids contents above 8 vol% are already too viscous. Deposits are several microns thick just by dipping the substrate into the suspension. The surface of the coatings is also inhomogeneous across the structure of the substrate, as the coating is thicker especially on the ground of the trenches.

In order to avoid cracks in the coating layers the drying process is a crucial step. Horizontal and vertical drying has been examined, both at environmental conditions and in a vacuum chamber. Best results are achieved when drying is carried out in a horizontal position in a vacuum chamber.

Figure 4 shows LIM-images of two coatings. The left has been produced from a 4 vol% suspension (pH-value 6.1), the right from a 5 vol% suspension (pH-value 5.5). A large quality difference between the two coatings can be observed. This verifies that the process is very sensitive to changes in parameters or to imprecise handling, as the solids content as well as the pH-value are very similar in both cases. The thickness of a coating deposited from a 6 vol% suspension at a pH-value of 5.5 with an applied potential of –0.5 Volts for 5 seconds is approximately 5 to 8 microns (see figure 6).

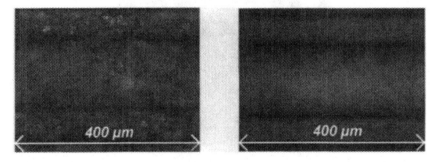

Fig. 4. Coated trench (left: from a 4 vol% suspension at pH-value 6.1,
right: from a 6 vol% suspension at pH-value 5.5).

Sintering is carried out at a temperature of 1000°C for a duration of 15 minutes and the result is shown in figure 5. Even though the substrate and the layer have different thermal expansion coefficients, there is no delamination of the ceramic layer. The right part of the figure also shows that the ceramic is not fully sintered, porosity is still remaining. To overcome this problem either a higher sintering temperature or a longer duration of sintering would be needed. But unfortunately both measures usually lead to a deformation of the substrate. Another way to increase sintering density would be the use of finer particles. As it was shown in figure 2, the dispersion of the particles is not yet complete, aggregates are still remaining. A better dispersion of the particles might already lead to a higher specific surface and hence a denser sintering of the layer.

Figure 6 shows a cross-section of a sintered zirconia layer. It can be seen that the coating levels out the roughness of the underlying steel substrate and therefore has a thickness ranging from about 5 to 8 microns.

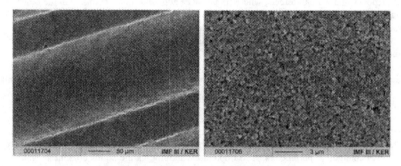

Fig. 5. SEM images of sintered layer.

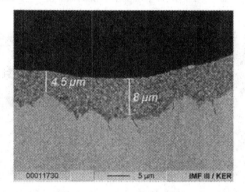

Fig. 6. Cross-section of a sintered layer.

CONCLUSIONS

The presented work shows that it is possible to electrophoretically deposit zirconia particles to form layers with a thickness of 5 to 8 microns on a microstructured steel surface. Therefore, aqueous ceramic suspensions with Unitec PYT05.0-001H zirconia powder and a solids content of ideally 6 vol% are prepared and stabilized with HCl at a pH-value between 5 and 6. This is the ideal range avoiding both instability of the suspension (pH-value > 6) and too high conductivity with negative effects on deposition (pH-value < 5).

The deposition takes place at a potential of –0.5 Volts for duration of 5 seconds. Drying is performed in a horizontal position in a vacuum chamber. The substrates are sintered for 15 minutes at a temperature of 1000°C. However, sintering does not lead to a fully dense ceramic and some porosity is still remaining. A solution could be the use of even smaller ceramic particles. Due to their higher sintering activity this would allow to reduce sintering temperature and time, avoiding deformation of the steel substrate.

The presented work demonstrates that the electrophoretic deposition of nanosized particles is a suitable processing way for the production of ceramic coatings with a thickness of a few microns on microstructured substrates.

ACKNOWLEDGEMENTS
The authors thank Dr. W. Benzinger, Institute for Micro Process Engineering at Forschungszentrum Karlsruhe, for providing the microstructured steel substrates used in this work.

REFERENCES
[1] P. SARKAR, P.S. NICHOLSON: *Electrophoretic Deposition EPD: Mechanisms, Kinetics and Application to Ceramics*, J. Am. Ceram. Soc., 79: 1987-2002, 1996

[2] S. APPEL, R. CLASEN, S. SCHLABACH, B. XU, D. VOLLATH: *Sintering behavior and grain structure development of ZrO_2- and Al_2O_3-compacts fabricated from different nanosized powders*, 26th Ann. Meeting Int. Conf. on Advanced Ceramics & Composites, Cocoa Beach, Florida, USA, 13. - 18.01.2002, in: Ceramic Engineering and Science Proceedings, 23:609-616, 2002

[3] Z. WANG, J. SHEMILT, P. XIAO: *Fabrication of ceramic composite coatings using electrophoretic deposition, reaction bonding and low temperature sintering*, J. Eur. Ceram. Soc., 22: 183-189, 2002

[4] X. WANG, P.XIAO, M SCHMIDT, L. LI: *Laser processing of yttria stabilised zirconi/alumina coatings on Fecralloy substrates*, Surface and Coatings Technology, 187: 370-376, 2004

[5] R.H. MÜLLER: *Zetapotential und Partikelladung in der Laborpraxis*, Wissenschaftliche Verlagsgesellschaft Stuttgart, 1996

[6] G. LAGALY, O. SCHULZ, R. ZIMEHL: *Dispersionen und Emulsionen*, Steinkopf Verlag, 1997

[7] J.N. ISRAELACHVILI: *Intermolecular & Surface Forces*, Academic Press, 2000

[8] MALVERN INSTRUMENTS: *Documentation for Zetasizer 1000/2000/3000: Principles of operation*, Manual Number MAN 0152, Issue 1.1, 1996

HIGH DAMPING IN PIEZOELECTRIC REINFORCED METAL MATRIX COMPOSITES

Ben Poquette, Jeff Schultz, Ted Asare, Stephen Kampe, Alex Aning
Virginia Polytechnic Institute and State University
Materials Science and Engineering Dept., 213 Holden Hall
Blacksburg, VA 24061

ABSTRACT

Piezoelectric-reinforced metal matrix composites (PR-MMCs) show promise as high damping materials for structural applications. Most structural materials are valued based on their stiffness and strength; however, stiff materials have limited ability to dampen mechanical or acoustical vibrations. PR-MMCs represent a potential material system capable of exhibiting increased damping ability, as compared to the structural metal matrix alone. In addition, the piezoelectric ceramic particles may also augment the strength of the matrix, creating a multifunctional composite. In this work, the damping behavior of PR-MMCs created by the addition of barium titanate ($BaTiO_3$) discontinuous reinforcement in bearing bronze (Cu- 10w% Sn) matrices has been studied. The degree of damping has been found to rely on the reinforcement volume fraction. This additional damping is thought to be attributed to twinning in the pyroelectric domains of the reinforcement. This research has been sponsored by the Army Research Office under contract no. DAAD19-01-1-0714.

INTRODUCTION

Most structural materials are valued based on their stiffness, and strength, but another inherent attribute of a stiff material is its ability to transmit vibrations. For most structural applications, vibrations are an unwanted side effect of mechanical motion. Great effort is made to eliminate vibrations, since they can cause mechanical failure, induce physical discomfort, and compromise stealth. To combat these undesirable effects, high damping, yet stiff material systems are needed.

In the literature, ferroelectric ceramics, having a piezoelectric crystal structure, have been shown to exhibit damping behavior below their Curie temperatures (T_c).[1] This damping ability can be lent to stiff structural materials through compositing. Piezoelectric Reinforced Metal Matrix Composites (PR-MMCs), utilizing structural metals as a matrix, form a material system that exhibits increased damping ability as compared to the structural metal alone. The addition of ferroelectric ceramic particles can not only improve damping capacity, but it can also augment the load bearing capacity of the matrix, creating a multifunctional composite.

A multifunctional material is designed to simultaneously address two or more engineering functions.[2] The concept of multifunctionality can be applied to many applications. These applications usually consist of structural materials that are at the same time self-interrogating, self-healing, stealth providing, or energy dissipating, to microscopic materials or systems that may exhibit some combination of sensing, moving, thinking, communicating, and acting. In all applications, the implementation of multifunctional materials can lead to improved system efficiency. The multifunctional material concept, studied here, employs a structural metal matrix composite, (MMC) reinforced with dispersed, isolated ferroelectric ceramic particles. This represents a marriage of two readily available material forms about which much is individually known.

This work demonstrates the ability to fabricate and characterize PR-MMCs. Here BaTiO$_3$, a ceramic having a piezoelectric crystal structure, is incorporated into a Cu-Sn (bearing bronze) matrix, and the damping properties of the resulting PR-MMC are investigated above and below the Curie temperature of the reinforcement.

EXPERIMENTAL PROCEDURE

The PR-MMCs were created using a bearing bronze matrix (Cu-10 w% Sn) reinforced with 0, 30, and 50 v% particulate BaTiO$_3$. To produce the bearing bronze, Cu and Sn -325 mesh powders of were purchased from Atlantic Equipment Engineers and combined in a roll mill. BaTiO$_3$ powder with a particle size of 25-45μm (-325 +500 mesh) was then be added to achieve the correct v %. This BaTiO$_3$ powder was obtained through the mechanical crushing and sieving of 3-12mm pieces purchased from Sigma-Aldrich. The Cu-Sn-BaTiO$_3$ powder mixtures were then uniaxially pressed at 280 MPa forming disk shaped powder compacts that were liquid phase sintered under argon at 820°C for six minutes. The composite samples tested were cut from these composite disks while the samples of bulk BaTiO$_3$ were cut from the as received 3-12mm pieces.

Differential scanning calorimetry (DSC) was used to demonstrate that the piezoelectric tetragonal phase was retained after sintering of the MMCs and dynamic mechanical analysis (DMA) was used to characterize the relative damping properties of the bulk BaTiO$_3$ and the PR-MMCs. All DMA tests were conducted using a three-point bend configuration.

RESULTS AND DISCUSSION

Barium Titanate has piezoelectric crystal structure, but is only piezoelectric as a bulk ceramic when poled. Bulk BaTiO$_3$ is however ferro- and thus pyroelectric, and it is these properties that are responsible for high damping. Pyroelectric materials are made up of several polarized domains which orient themselves such that the sum of their polarizations approaches zero. When a pyroelectric material is stressed the magnitude of polarization within each of its domains is changed, so they must reorient themselves to bring the overall polarization back to zero.[3,4] This reorientation requires energy. If the stress is cyclic, as in the case of vibrations, this domain reorientation is constantly occurring and thus constantly using energy which tends to damp the system.

Figure 1 is a plot of tan delta (tan δ) as a function of temperature for bulk BaTiO$_3$. Tan δ is the loss modulus divided by the storage modulus, and is a measure of a material's ability to store or dissipate mechanical energy. Some example values are: Tungsten 0.0001-0.0007, cast iron 0.007-0.03, and rubber 0.2- 4.0.[5]

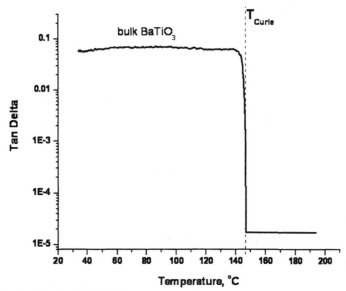

Figure 1. DMA results for bulk BaTiO$_3$

The sharp drop in the curve represents the phase transition from the lower temperature piezoelectric tetragonal phase to the higher temperature paraelectric cubic phase. This point is also known as the Curie temperature (T_c). Below the T_c the tetragonal phase exhibits a relatively high tan δ value due to the dissipation of energy through domain reorientation as previously mentioned. Above T_c the cubic phase exhibits a low tan δ value typical of a structural ceramic.[5] This is especially important since it allows for the distinction between added damping specific to domain reorientation from the damping due to mechanisms inherent to particulate reinforced composites.

Since the piezoelectric tetragonal phase is necessary for high damping in BaTiO$_3$, DSC was used to verify its existence after sintering of the composite. Figure 2 is a DSC plot of heat flow as a function of temperature for BaTiO$_3$ powder and a 50v% BaTiO$_3$ PR-MMC. The peaks occurring at 131°C for the powder and 128°C for the composites are the endothermic energies associated with the first-order tetragonal-to-cubic phase change of BaTiO$_3$, at T_c.

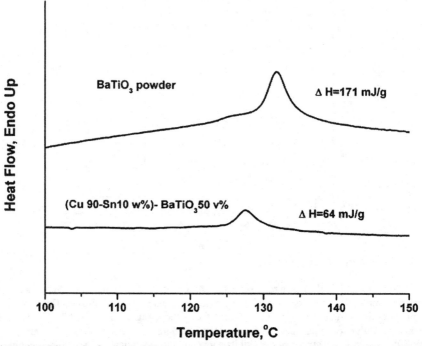

Figure 2. DSC results for the BaTiO₃ tetragonal-to-cubic phase transition in BaTiO₃ powder and a 50v% composite

The enthalpy of the tetragonal-to-cubic phase change for the powder was calculated to be 171mJ/g (40 J/mol). The enthalpy calculated for the tetragonal-to-cubic transition for the reinforcement in the composite is 64 mJ/g (15 J/mol), which is approximately equal to the product of the weight fraction of BaTiO₃ in the composite, 0.40, and the enthalpy for the BaTiO₃ powder. Thus, the chemical composition and crystal structure of the BaTiO₃ in the composite does not appear to be affected by the liquid phase sintering process used to densify the composite.

Figure 3 is a plot of tan δ as a function of temperature for PR-MMCs containing 30 and 50 v% BaTiO₃ as well as the matrix metal alone.

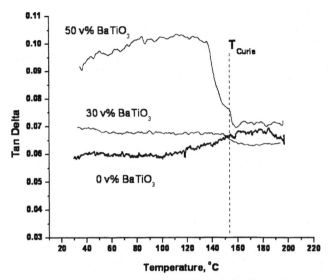

Figure 3. DMA results for Cu- 10w%Sn containing 0, 30, and 50 v% BaTiO₃

As was noted for the bulk BaTiO₃, a decrease in the tan δ is observed at the Curie temperature. The upward trend in tan δ exhibited by the matrix metal alone is a trend shown by most metals and alloys near room temperature.[6] In the composites below the Curie temperature, damping is due to three mechanisms: ferroelectric domain reorientation in the reinforcement, composite-damping due to interfacial matrix/reinforcement interactions, and inherent damping in the matrix. Above T_c however, only the later two mechanisms contribute.

Previous work has also shown success in this area.[7-10] In these studies as in this current work, MMCs were fabricated by the consolidation of mixtures of metal and ferroelectric ceramic powders. These consolidation methods consisted of pressing or rolling of the powders into a green body followed by high temperature sintering. Consolidation in this manner though, has some severe implications. It can result in unwanted porosity which tends to form around the reinforcing particles degrading the metal/ceramic interface. This reduces the amount of mechanical stress that can be transferred to the reinforcement particles which reduces their impact on the composite behavior. In severe cases, no stress might be transferred to the ceramic particles and they would simply behave as flaws in the matrix.

Currently new fabrication methods are being studied which could improve the overall quality of PR-MMCs as compared to work previously done in this area. These improved methods include secondary pressing and sintering (to decrease porosity) along with electroless copper plating of the ceramics (to improve the interface).

CONCLUSION

Vibration control is recognized as an essential means of ensuring stable and effective operation of structural materials. The use of multifunctional, self-damping, structural members could lead to significantly increased efficiency in many systems. In this work, such a material has been realized in the form of a metal matrix composite. This MMC was created by embedding dispersed, isolated ferroelectric particles into a matrix of bearing bronze. The resulting PR-MMC has become a novel, multifunctional, high damping material that with the use of improved fabrication methods could benefit many structural and mechanical systems in use today.

ACKNOWLEDGEMENT

This work was sponsored by the Army Research Office (ARO) under contract no. DAAD19-01-1-0714; Dr. William Mullins, project manager.

REFERENCES

[1]B. Cheng, M. Gabbay, M. Maglione, and G. Fantozzi, "Relaxation Motion and Possible Memory of Domain Structures in Barium Titanate Ceramics Studied by Mechanical and Dielectric Losses" *Journal of Electroceramics,* **10**, 5-18 (2003).

[2]L. Christodoulou, and J. Venables, "Multifunctional Material Systems: The First Generation" *JOM,* **55**, 39-45 (2003).

[3]J. Forrester, E. Kisi, and A. Studer, "Direct Observation of Ferroelastic Domain Switching in Polycrystalline BaTiO$_3$ Using in situ Neutron Diffraction" *Journal of the European Ceramic Society,* **article in press,** (2004).

[4]J. Calderon-Moreno, and M. Popa, "Stress Dependence of Reversible and Irreversible Domain Switching in PZT During Cyclic Loading" *Materials Science and Engineering A,* **A336,** 124-128 (2002).

[5]M. Ashby. *Materials Selection in Mechanical Design,* 2nd ed., Pergamon Press Ltd, (1999).

[6]B. Lazan. *Damping of Materials and Members in Structural Mechanics,* First ed., Pergamon Press Ltd.: Oxford, (1968).

[7]A. C. Goff, A. O. Aning, and S. L. Kampe, *TMS Letters,* **1,** 59-60 (2004).

[8]S. L. Kampe, A. O. Aning, J. P. Schultz, T. A. Asare, and B. D. Poquette. In *ICCE,* 2004.

[9]S. L. Kampe, A. O. Aning, J. P. Schultz, A. C. Goff, and J. S. Franklin. U. S. Patent Pending, 2004.

[10]I. Yoshida, M. Yokosuka, D. Monma, T. Ono, and M. Sakurai, "Damping Properties of Metal-Piezoelectric Composites" *Journal of Alloys and Compounds,* **355,** 136-141 (2003).

Carbon/Carbon and Ceramic Composite Materials in Friction

PREPARATION OF LARGE-SCALE CARBON FIBER REINFORCED CARBON MATRIX COMPOSITES (C-C) BY THERMAL GRADIENT CHEMICAL VAPOR INFILTRATION (TGCVI)

Jinyong Lee
Agency for Defense Development
Yuseong P.O. Box 35-5
Daejeon, 305-600
Korea

Jong Hyun Park
DACC Co. Ltd.
24-4 Seongju-dong, Changwon
Kyungnam, 641-120
Korea

ABSTRACT

In this study, C-C composites brake disks with density of 1.79~1.81 g/cm^3 are obtained in single TG-CVI processing. Density including porosity, microstructure, thermal, mechanical properties are tested, and evaluated that the C-C composites have a uniform properties in the disk radial and thickness directions. Friction-wear tests were carried out and compared with the existing carbon brake C-C materials. Dynamic torque tests reveal that C-C brakes by TG-CVI technique have satisfying braking-wear performance for application.

INTRODUCTION

Carbon fiber reinforced carbon matrix composites (C-C) have been used for the past 35 years as aircraft brakes due to their low density, high wear resistance performance and ability to withstand the high temperatures associated with aggressive braking conditions. In the C-C brake manufacturing industry, two main processing routes, so called liquid polymer and CVI (chemical vapor infiltration) processing, have been commonly applied. In the liquid polymer processing, high charring polymers are being used for the green body manufacturing and carbon matrix precursor, and additional densification is achieved by CVI until required density is obtained. So the matrix in C-C composites is consisted with resin char and pyrolytic carbon. In the liquid process, polymers undergo high shrinkage during pyrolysis, which results in severe carbon fiber damages. Also, the resin char matrix has a relatively low graphitizability even heat treated at more than 2000℃. These allow polymer-based C-C lower thermal and wear properties comparing with CVI C-C. So, C-C brake disk market is consciously shifting toward composites formed with a matrix derived entirely from CVI processing.[1]

In the CVI processing, specially prepared dry carbon fiber preforms, usually by needle-punching technique, are densified by the pyrolytic carbon. In this process, steady and slow build-up of carbon matrix around carbon fibers renders resulting C-C as high-performance composites with superior thermo-mechanical and friction-wear properties. Also, CVI processing offers easy control of matrix microstructure. Therefore, CVI is the most widely used commercial process nowadays. There are several types of CVI, such as isothermal, thermal gradient, pressure gradient and pulse CVI. Among these techniques, isothermal with low pressure is still the most prevalent process for mass production. Its main advantage is good parameter control, in particular for large furnaces where a large number of complex preforms can be densified simultaneously. The drawback is that carbonaceous materials are preferentially deposited on and near the surface region of the preforms. Surface bottleneck pores are gradually closed off prior to densification of inner part of the preform as the process continues. Consequently, open porosity of preform decreases with processing time. So this process would require 600~2000 hours to achieve the desired density with 2~3 intermittent surface grindings to open up porosity under the surface carbon crust.[2]

The thermal gradient CVI process allows considerably higher densification rate comparing with isothermal CVI technique.[3,4,5] However, TG-CVI has been considered that it is not suitable to production scale-up. This is may be true in case of using preforms with irregular shapes. But regular and symmetric preforms may give different results.

In partnership with DACC Co. Ltd. Agency for Defense Development has been developed C-C brake disks using TG-CVI technique. This paper describes C-C manufacturing process, properties, and the dynamic torque test results of C-C brake disks prepared by TG-CVI.

C-C MANUFACTURING
Carbon Fiber Preforms

Carbon fiber 3-D preforms were prepared by the needle-punching technique using oxidized PAN (OXI-PAN) tows. The OXI-PAN preforms were prepared by needle-punching and then carbonized in an inert atmosphere at more than 1600℃. Typical properties of the preforms were: 1.74 g/cm³ fiber density, 28% fiber volume fraction, and 0.51 g/cm³ bulk density. Preform dimensions were 350 mm (OD), 120 mm (ID), and 35 mm (thickness).

TG-CVI Densification

Figure 1 shows the schematic representation of TG-CVI used in this project. An electrical resistance furnace was used. In order to obtain thermal-gradient along the radial direction of the carbon fiber preform, a graphite-heating element was located at the center of the preforms. So, the carbon matrix infiltration into the preforms proceeded with radial direction from the center of the preform to the outer area of the preform. Hydrocarbon gases with hydrogen carrier gas, such as methane and propane were supplied into the TG-CVI furnace. The pressure of the densification furnace was maintained at slightly higher than 1.01 kPa. The processing temperature and densification rate controlled by the moving-out rate of a thermocouple from the heating element surface were varied with from 900℃ to 1000℃ and 0.2 to 1.0 mm/hour in order to find out the optimum densification parameters. In order to obtain the desired density of C-C composites, a single TG-CVI densification taking 1~3 weeks was carried out dependent on the processing rates.

TEST AND EVALUATION
Density Measurement

Bulk density and porosity of C-C composites were measured in accordance with ASTM C 373-88. Figure 2 shows the results of C-C composites prepared at 1000℃ and processing speed at 0.75 mm/hour. It is found that C-C composites have very uniform density and porosity distribution along the disk radial direction, which are around 1.8 g/cm³ and 6.6~8.1%, respectively. The obtained density seems to be slightly higher than that of typical carbon brake C-C composites, which are typically around 1.75 g/cm³.[4,6]. Also, C-C composites are seen to have very uniform density distribution along the radial direction. Density in the disk thickness direction, which is not shown in this paper, was measured, and found to have a uniform distribution. From the density point of view TG-CVI technique used in this project seems to be very effective for C-C brake manufacturing.

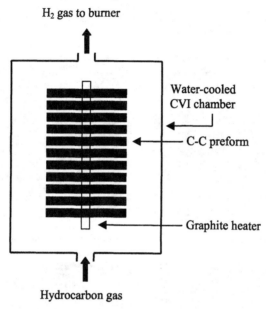

Figure 1. Simplified schematic representation of TG-CVI technique used.

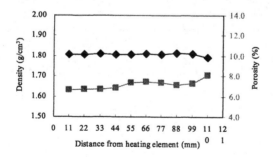

Figure 2. Bulk density and porosity distribution of C-C composites manufactured at 1000 ℃ and 0.75mm/hour processing rate.

123

Microstructural Observations

SEM microstructures of C-C composites along the C-C disk radial direction (from the heating element surface) were evaluated, and the results are shown in Figure 3. From the SEM observation, there seems to be little difference in microstructure from the inner portion (inside of 50mm from the heating element surface) to the outer portion (area from 65mm to 115mm). It is seen that large pores are still existed after TG-CVI between the carbon fiber yarns. On the other hand only small porosity is observed inside of the carbon fiber tows. An area with high density of Z-direction carbon fibers by the needle-punching showed improved microstructure by reducing delamination cracks between the tows.

20mm   80mm

50mm   110mm

Figure 3. Typical SEM Microstructures of C-C composites manufactured at 1000 ℃ and 0.75mm/hour processing rate.

Thermal Conductivity

Thermal conductivity was measured with a homemade testing machine at room temperature. In order to evaluate the uniformity of thermal conductivity, testing samples were taken from the inner and the outer portions of C-C disk in the thickness direction, as described in the microstructural observation section. Also, the specimens were obtained from the disk surface and middle portions in the radial direction. Table 1 summarizes the thermal conductivity. It can be seen that the thermal conductivity of the thickness direction of the inner portion (AT specimen) is slightly higher than that of the outer portion (BT specimen). Samples taken from the disk surface (ARS, BRS) showed marginally higher thermal conductivity than those of the middle of the disk.

Table 1. Thermal conductivity (TC) of C-C composites manufactured at 1000°C and 0.75mm/hour processing rate.

	Specimen positions		Thermal conductivity (W/mK)
Thickness direction	Inner part (AT)		14.19 (±0.22)
	Outer part (BT)		12.63 (±0.09)
Radial direction	Inner part (AR)	Middle (ARM)	28.44 (±0.66)
		Surface (ARS)	29.59 (±0.35)
	Outer part (BR)	Middle (BRM)	22.21 (±0.85)
		Surface (BRS)	22.40 (±0.40)

Mechanical Properties

For the mechanical property observation, compressive strength, flexural strength, and fracture toughness (K_{IC}) were determined. The mechanical tests were conducted according to ASTM C 695-91 for compressive, ASTM D 790-03 for flexural, and ASTM C 1421-01b for fracture toughness. It is seen that flexural and compressive strengths of the inner area are marginally higher than those of outer portion of C-C brake disks. And these values are determined to have similar strength level when compared with C-C composites brakes produced current carbon brake company.[7] In the case of fracture toughness, the values seem to be slightly lower than those of commercial C-C composites brake disks.[6]

Table 2. Mechanical properties of C-C composites brake disk prepared at 1000°C and 0.75mm/hour processing rate.

Properties	Specimen positions	
	A (inner)	B (outer)
Flexural strength (MPa)-Radial direction	157 (±5.8)	154 (±1.9)
Compressive strength (MPa)-Thickness direction	186.3 (±7.7)	180.0 (±2.1)
Fracture toughness (MPa m$^{1/2}$)-Radial direction	7.0 (±0.46)	7.2 (±0.32)

Friction and Wear Properties

To evaluate the friction and wear performances, C-C specimens with dimensions of OD 78mm x ID 53 mm x t15 mm were tested with a sub-scale dynamometer. 100 stops of friction and wear tests were conducted at testing kinetic energy 108 kJ, and the properties were evaluated at after every 25 stops. Table 3 is friction and wear properties of C-C composites manufactured in this project. And these values are evaluated to have similar level of friction coefficient and wear rate compared with existing C-C composites brake disks produced current carbon brake company.

Table 3. Friction coefficient and wear rate of C-C composites prepared at 1000 ℃ and 0.75mm/hour processing rate.

Properties	25	50	75	100	Average
Friction coefficient	0.24	0.25	0.25	0.24	0.25
Wear rate (μm/stop)	0.63	0.25	0.30	0.33	0.38

CONCLUSION

In order to reduce the manufacturing time, TG-CVI technique for the densification of C-C composites was applied. Properties including density, microstructure, thermal, mechanical, and friction-wear of C-C were evaluated. The results indicate that TG-CVI technique can produce C-C brake disks with diameter 350 mm within 1~3 weeks. And their properties are comparable with those of commercial C-C composites brake disks.

REFERENCES

[1]Christopher Byrne, "Modern carbon composite brake materials,"*J. Comp. Mater.*, **38**, 1837-1850 (2004).
[2]I. Goleki, "Industrial carbon chemical vapor infiltration (CVI) processes,"P. Delhaes (ed) in *World of Carbon*, Vol. 3, 112-138 (2003).
[3]Zhong-hua, Dian-ning Qu, Jie Xiong, Zhi-qiang Zou, "Effects of infiltration conditions on the densification behavior of carbon/carbon composites prepared by a directional-flow thermal gradient CVI process," *Carbon*, **41**, 2703-10 (2003).
[4]Ruiying Luo, "Friction performance of C/C composites prepared using rapid directional diffused chemical vapor infiltration processes," *Carbon*, **40**, 1279-85 (2002).
[5]Kent J. Pbobst, Theodore M. Besmann, David P. Stinton, Richard A. Lowden, Timothy J. Anderson, Thomas L. Starr, "Recent advances in forced-flow, thermal-gradient CVI for refractory composites," *Surface and Coating Technol*, **120-121**, 250-58 (1999).
[6]Sundar Vaidyaraman, Mark Purdy, Terence Walker, Susan Horst,"C/SiC Material Evaluation for Aircraft Brake Applications," *Proceedings of High Temperature Ceramic Matrix Composites,* 802-808 (2001).
[7]Jinyong Lee, "Final Report on the Development Test of KF-16 Block-52 Brake Disks," Agency for Defense Development, Korea (2000).

FRICTIONAL PERFORMANCE AND LOCAL PROPERTIES OF C/C COMPOSITES

Soydan Ozcan, Milan Krkoska, Peter Filip
Center for Advanced Friction Studies
Southern Illinois University at Carbondale
Carbondale, IL, 62901-4343, USA

ABSTRACT

The microstructure and mechanical properties of two-dimensional (2D) C/C, C/C-SiC composites with randomly chopped carbon fibers, and three dimensional (3D) non-woven C/C composites subjected to different friction experiments were investigated. The microstructure was characterized using polarized light microscopy (PLM), scanning electron microscopy (SEM), and transmission electron microscopy (TEM). Different texture degrees of pyrolitic carbon matrices were determined utilizing the polarized light measurement technique and measuring the orientation (opening) angle (OA) of the carbon (002) electron diffraction arcs. The local properties (hardness, elastic modulus, and stiffness) of individual structural components of C/C composites were characterized using nanoindentation. Nanoindentation results showed that increasing the degree of texture of pyrocarbon leads to a decrease in elastic modulus and hardness. The level and stability of the coefficient of friction as detected in subscale aircraft dynamometer simulations is influenced by both the characteristics of the bulk of C/C composite. It is possible to optimize the wear rate as well as the frictional performance of C/C composites by controlling the microstructure and mechanical properties of individual components of C/C composites.

INTRODUCTION

Carbon/Carbon composites exhibit an attractive combination of mechanical, thermal and electrical properties and are being successfully used in industries ranging from aerospace to sporting goods[1,2]. One of the major application areas includes friction materials such as high performance clutches and brakes. The characterization and prediction of the properties of C/C composites fabricated by chemical vapor infiltration and establishing the structure/property relationships are important for effective industrial applications[3]. Structure and property relationships, such as mechanical, wear and friction properties, are of interest in the field of carbon/carbon composite clutches and aircraft brakes.

Nanoindentation is a powerful technique of characterizing the near surface mechanical properties of the materials[4-6]. Most recent research on carbon materials using nanoindentation has been focused on amorphous carbon films[7-9], however, relatively few studies addressing pyrocarbons or fiber in brake composites have been published[10]. The goal of this paper is to determine the relationship between microstructural characteristics, corresponding properties of individual elements of C/C composites and frictional performance.

EXPERIMENTAL

Table 1 lists the information on the three composites which were involved in this research and Fig. 1 represents a schematic view of the C/C disc used in friction tests.

Table 1.Lists characteristic features of on three materials involved in this research

Sample	Fiber type	Matrix	Fiber orientation
C/C-2D	randomly chopped pitch	Charred Resin- CVI	2D random
C/C-SiC-2D	randomly chopped pitch	Charred Resin-SiC- CVI	2D random
C/C-3D	non-woven PAN	CVI	3D

The "2D randomly chopped" disc samples have the fiber preferentially oriented in x-y plane, while the "3D non-woven" materials contained also carbon fiber oriented perpendicular to the friction surfaces of disc (in the z direction). The nominal outer and inner diameters of the disc specimens were 92.25 mm (3.75") and 69.85mm (2.75"), respectively.

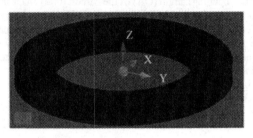

Fig. 1. A Schematic view of C/C disc. Tangential (X), Radial (Y) and Normal (Z) directions are shown.

A Nikon Microphot-FX polarized light microscope (PLM) was used for the characterization of the microstructure and measurement of the extinction angles of the different textures of pyrocarbons. Three specimens corresponding to different directions (X, Y, Z) were prepared from each of the three composites. They were mounted in epoxy resin ground and polished. Different diamond polishing slurries with the grain size from 6μm down to 0.25μm were used, subsequently. The size of the isochromatic regions was determined and the study followed with the evaluation of the anisotropy degree of each carbon type by measuring the extinction angle (Ae)[11].

Thin foils for TEM were prepared from 300 μm thick cuts (Linear precision diamond saw, Isomet 4000). Discs with 3mm diameter were core-drilled from these cuts (Core Drill, VCR Group, Model V 7110). The discs were dimpled down to a thickness of about 5μ (Dimpler, D 500i, VCR Group). The ion mill (Gatan, 691 Precision Ion Polishing System) was employed for the final thinning and polishing stage with beam at 4° angle. The TEM studies were carried out using a 200kV JEOL 2010F transmission electron microscope.

The indentation experiments were conducted at room temperature using a Nano Indenter® XP system (MTS Nanoinstruments, Knoxville, TN) with Berkovich type diamond indenter. Samples were prepared in identical procedures as described for PLM studies. Before beginning the nanoindentation tests, the system was calibrated using the fused silica. The continuous stiffness mode (CSM) was used for the tests. The method of Oliver and Pharr was employed in evaluation of nano-hardness and modulus of elasticity[4]. The reduced modulus, E_r, is defined as[4]:

$$E_r = \frac{\sqrt{\pi}}{2} \frac{S}{\sqrt{A}} \qquad (1)$$

where S is the stiffness and A is the projected contact area. Poisson's ratio that used for the calculations is 2.2. The hardness was calculated as a ratio of the maximum load P_{max} observed during the penetration of to 300nm and the projected area A:

$$H = \frac{P_{max}}{A} \qquad (2)$$

The "disc-on-disc" configuration and Link Engineering subscale aircraft dynamometer were chosen for frictional tests which simulated different aircraft landing conditions. The schematic picture of the subscale dynamometer is illustrated in Fig. 2. The subscale aircraft dynamometer was connected with PC and particular stops (landing or taxi) were monitored. The coefficient of friction, temperature, torque, decelerating speed and normal force are measured in real time.

The dynamometer was operated in constant torque mode. The energy levels simulating 12.5,25,50 and 100% normal landing energy conditions (NLE) were selected[12]. To understand how the moisture affects the frictional performance of tested materials, three environments were used: 2, 50, and 90% relative humidity (RH). An advanced test-chamber humidity control system was installed to maintain the relative humidity within 1% of the desired value. Three taxis and one landing stop sequences were used to simulate operations of brake. The sequences were repeated 50 times for each energy condition. The initial temperature of each test stop was set up on $T_0 = 49\,°C\ (120\,°F)$ and the applied normal force ramp rate was set up on value $dF/dt = 1780$ $N/s\ (400\ lb/s)$.

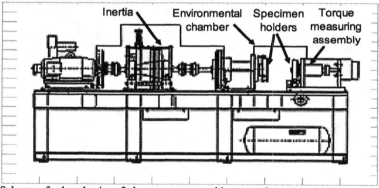

Fig. 2. Scheme of subscale aircraft dynamometer used in research

RESULTS and DISCUSSION

Figure 3 shows the PLM images of polished surface of investigated samples. 2D composites exhibited similar microstructure of pitch fiber and charred resin matrix. The final densification was achieved by applying all sequences. In contrast, the 3D composite is typified by the presence of the ex-PAN carbon fiber and dry CVI matrix.

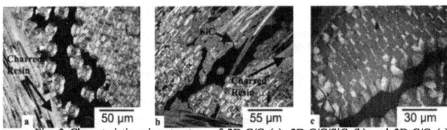

Fig. 3 Characteristic microstructure of 2D-C/C (a), 2D-C/C/SiC (b) and 3D-C/C (c) composites.

Based on measurement of extinction angles (A_e), the optical texture of CVI carbon in 3D-C/C is rough laminar (A_e=21°) and CVI carbon of 2D-C/C is smooth laminar (A_e=16°). The microstructure of CVI carbon in 2D-C/C was found slightly better organized than that of CVI carbon in 2D-C/C/SiC with a 17° extinction angle.

High magnification TEM bright field images and inserted SAED patterns of CVI carbon from different studied composites are given in Figs. 4a-b-c. The inserted SAED patterns clearly illustrate the variation of CVI carbons. The corresponding orientation angles obtained from SAED patterns of 2D-C/C and 2D-C/C/SiC indicate that they are both smooth laminar with an OA=70° and OA=80°, respectively (Fig 4a and 4b). However, 3D composite has a rough laminar CVI carbon with an OA=50° (Fig 4c).

Fig. 4. High magnification TEM image and inserted SAED patterns of 2D-C/C (a), 2D-C/C/SiC (b) and 3D-C/C (c) CVI carbons.

The nanostructure of fibers in each composite was also investigated using TEM and measured OA from SAED pattern for pitch fibers in 2D-C/C and 2D-C/C/SiC composites are 30° and 40°, respectively. The PAN fiber in 3D-C/C is less ordered (OA=45°) than pitch fibers as could be expected.

Local Mechanical Properties

Each indentation was performed in the semiparallel direction with the carbon fibers longitudinal axis. Characteristic SEM images of the indents and the loading/unloading plots for fiber and matrix, respectively are given in Figs. 5a to d. The presence of indents indicates that plastic deformation of C-fibers and CVI matrix occurred during the indentation. Hardness (H) and elastic modulus (E_r) for both samples were calculated from the linear part of the unloading curve as shown in Fig 5b and d. The average value of H and E_r and standard deviation were calculated for 7 indentations. The results for fiber and matrix for all three samples are given in Table 2.

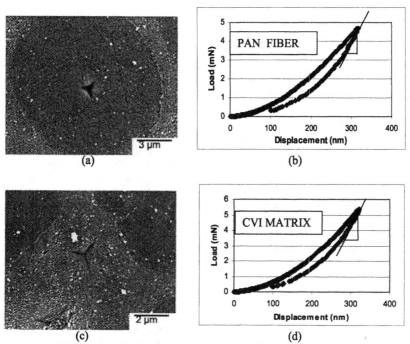

(a)   (b)

(c)   (d)

Fig. 5. a) SEM images of 300nm depth indent and (b) load versus displacement graphs for PAN fiber (a) and SEM (c) and load vs. displacement (d) as detected for CVI carbon.

Table 2. Mechanical properties of composite components determined from nanoindentation.

	2D-C/C			2D-C/C/SiC			3D-C/C	
	Pitch Fiber	CVI-carbon	Charred-resin	Pitch Fiber	CVI-carbon	Charred-resin	PAN fiber	CVI-carbon
Hardness,H(GPa)	1.90	2.24	3.32	2.49	2.83	3.52	5.78	1.36
Modulus,Er(GPa)	21.45	14.55	19.38	24.74	22.03	21.65	38.70	14.38

131

Measured anisotropy degree reveals that the CVI-carbon of the 3D-C/C is better organized (rough laminar) than that of the 2D-C/C (smooth laminar). Table 2 shows that the hardness of CVI carbon in the 3D-C/C (1.36 GPa) is lower than the hardness of CVI carbon in 2D-C/C (2.24 GPa) and 2D-C/C/SiC (2.83 GPa), respectively. However the moduli of elasticity of the CVI carbons are comparable for the 2D-C/C and 3D-C/C. The 2D-C/C/SiC exhibited higher modulus. This leads to the assumption that the individual composites were exposed to different heat treatment temperatures (HTT). The 2D-C/C/SiC composites were heat treated at slightly higher temperatures while the 2D-C/C and 3D-C/C composites are heat treated at almost the same temperature. At temperature below 2000°C, higher modulus for PAN fibers can be achievable compared to the pitch fibers[13]. Based on the data shown in table 2, the hardness of PAN fibers in the 3D-C/C is considerably higher than pitch fibers in the 2D composites. This is possible with a final HTT lower than 2000°C.

The Coefficient of Friction

For the engineering evaluation of the coefficient of friction (COF), the individual stops were averaged and plotted against the normal landing energy (NLE). The plots are shown in Figs. 6, 7, 8. Each landing shown in these figures represents the average of 50 landing stops and each taxi represents an average of 150 stops. Although the taxi stops are plotted for different energy levels, all taxiing was performed at 4.5% NLE. The purpose, why they are plotted against different NLE, is to show the previously and subsequently applied landing energy.

The average "landing" coefficient of friction (α(L)) of the 2D-C/C material showed significant sensitivity to the energy level and almost no sensitivity to the humidity level applied during braking (see Fig. 6). The average values of the coefficient of friction μ(L) detected at lower energy levels (12.5, 25, 50 % NLE) are very similar at all humidities. At these low normal landing energy conditions, the average COF was never lower than $\mu \approx 0.45$. At the higher energy simulations (100% NLE), the average coefficient of friction oscillates around the average value of $\mu \approx 0.3$. The peaking of COF was detected at 25% NLE. If the temperature on the friction surface is insufficient to evaporate the adsorbed moisture (all taxi stops), the sensitivity of the 2D-C/C material to moisture is clearly visible. As the level of the humidity in environment increases, the average coefficient of friction decreases. Only at "dry" conditions (2% RH), the average coefficient of friction is comparable to the landing situation, where the temperature on the rubbing surfaces is significantly higher.

The 2D-C/C/SiC material shows a similar sensitivity to the applied normal landing energy as the 2D-C/C material, but the average coefficient of friction is always slightly higher when compared to the 2D-C/C material (Fig.7). This can be ascribed to the presence of SiC in the structure. At the lowest normal landing energy (12.5 % NLE), the average "landing" COF μ(L) is higher than $\mu \approx 0.45$. Once again, the peaking of the average "landing" COF μ(L) was detected at 25% NLE ($\mu \approx 0.65$) and the COF remains higher than 0.5 also for 50% NLE at all humidity levels applied. The humidity impact during all landing conditions was low, but stronger when compared to the 2D-C/C material without SiC in the microstructure. Filip et al.[13] reported lower sensitivity of COF to moisture, which is in contrast with these results. Apparently, a critical amount of SiC is required in order to minimize the environmental impact on frictional performance. The average "taxi" COF μ(T) of the 2D-C/C/SiC samples shows a similar behavior as described for the 2D-C/C material.

The average coefficient of friction of the 3D-C/C material did not exhibit remarkable sensitivity to the applied testing energy level (Fig. 8). Nevertheless, the average values μ(L) that were detected at lower energy conditions (12.5 and 25% NLE) are slightly lower when compared to the μ(L) data detected at 50 or 100% NLE. As the humidity increases, the average coefficient of friction decreases for 3D composite. During taxing, the trend is similar as shown for the 2D-C/C and the 2D-C/C/SiC materials. Increasing the humidity causes decreasing of the average coefficient of friction μ(T).

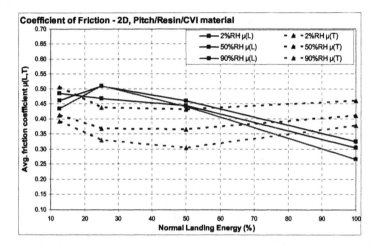

Fig. 6. The average coefficient of friction for taxi μ(T) and landing μ(L) stops performed at all humidity levels, 2D-C/C material.

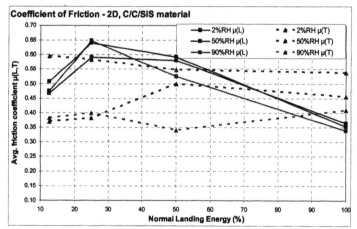

Fig. 7. The average coefficient of friction for taxi μ(T) and landing μ(L) stops performed at all humidity levels, 2D-C/C/SiCmaterial.

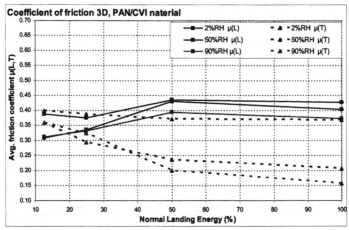

Fig. 8. The average coefficient of friction for taxi μ(T) and landing μ(L) stops performed at all humidity levels, 3D-C/C material.

Wear Properties

After each sequence 50 landing and 150 taxi stops, the discs were weighed and the wear was calculated as mass loss. The wear of the 2D-C/C, 2D-C/C/SiC and 3D-C/C samples, tested at different energy and humidity condition is shown in Figs. 9, 10, 11, respectively.

Wear of the 2D-C/C material is sensitive to the applied humidity level. As the humidity increases, the wear increases proportionally at all normal landing energy simulations (Fig. 9). Wear of 2D-C/C material showed only a slight sensitivity to the different applied energy levels. In general, the wear is slightly lower at low NLE and higher for 100% NLE.

Wear of the 2D-C/C/SiC material (Fig. 10) showed similar tendencies as detected in the 2D-C/C material. As the humidity in environment increases, the mass loss of the 2D-C/C/SiC samples increases too. Similarly, as seen for the 2D-C/C material, wear increases with increasing applied NLE, but the absolute values of wear are as high for the 2D-C/C/SiC material compared to wear measured for the 2D-C/C samples. When hard, abrasive SiC particles were added to the C/C composite, the wear rate of the C/C/SiC composites was twice as high as in the C/C composite at higher NLE and RH conditions.

Wear of the 3D-C/C material (Fig. 11) showed different development when compared to 2D samples. While the 2D materials demonstrated increased wear as the applied energy increased, the 3D-C/C material exhibited an opposite tendency. A scatter of data was particularly observed at very low energy levels (12.5 and 25% NLE). It is easily visible that moisture has the highest impact on wear at the extreme testing conditions. For 90% RH, the wear is highest at 12.5% NLE, and decreases with increasing landing energy. The lowest wear was detected for 100% NLE. At the "dry" conditions (2% RH), the wear was relatively stable.

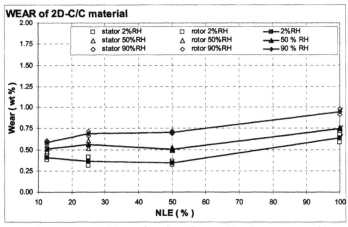

Fig. 9. The wear of 2D-C/C material as a function of normal landing energy and humidity.

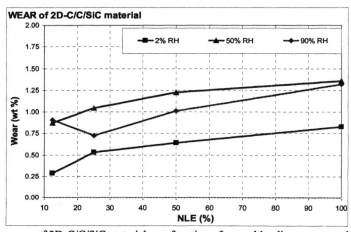

Fig. 10. The wear of 2D-C/C/SiC material as a function of normal landing energy and humidity.

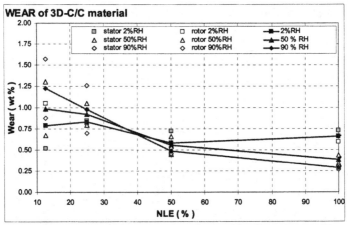

Fig. 11. The wear of 3D-C/C material as a function of normal landing energy and humidity.

The frictional performance can be related to microstructure and mechanical properties (hardness) of investigated materials. At low landing energies, the smooth laminar CVI carbon in 2D composites exhibited more than 100% larger hardness values compared to the rough laminar CVI carbon in 3D-C/C composites. Although the hardness of PAN fiber in 3D composite is higher, the lower fiber content, compared to matrix. The harder 2D composite material wears less when compared to the 3D samples. However, at 100 %NLE simulations, the tougher 3D samples exhibited significantly lower wear rate. It is often demonstrated that the wear is inversely proportional to the hardness. Apparently, the increased level of fracture toughness in softer 3D-C/C is a relevant factor when a more dynamic type of loading with an intense mechanical effect is applied at high speed and energy friction process.

In contrary to high landing energy conditions, 2D composites with a smooth laminar CVI carbon display a better frictional performance than softer 3D-C/C composite at low landing energy conditions. Taxi conditions performed after high energy landing, 2D composites display higher frictional performance compared to the 3D-C/C composite which display low frictional performance. Thermal properties of the bulk of the brake materials should be taking into account. The orientation of the fibers in the composite play an important role on the thermal diffusivity. Fibers generate a direct thermal pathway and provide as a pattern for the directional deposition of CVI carbon. While 3D-C/C includes 64 vol % of PAN fibers 2D composites includes 40 vol% of pitch fibers. Thermal conductivity increases with an increased crystallinity and with an increased amount of fibers. So, greater thermal diffusivity is expected with increased fiber content where 3D sample has. The temperature achieved on the surface of 2D composites might always be higher than the 3D-C/C composite which may contribute the particular character of wear, fracturing and friction layer formation. This leads to the assumption that better organized CVI matrix and well bounded interface[3], known from microstructural characterization, contribute to achieve high thermal conductivity directly related to frictional performance.

CONCLUSIONS

The investigated 2D composites exhibited significant dependencies of the average coefficient of friction (COF) on applied energy levels. Generally, as applied normal landing energy increased, the average coefficient of friction decreased. The peaking of the COF at 25% NLE and all humidity levels was recorded for both, the C/C and C/C/SiC composites.

The 3D composite a demonstrated lower influence of applied normal landing energy on the average coefficient of friction. This phenomenon probably occurs because of the higher thermal conductivity of the 3D C/C composite in the direction normal to the friction surface.

The microstructure and local mechanical properties of the C/C composites play a major role on the wear and friction. It was shown that increased fracture toughness may eliminate the amount of wear, particularly at high energy simulations.

REFERENCES

[1]L.M. Manocha, A. Warrier, S. Manocha, D.D. Edie, A.A. Ogale, Microstructure of carbon/carbon composites reinforced with pitch-based ribbon-shape carbon fibers, *Carbon* **41**, 1425-36 (2003).

[2]B.Reznik, D. Gerthsen, Microscopic study of failure mechanisms in infiltrated carbon fiber felts, *Carbon* **41**, 57-69 (2003).

[3]S. Ozcan, P. Filip. Microstructure and Wear Mechanisms in C/C composites. In Press, *Wear* (2005)

[4]W.C. Oliver and G.M Pharr, An improved technique for determining hardness and elastic modulus using load and displacement sensing indentation experiments, *J. of Mat. Res.* **7**, No.6 1564-83 (1992).

[5]B. Wolf, A. Richter, V. Weihnacht, Differential and integral hardness—new aspects of quantifying load–depth-data in depth-sensing nanoindentation experiments, *Surface and Coating Technology* **183**, 141-150 (2004).

[6]Q. Wei, A.K. Sharma J. Sankar, J. Narayan, Mechanical properties of diamond-like carbon composite thin films prepared by pulsed laser deposition, *Compoite B* **30**, 675-684 (1999).

[7]S. Dub, Y. Pauleau and F. Thièry, Mechanical properties of nanostructured copper-hydrogenated amorphous carbon composite films studied by nanoindentation, *Surface and Coating Technology,* **180-181**, 551-55 (2004).

[8]S. Logothetidis, S. Kassavetis, C. Charitidis, Y. Panayiotatos and A. Laskarakis, Nanoindentation studies of multilayer amorphous carbon films, *Carbon* **42**, no.5-6, 1133-36 (2004)

[9]E. Martínez, J. L. Andújar, M. C. Polo, J. Esteve, J. Robertson and W. I. Milne, Study of the mechanical properties of tetrahedral amorphous carbon films by nanoindentation and nanowear measurements, *Diamond and Related Materials* **10**, no.2, 145-52, (2001)

[10]Greg Hofmann, Martin Wiedenmeier, Matt Freund, Al Beavan, Jennifer Hay and G. M. Pharr, An investigation of the relationship between position within coater and pyrolytic carbon characteristics using nanoindentation, *Carbon* **38**, 645-53 (2000)

[11]P. Dupel, X. Bourrat and R. Pailler, Structure of pyrocarbon infiltrated by pulse-CVI, *Carbon* **33**, no.9 1193-1204 (1995)

[12] T. Policandriotes, T. Walker, P. Filip, "Subscale Aircraft Brake Dynamometer Testing on C/SiC and C/B4C Ceramic Friction Materials", *28th Annual Conference on Ceramics, Metal & Carbon Carbon Composites, Materials and Structures*, Cocoa Beach, FL, January 26-30, 2004 (ZAI 1815 N. Forth Myer Drive, Arlington, VA 22209).

[13] J. Donnet, T. K. Wang, J.C.M. Peng, S. Rebouillat, *In Carbon Fibers,* Chapter 5-Mechanical Properties of Carbon Fibers, Marcel Dekker Inc. New York (1998).

[14] D.E. Wittmer and P. Filip, "Thermal Shock Impact on C/C and Si Melt Infiltrated C/C Materials", *29th International Conference on Ceramics and Composites, Heat Treatment, Microstructure and Properties of Ceramics and C/C Composites.* Cocoa Beach, FL, January 23-28, 2005

ACKNOWLEDGEMENT

The authors acknowledge the contributions of Adam Pawliczek for the performing the friction tests for 2D-C/C/SiC composites. This research was performed within core research program of the Center for Advanced Friction Studies (CAFS) and, sponsored by National Science Foundation (Grant EEC 9523372), State of Illinois and consortium of nine industrial partners (Aircraft Braking Systems, Boeing, Composite Innovations, Climax Molybdenum, Honeywell International, Hyper-Therm, Starfire Systems, Carbon NanoTechnologies, Tribco Incorporated). Microscopic characterization part of the research was carried out in the Center for Microanalysis of Materials, University of Illinois, which is partially supported by the U.S. Department of Energy under grant DEFG02-91-ER45439.

HUMIDITY AND FRICTIONAL PERFORMANCE OF C/C COMPOSITES

Milan Krkoska, Peter Filip

Southern Illinois University at Carbondale,
SIUC
1230 Lincoln Dr.
Carbondale, IL
62901-4343

ABSTRACT

This paper focuses on friction and wear of the carbon-carbon composite materials tested at different simulated landing energy and humidity conditions. The strong influence of humidity was visible at the low landing energy conditions. Temperature on the friction surface is insufficient for the evaporation of moisture, and the average coefficient of friction is low ($\mu \approx$ 0.15). At higher energy, the average μ increases to 0.4. However, the friction becomes unstable and depending on time. Similar frictional response was detected for both, 2D and 3D materials, however the effect is more pronounced in 3D composites. Wear of 2D material is sensitive to the level of the moisture and less sensitive to energy dissipated in friction process. As the humidity increases, wear of 2D material decreases proportionally. Wear data of 3D material demonstrated considerable sensitivity to humidity and to braking energy levels. At low energy braking simulations, wear of 3D composite is significantly higher compared to 2D samples. Opposite tendency was observed at 100% normal landing energy (NLE) simulations and wear is similar for 3D and 2D tested at 50% NLE. Microstructural studies showed that the friction layer was developed on the rubbing surface during all braking simulations. The proportion of the area covered by the friction layer is markedly influencing the frictional properties of C/C composite materials. TEM analysis revealed that at small energies, the friction layer was always amorphous. At highest energy simulations, a mixture of amorphous and highly crystalline carbons was detected in the friction layer. No direct correlation between crystal order of friction film and frictional performance was found.

I. INTRODUCTION.

The frictional performance of carbon-carbon composites may change dramatically when moisture and other condensable vapors are present[1,2,3]. The "morning sickness" of plane brakes is well known. The minimum amount of water sufficient for lubricating of carbon-carbon surfaces may be as low, as 100 ppm[2]. The presence of oxygen in the environment also influences frictional performance. The oxygen may act as a lubricant. Under dusting conditions, oxygen can reduce the coefficient of friction and wear of carbon materials. The covalent bonds on the edge-faces are highly reactive[2]. The carbon will easily react with available gases and produce the complex groups on the surface. Chemisorbtion of oxygen and adsorbtion of vapors on the surface leads to passivation of covalent bonds which allows maintaining a low friction. Controversially, at non-dusting conditions the presence of oxygen may increase the friction level of carbon materials. Marchon at al.[4] have shown that adsorbtion of water vapors occurs even at room temperature and chemisorbtion of O_2 was observed above 227°C. When braking takes place, the frictional surfaces sliding against each other exhibit significant damage[5-12]. As shown by Harker at al.[13], when the structural defects are introduced into the carbon, the chemical reactivity of carbon increased abruptly.

This paper summarizes the data obtained from the subscale aircraft dynamometer testing of commercially available 2D and 3D carbon/carbon brake materials. The goal was to compare the frictional performance and to contribute to better understanding of the relationship between microstructure and braking effectiveness.

II. EXPERIMENTAL.
Tested materials.

The carbon-carbon composite materials used in this research were kindly provided by Honeywell Aircraft Landing Systems. The first type is a 2D randomly chopped pitch fiber reinforced carbon with charred resin matrix which was densified by CVI process in the final stage. The second type is a 3D non-woven "PAN fiber" reinforced carbon with dry CVI matrix. The 2D material has fibers oriented parallel to the frictional surface while the 3D material also has fibers more or less oriented perpendicular to the frictional surface. The characteristic structure of the 2D and 3D materials is shown in Fig. 1.

a)   b)

Figure 1. Light microscopy images of a) 2D pitch/resin/CVI and b) 3D PAN/CVI materials.

The subscale-dynamometer testing.

In this research, the "disc-on-disc" configuration and Link Engineering subscale aircraft dynamometer were used for frictional tests simulating the several landing and taxi conditions. The schematic picture of the subscale aircraft dynamometer is illustrated in Fig.2. Applied dynamometer testing conditions represent three taxi stops followed by one landing simulation repeated 50 times (200 stops)[14]. The simulated landing was performed at 12.5, 25, 50, 100 and 200% of normal landing energy (NLE). The nominal outer and inner diameters of the ring specimens were 92.25 mm (3.75") and 69.85mm (2.75"), respectively. The kinetic energy of 128.8 kJ was taken as the 100% Normal Landing Energy. The relevant equations were taken in consideration for the calculation and setting of the testing conditions:

$$KE = \tfrac{1}{2}I\omega_0^2 \qquad (1)$$

$$\tau = I\alpha = I\frac{\omega_0}{t_s} \qquad (2)$$

where the KE stands for the Kinetic Energy, I stands for the rotational inertia of the rotating system, ω_0 is the initial rotational speed, α is the angular deceleration (which is assumed to be constant), and t_s is the stop time. Therefore, for a given torque τ, rotational inertia, initial rotational speed, and the stop time can be calculated using Equation (2). This method was used to determine the test conditions. The subscale aircraft dynamometer was operated in constant torque mode, the normal force varied in response to achieve the desired torque set point. The initial temperature of each test stop was set up on $T_0 = 49\,^{\circ}\mathrm{C}$ ($120\,^{\circ}F$) and the applied normal force ramp rate was set up on value $dF/dt = 1780\ N/s$ ($400\ lb/s$).

Three relative humidity environments were used in frictional tests: 2, 50, and 90% RH. An advanced test-chamber humidity control system maintained the relative humidity to within ±1% of the desired value. The system includes: Holmes ultrasonic humidifier with 8.7 liter capacity, JC Systems Model 600A relative humidity and temperature controller, General Eastern Model 1211H optical dew point sensor, General Eastern Hygro M4 Dew Point Monitor. The subscale aircraft dynamometer was connected with PC and particular stops (landing or taxi) were recorded. The coefficient of friction, temperature, torque, deceleration speed and normal force is measured in real time.

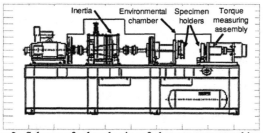

Figure 2. Scheme of subscale aircraft dynamometer used in research

Wear measurement.

After each sequence (50 landings + 150taxi stops), the discs were weighed and the wear was calculated as mass loss. The analytical balance Metler-Toledo AB135-S was used. The thickness of the discs was measured too, but due to extreme waviness of the discs, the measurements were not taken in consideration.

Profilometry measurements.

Roughness parameters of the rubbing surface were measured after each testing sequence, so the evolution of the surface roughness, depending on the energy and humidity, could be recorded. A computer-based, stationary surface measurement device produced by MAHR GmbH Company was used. The statistical parameters as arithmetical mean deviation Ra, root mean square deviation Rq, skewness Rsk or kurtosis Rku remain stable when taken at random locations on an isotropic surface. The worn surfaces are usually anisotropic, which means they have characteristic directionality or lay associated with them. Moreover, statistic techniques may be often out of approximations of the natural forms (inherently complex or irregular)[15,16]. This was the reason why the fractal theory was chosen as a tool for characterization of the robbed surfaces. The rough surface may be characterized by parameters D and G. The value of parameter D characterizes the amount of roughness, while the value of parameter G characterizes how wavy is the surface. The Hurst analysis method for the calculation of D and G parameters was used[15]. Technical data of the profilometry device and the measured area are summarized in Tab. 1.

Table 1. Configurations of profilometry equipment and the measured surface

	profilometer	MAHR GmbH, Ger
Equipment	Measur.range	500 mm
	Measur. Speed	0.5mm per second
	Tip radius	5 µm
	Tracelenght T	10.00 mm
	Width W	5.6 mm
Measured area	Point spacing of T	11200 pts
	Number of T	128
	Point spacing of W	44.09 µm

Macro and microstructure observations.

The scanner Agfa Arcus 1200 was used for scanning the surfaces of the discs (macrostructure), after each aircraft sub-scale dynamometer sequence test at specific energy and humidity condition. The Nikon Microphot-FX polarized light microscope was used for microstructure observation. Simultaneously, the samples for TEM analysis were taken from the friction layer. The Hitachi H-7100 transmission electron microscope (TEM) was used for microstructure observation.

III. RESULTS AND DISCUSSION.

Frictional performance.

The characteristic dependences of the coefficient of friction, temperature, speed and normal force during individual landing stops for 3D (PAN fiber, CVI matrix) material are shown in Fig. 3. At lowest energy (12.5% NLE) and at all humidity levels, the coefficient of friction is stable (without any transients). The average value of the coefficient of friction (Fig. 4) is slightly higher for "dry" (2% RH) humidity conditions when compared to "medium" (50% RH) or "wet" (90% RH) conditions. When the applied NLE increased to 25%, the changes of the coefficient of friction were registered. The stops are relatively stable at a 2% relative humidity conditions. With increasing of the moisture in environment, the coefficient of friction becomes unstable and transients (sudden increases) were recorded after a few seconds, while the discs are in full contact. At 50% NLE, the transient behavior is obvious and drastic changes in the coefficient of friction are accompanied by vibrations and noise. The transients at higher humidity are longer in duration when compared to dry conditions.

At 100% NLE, the coefficient of friction is unstable too, and two transients are detected. The first transient occurs at the beginning of the stop and the second transient was observed at the final stage of the test. The coefficient of friction is increasing rapidly from the value of $\mu \approx 0.2$ to the highest value of $\mu \approx 0.6$ in the first transient. The temperature measured near the surface increases dramatically at this stage. The maximum peak level of friction is followed by its continuous decrease to the value of $\mu \sim 0.35$. Subsequently, the friction increases again to the value of $\mu \approx 0.5$ (second transient). If the first peak of the coefficient of friction is wider for some reason, (roughness, real contact area, etc.), temperature measured near the friction surface exhibits yielding during the time when peak occurs. This may indicate an endothermic reaction associated with the chemical changes on the surface, most likely the desorbtion of physisorbed water. It is interesting that at the highest humidity level (90% RH), the peak is mostly spread out and the associated yielding of the temperature is larger, probably due to the highest value of the moisture adsorbed on the surface (see Fig. 3). Figure 5 summarizes the frictional characteristics of the 2D material.

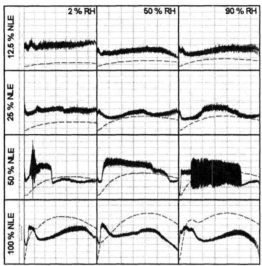

Figure 3. Characteristic selected individual stops for varying
energy and humidity levels, 3D PAN/CVI material

Typical examples are given for the different energy and humidity levels applied during testing. The most characteristic attributes of the 2D material is the number of transients when compared to the 3D samples and oscillation of μ, which starts at the lowest energy conditions (12.5 % NLE). While the coefficient of friction was relatively stable in the 3D material tests, the 2D material exhibits transient behavior at the lowest applied 12.5 % NLE conditions. Similar behavior with one transient of μ was detected for 25% NLE and at all humidity levels. At the very beginning of the 25% NLE stops, the coefficient of friction is unstable. After a short time (2-3 seconds), the coefficient of friction increases with the temperature up to the moment when the peaking (plato) occurs at high value of μ ≈ 0.65. At the very end of the landing stop, the coefficient is usually lower. This behavior corresponds to the "one transient" behavior. At 50% NLE, the transition occurs at the very beginning of the braking process. The coefficient of friction is relatively high and oscillates during almost the entire stop, particularly at dry conditions. Dependence of the coefficient of friction detected for the 100% NLE is similar to the dependences of the 3D material described earlier. The coefficient of friction is unstable from the very beginning of the stop. The surface temperature increases rapidly, but without the "yielding" observed for the 3D materials. There are also two transients detected in the 2D samples, but timing of the peaks is different. While in the 3D material, the first peak was detected at the very beginning of the stop and the second close to the end of stop, the peaks occurred immediately after brake engagement in the 2D material. During the peaking of μ, temperature increases rapidly and the coefficient of friction remains stable during the rest of the stop. As shown elsewhere[17], the transients of the coefficient of friction can be ascribed to the consecutive release of physisorbed and chemisorbed species.

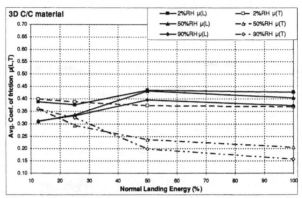

Figure 4. The average coefficient of friction (a)" taxi" and (b)" landing", calculated from repeated (150 taxi, 50 landing) stops performed at all humidity levels, 3D PAN/CVI material

The bulk temperature plays a crucial role in friction[3,18,19]. The thermal conductivity of the 2D material is low in the "z" direction (perpendicular to the frictional surface) due to the given orientation of the fibers and the less organized matrix. Figure 6 shows the measured temperature across the entire disc thickness. Obviously, the surface of the 2D material with significantly lower thermal conductivity in "z" direction will reach the higher temperature which facilitates the release of chemisorbed species. When compared to the 3D samples both, physisorbed and chemisorbed species are released faster from the 2D samples. Correspondingly, the peaking of μ is observed earlier in the 2D samples. The oxidation of the bulk material may take place at relatively low temperature[19]. The oxygen adsorbed from the air and possibly hydrogen released from the moisture can react with the carbon and form carbon dioxide and carbon monoxide. As shown in Fig.5, the coefficient of friction of the carbon samples at low moisture levels increases. Vice versa, when moisture, oxygen and hydrocarbons (product of reactions) are available as a consequence of the chemisorbed species, the coefficient of friction decreases. Simultaneously, the accompanying oxidation reactions increase wear (discussed later). Both investigated, the 2D and 3D samples are sensitive to moisture levels. When the 2D and the 3D materials are compared, it is obvious that the 2D material (lower thermal conductivity) exhibits transients at all energy levels, which contrasts with the 3D material, where transients occur at higher energy levels only (25% NLE and higher). However, comparatively larger differences in μ during one stop were detected in the 3D samples.

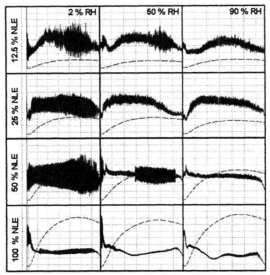

Figure 5. Characteristic selected individual stops for varying energy and Humidity levels, 2D pitch/resin/CVI material

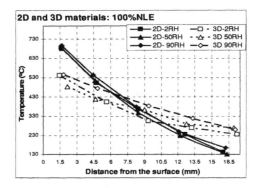

Figure 6. The temperature measured in different distance from the friction surface, 100% NLE all humidity, 2D and 3D materials

When the average values of the coefficient of friction detected for taxi and landing stops are averaged for one energy condition, the plots shown in Fig. 4 (3D material) and Fig. 7 (2D material) are obtained. These dependencies are typically used for casting the opinions on the performance of individual materials. It is necessary to note that all taxi stops were performed at 4.5% NLE, however, the plots in Figs. 4 and 7 shows the dependence on previously applied landing stops energy.

In the case of 3D PAN/CVI composite, the average coefficient of friction detected for repeated landing stops μ(L) did not show significant sensitivity to the energy level (Fig.4). Nevertheless, the average values μ(L) detected at lower energy conditions (12.5 and 25% NLE) are slightly lower compared to μ(L) data detected at 50 or 100% NLE. It is clear that one transient behavior (detected at 50% NLE) or two transients (observed during 100% NLE stops) do not have significant impact on these average values. During taxiing, the average coefficient of friction μ(T) remains stable only at the lowest humidity conditions (2% RH). At elevated humidity levels of 50 and 90 % RH, the average coefficient of friction μ(T) decreased with increased amount of energy spent during the simulated landing preceding the actual taxiing. The most dramatic change occurred when taxiing after 50% NLE stops.

In the case of 2D pitch/resin/CVI composite, the average coefficient of friction μ(L) detected at different landing conditions is very sensitive to the energy level applied during braking and is less sensitive to the applied humidity level. While μ(L) values are high and similar for 2, 50 and 90% RH (μ(L) ≥ 0.45), they drastically decrease when the percentage of NLE dissipated in friction process increases (100% NLE). The data detected for 12.5, 25 and 50 % NLE are very similar, ranging between 0.45 and 0.50, however a significant lower level of friction (μ(L) ~ 0.3) was detected at 100% NLE. The detected average μ(T) is not very sensitive to the level of energy applied during the preceding landing. The values of the average coefficient of friction are similar for taxis performed after landing at 12.5, 25, 50, 100 and 200 %NLE. The increased humidity level causes a drop in the average μ(T). It can be easily deduced from Fig. 7 that the testing energy is not sufficiently high to be able to drive off the moisture from the surface. Hence, the level of friction decreased with increased humidity level.

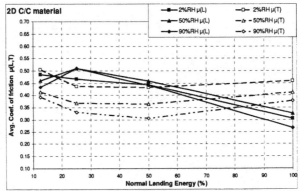

Figure 7. The average coefficient of friction taxi and landing, calculated from repeated (150 taxi, 50 landing) stops performed at all humidity levels, pitch/resin/CVI material

Wear.

The wear of the investigated materials was measured as the percentage of mass loss related to the original mass of the machined sample. The wear of the 2D and 3D samples tested at different humidity and energy levels is shown in Figs. 8 and 9, respectively. A scatter of data in 3D material (Fig. 8) was particularly observed at very low energy levels (12.5 and 25% NLE). It is easily visible that moisture has the highest impact on wear at extreme testing conditions. For

90% RH, the wear is highest at 12.5 % NLE, and decreases with increasing landing energy. The lowest wear was detected for 100 % NLE. The wear of the 2D sample is the lowest at low energy conditions and increases slightly with increased normal landing energy. At 12.5 and 25% NLE conditions, the wear levels detected in the 2D material represent only 50% of those detected for the 3D samples at identical conditions. Vice versa, at 100% NLE, the wear of the 2D material is doubled when compared to the 3D material sample. In all probability, the major factor is the different temperature (significantly higher for 2D compared to 3D materials) and corresponding heat related damage of the 2D sample. It is commonly believed that when the braking energy is low, the particular debris is generated during friction[3,7,8,20,21]. The small debris particles are very tough, they are moving between the stator and the rotor and abrasive wear is taking place. At higher energy conditions, these particles are transformed into a friction layer due to their exposure to higher temperature and shear stresses. Since the friction layer remains on the friction surface, the wear rate decreases. Moreover, it is often suggested that the friction layer becomes a barrier for the oxygen and the moisture transport[7]. However, if the energy is too high, the oxidation may take place, and the wear increases. The wear data detected for the 2D material samples (Fig.9), particularly those for low energies, do not support the above mentioned theory addressing the wear of the C/C composites.

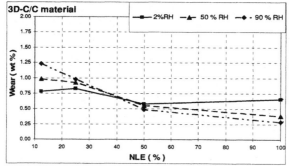

Figure 8.   Wear of investigated 3D PAN/CVI material as a function of normal landing energy and humidity.

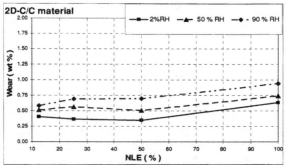

Figure 9. Wear of investigated 2D pitch/resin/CVI material as a function
of normal landing energy and humidity.

Hutton, Johnson and McEnamy[8] showed that the fiber orientation has a significant effect on the tribology of the C/C composites. According to their results, if the carbon-fibers are parallel to the friction surface, the wear is relatively low and higher at low energy than at high energy landing. The presented results are similar to those reported in [8]. According to[12], if a test is performed on the discs with fibers oriented perpendicular to the friction surface (z-direction), the wear rate at low energy level is almost doubled when compared to parallel-parallel configuration. This is in excellent accord with the data generated in this research. However, according to[12], at higher energy conditions, wear should be identical for parallel-parallel, parallel-perpendicular and perpendicular-perpendicular carbon-fiber configurations. This conclusion contrasts with our results. Apparently, the architecture of fiber has a significant impact on frictional performance, however it is not possible to generalize in the way authors[12] did. The microstructure of the fiber and matrix plays a significant role.

In our PAN/CVI material samples, the shear modulus of the graphitic matrix is very low[22], so shear deformation occurs easily and the friction layer is easily generated. Moreover, the transverse shear modulus of the PAN fiber is much lower when compared to the parallel shear modulus and this is a possible explanation to why the friction layer quickly forms on the friction surface of samples with parallel fibers. If the fibers emerge perpendicularly to the surface, more difficulties may occur in the formation of the frictional layer so the very aggressive abrasive wear is observed in the PAN/CVI material samples at low energies. When the energy is increasing, the frictional layer forms easily, no matter of fiber orientation. Correspondingly, the wear is decreasing.

Surface characteristics.

The study of friction surface is fundamental and inseparable part of the investigation of the frictional phenomena. There is practically no published paper addressing friction in carbon-carbon composites, which would not discus the effect of frictional debris or lubricating friction layer. In the present study, the friction layer was always observed. Figures 10a and 10b shows the structures of the 3D PAN/CVI material at low energy (taxi condition) and at 50% RH. The friction layer is clearly visible. On the other hand, at high energy, the friction layer was observed too. Figure 10b is showing the friction layer at 100% NLE at same humidity level.

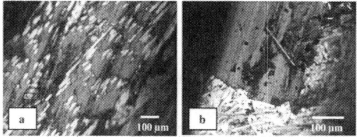

Figure 10. Light microscopy of 3D PAN/CVI material ant (a) taxi at 50% RH and (b)
100% NLE at 50% RH.

Figures 11 and 12 show the surface scans taken after the friction tests of the 2D pitch/resin/CVI and 3D PAN/CVI materials. Both materials have darker looking surfaces after the friction tests performed at low NLE when compared to surfaces performed at higher energies. This indicates that, at low energies, the abrasive wear dominates in spite of the fact that friction layer was developed. The surfaces, scanned after friction tests performed at high NLE are polished which indicates the non-abrasive wear. Figure10 demonstrates that the friction layer surface covered a significantly larger surface area. Obviously, at higher energies, a stable and larger friction layer was generated and prevented the friction surface from wearing. The 2D material has the fibers oriented reverently parallel to the friction surface and wears uniformly at low energies and at dry conditions (2% RH). The grooves were detected more frequently when humidity and normal energy increased. This observation perfectly corresponds to the detected wear. As it was discussed earlier, the wear of the 2D pitch/resin/CVI material is increasing equally as the humidity level is increasing. The correlation between the wear and developing of surface behavior in the 3D material is not as clear as for the 2D material. The frequent grooves were detected at the lowest normal energy levels and at all humidity conditions. In spite of the fact that the friction surface at 2% RH and at 90% RH (12.5% NLE) looks very similar, the wear detected at highest humidity level was considerably greater. Equally, at 100% NLE the friction surface was scanned after the test and looks very similar to that detected after testing at 2% RH and at 90% RH. However, wear observed at dry humidity condition is twice as high as wear at 2% RH. Complexity of possible microstructure variations needs to be taken into consideration and generalization is not easily achieved.

The main idea of measuring the topography after tests and calculating the roughness parameters was to determinate how the surface roughness effects the frictional performance of the C/C composites. Figure 13 shows examples of the detected dependencies. The friction and wear data are also plotted along the roughness parameters. Unfortunately, the correlations between coefficient of friction, wear and roughness parameters were not found. It is suggested that fuzzy logic and/or neural network modeling will be employed in future to solve this problem.

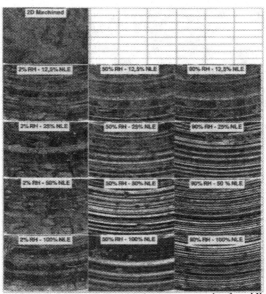

Figure 11. Scan of the surfaces after friction test at varying humidity and NLE,
2D pitch/resin/CVI material

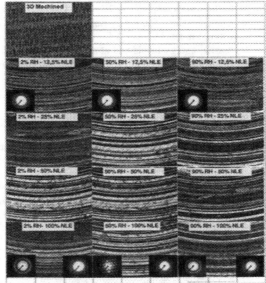

Figure 12. Scan of the surfaces after friction test at varying humidity and NLE,
3D PAN/CVI material

150

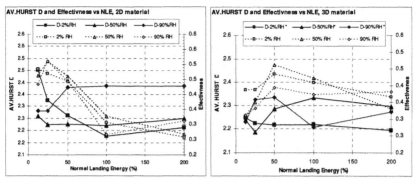

Figure 13. Roughness data (HURST "D" parameter) and coefficient of friction plotted against the Normal Landing Energy, 2D and 3D materials

TEM studies of the friction layer.

Figure14 shows the "spots" (circles) where the stops were interrupted and the TEM samples were taken from the friction surface. In the case of taxi stops, the samples were collected after the third taxi (following landing stop), where the influence of the previous landing on the microstructure of the friction layer is minimal. Only the results from the 3D PAN/CVI material are presented in this paper. TEM samples were taken only after the termination of braking performed at taxi and 25% NLE simulations, since no significant changes in μ were detected during the stop. However, for the 100% NLE conditions, the subscale dynamometer tests were interrupted at extreme values of μ.

The microstructure of the friction layer observed after the taxi stops and after stops performed at 25% NLE exhibited amorphous nature. A characteristic example of a friction layer observed after a 25% NLE stop performed at 50% RH level is shown in Fig.15. The bright field image shows the morphology of the carbon forming the friction layer (Fig.15a), the "fuzzy" diffraction pattern indicates the lack of long range order of C atoms typical for amorphous state. The dark field image (Fig.15c) demonstrates that the sample is transparent and the character of diffraction pattern is not influenced by lower transmissivity of electrons. At the very beginning of testing at 100% NLE, the initial decrease of the coefficient of friction caused by the release of physisorbed moisture is followed by first transition accompanied by a sharp and sudden increase of the coefficient of friction. At this peaking of μ, the microstructure of the friction layer is highly crystalline as shown in Fig.16. Figure 16 is an example, well representing the surface observed during testing at all humidity conditions. The release of chemisorbed moisture and species following the first peaking of friction causes the subsequent decrease of the coefficient of friction. After the chemisorbed species are removed from the friction surface, the second transition (increase of μ) occurs. The second peaking of the coefficient of friction well correlates with the highest temperature measured during the stop (see Fig.18). The microstructure of the friction layer exhibited amorphous character (Fig.16) at this second peaking point. Based on earlier published results, this is a surprise because it was speculated that microstructure of carbons tends to be graphite-like at higher energies and temperatures[6]. Also, it is clear that the microstructure of carbon itself does not dictate the level of μ. In the discussed situation, the coefficient of friction at its two peaking is very similar (μ ≈ 0.6). However, the microstructure of the carbonaceous friction layer is different (crystalline at the first and amorphous at the second transition).

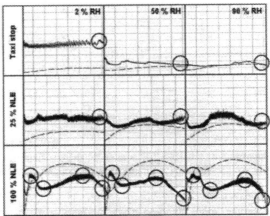

Figure 14. The spots of taxi and landing stops where the testing was interrupted and TEM samples were taken from the friction surface of PAN/CVI material samples. Coefficient of friction, temperature measured near the friction surface (dashed line).

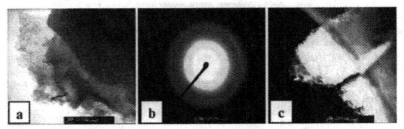

Figure 15. Amorphous carbon detected on surface of sample3D PAN/CVI material tested at 50% RH and 25% NLE (end of the stop). Bright field TEM image (a) diffraction pattern (b) and dark field (c).

Figure 16. Highly crystalline carbon detected in at 50% RH and 100% NLE during subscale dynamo testing (4sec) of 3D PAN/ CVI material. Bright field TEM image (a) diffraction pattern (b) and dark field (c).

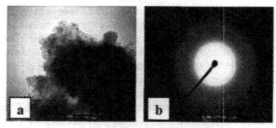

Figure 17. Amorphous carbon detected in at 50% RH and 100% NLE during subscale dynamo testing (20sec) of 3D PAN/CVI material. Bright field TEM images (a, diffraction pattern (b).

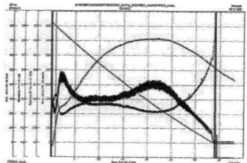

Figure 18. Characteristic stop for 3D material,
50% RH and 100% NLE

IV. CONCLUSION.

The 3D, PAN fiber/CVI matrix and 2D, pitch fiber/ resin-CVI matrix materials showed different frictional performance when tested at similar kinetic energy and humidity levels.

The 2D material exhibited transients during braking simulations at all energy and humidity levels. At the highest energy level (100% NLE), two transients were detected. It was suggested before that transients of the coefficient of friction are influenced by release of physisorbed and chemisorbed species from the surface[17]. Oscillations of the COF and noise were recorded during braking simulations, particularly at "dry" 2% RH conditions. The volume of oscillation and noise decreased when the level of humidity increased in testing environment. However, the average coefficient of friction did not exhibit significant sensitivity to humidity level. The average coefficient of friction for 2D material is highly sensitive to the energy level applied during braking. While the average coefficient of friction is high (μ(L) = 0.45 to 0.5) at lower and intermediate normal landing energy (12.5,25 and 50%), it is significantly lower (μ(L) ~ 0.3) at 100% NLE. Wear of 2D material was found very low and similar for all energy levels. The humidity level has stronger affect on wear of 2D material. When the humidity in the environment was increasing, the wear of 2D material was proportionally increasing.

The 3D material did not exhibit transients at the lowest (12.5%) normal landing energy level. Transients detected at higher humidity levels ware stronger and longer in duration. No significant oscillation or noise was recorded at any energy and humidity level, with exception of landing stops performed at 50% NLE and 90% RH. The average coefficient of friction showed higher sensitivity to humidity level but only low sensitivity to applied energy level. The average coefficient of friction of 3D material is slightly lower at lower normal landing energy and higher humidity level. Wear data exhibited the scatter. At 2% RH, wear of the 3D material is low, but increases with increasing humidity. At 100% NLE the situation is reverse and the wear detected at higher humidity is lower when compared to wear detected at 2% RH.

Correlations between coefficient of friction or wear and roughness parameters were not found. Similarly the correlation between the microstructure of the friction layer forming on the rubbing surface during braking. It is clear that the microstructure itself does not dictate the frictional performance of C/C composites. All factors, including environmental impact play important role. It is suggested that fuzzy logic or neural network models are used for prediction of frictional performance.

V. ACKNOWLEDGEMENT.

This research was performed within core research program of the Center for Advanced Friction Studies and sponsored by the National Science Foundation (Grant EEC 9523372), State of Illinois and following industrial partners: Aircraft Bracing System, Boeing, Composite Innovations, Climax Molybdenum, Honeywell International, Hyper-Therm, Starfire System, Carbon Nanotechnologies, and Tribco Incorporated.

References.
[1]C. Blanco, J. Bermejo, H. Marsch, R. Menendez: "Chemical and physical properties of carbon as related to brake performance, carbon and carbonaceous composite materials," *World Scientific Publishing Co.Pte.Ltd.* ,331-363, (1996)

[2]B.K. Yen: "Influence of water vapor and oxygen on the tribology of carbon materials with sp^2 valence configuration," *Wear* **192**, 208-215, (1996)

[3]B.K. Yen, T. Ishihara: "An investigation of friction and wear mechanism of carbon-carbon composites in nitrogen and air at elevated temperatures," *Carbon Vol.* **34**, No. 4, 489-498, (1996)

[4]Marchon, J. Carrazza, H. Heinemann, G.A. Somorjai : "TPD and XPS studies of O2, CO2 and H2O adsorbtion on clean polycrystalline graphite," *Carbon Vol.*26, 507-517, (1988)

[5]W.T. Clark, J.K. Lancaster: "Breakdown and surface fatigue of carbon during repeated sliding," *Wear vol.*6, 467-482, (November-December 1963)

[6]K.J. Lee, J.H. Chern Lin, C.P. Ju : "Electron microscopic study of worn PAN-pitch based carbon-carbon composite," *Carbon Vol.* 35, 613-620, (1997)

[7]J.D. Chen, C.P. Ju : "Friction and wear of PAN/pitch, PAN/CVI and pitch/resin/CVI based carbon-carbon composites," *Wear* 174, 129-135, (1994)

[8]T.J. Hutton, B. McEnaney, J.C. Crelling : "Structural studies of wear debris from carbon-carbon composite aircraft brakes," *Carbon* 37, 907-916, (1999)

[9]B. K. Yen, T. Ishihara: "The surface morphology and structure of carbon-carbon composites in high-energy sliding contact," *Wear* 174, 111-117, (1994)

[10]J.K.Lancaster: "Transitions in the friction and wear of carbons and graphites sliding against themselves," *ASLE Vol.*18, 187-201, (1975)

[11]A. Samah, D. Paulmier, M.E. Mansori : "Damage of carbon-carbon composite surface under high pressure and shear strain," *Surface and coatings technology* 120-121, 636-640, (1999)

[12]C.P. Ju, K.J. Lee, H.D. Wu, C.I. Chen: "Low-energy wear behavior of palycrylonitrile, fiber-reinforced, pitch-matrix, carbon-carbon composites," *Carbon Vol.*32, No.5, 971-977, (1994)

[13]H. Harker, J.B. Horsley, D. Robson : "Active centers produced in graphite by powdering," *Carbon Vol.*9, 1-9, (1971)

[14] T. Policandriotes, T. Walker, P. Filip: "Subscale Aircraft Brake Dynamometer Testing on C/SiC and C/B4C Ceramic Friction Materials", 28 *th Annual Conference on Ceramics, Metal & Carbon Carbon Composites*, Materials and Structures, Cocoa Beach, FL, January 26-30, (2004)

[15]T. Policandriotes: "Surface characterization and evolution of sub-scale brake materials," *B.S. Thesis*, Department of physics, SIUC,IL, (1995)

[16] B. Bhushan and co. "Modern tribology handbook," *CRC Press LLC*, New York, (2001)

[17] K.Peszynska-Bialczyk, A. Pawliczek, P. Filip, K. Anderson, M. Krkoska : "Study of „adsorbtion/desorbtion" phenomena on friction debris of aircraft brakes," 29*th International conference on advanced ceramics and composites*, Adsorbtion/desorbtion phenomena and frictional performance, Cocoa Beach FL, January 23-28, (2005)

[18]B. Venkataraman, G. Sundararajan : "The influence of sample geometry on the frictional behavior of carbon-carbon composites," *Acta materialia* 50, 1153-1163, (2002)

[19]M.Goudier, Y. Berthier, P. Jacquemard, B. Rousseau, S. Bonnamy, H. Estrade-Szwarckopf :"Mass spectrometry during C/C composite friction: carbonoxidation associated with friction coefficient and high wear rate," *Wear* 256, 1082-1087, (2004)

[20]B.K. Yen, T. Ishihara: "The surface morphology and structure of carbon-carbon composites in high-energy sliding contact," *Wear* 174, 111-117, (1994)

[21]J.R.Gomes, O.M.Silva, C.M.Silva, L.C.Pardini: "The effect of sliding seed and temperature on the tribological behavior of carbon-carbon composites, "*Wear* 249, 240-245, (2001)

[22]S. Ozcan, P. Filip, M.Krkoska : "Microstructure and wear mechanisms in C/C composites," *Wear*, submitted for publication, (2005)

STUDY OF "ADSORPTION/DESORPTION" PHENOMENA ON FRICTION DEBRIS OF AIRCRAFT BRAKES

Katarzyna Peszynska-Bialczyk, Milan Krkoska, Adam Pawliczek, Peter Filip
Center for Advanced Friction Studies, Southern Illinois University Carbondale
1230 Lincoln Drive, Engineering A 108
Carbondale, IL, 62901

Ken Anderson
Department of Geology, Southern Illinois University Carbondale
1259 Lincoln Drive
Carbondale, IL, 62901

ABSTRACT

This paper discusses the influence of humidity and landing energy on the "adsorption/desorption" phenomena of friction debris obtained from carbon-carbon composite brake materials. Debris samples were collected from three discrete C/C materials during subscale dynamometer disc on disc type testing at various energy and humidity conditions (25 % and 100 % of normal landing energy (NLE); 2, 50 and 90 % relative humidity (RH)). Samples have been characterized using Temperature Programmed Reduction (TPR), Thermogravimetric Analysis coupled with Fourier Transform Infrared Spectroscopy (TGA/FTIR) and specific surface area (Brunauer-Emmet-Teller – BET) methods. The amount of physisorbed, chemisorbed and decomposed species strongly depends on the applied testing (brake dynamometer) conditions. Specific surface areas of debris obtained from different samples of materials are ranging from 800 to 1300 m^2/g. TPR analysis indicates two reaction ranges, the first reaction range between 350 °C and 600 °C, and the second from 450 °C to 800 °C (maximum at 475 °C and 680 °C, respectively). CO_2 and CO peaks detected using TGA/FTIR methods, occur as a result of decomposition of oxygen-containing functional groups that are formed by oxidation of the C/C during braking. The relative amount of these groups indicates the resistance of the brake materials to chemical reactivity.

Keywords: carbon-carbon composite, friction debris, "adsorption/desorption" phenomena, TGA, FT-IR, TPR, BET

INTRODUCTION

Carbon-carbon composites are widely used in numerous demanding applications, including aircraft brake systems. Their low specific mass, high thermal conductivity, high shock resistance and low thermal stress generated during rapid heating up to temperatures above 2000 °C make them an ideal candidate for high energy, dry friction applications[1, 2]. The dominant advantage of carbon fiber reinforced matrix composite is the stability of structure at high temperatures[3].

The C/C composites currently used as aircraft brake discs combine a variety of fiber types, architectures and carbonaceous matrices. Oxidized PAN fiber/CVI matrix or pitch fiber charred resin matrix, subjected to specific heat treatment are the two most typical friction material candidates[4]. The mechanical and tribological properties of these composites vary with fiber architecture, density, porosity, and the type of heat treatment applied. Different microstructures of matrices and fibers regulate the wear mechanism and friction coefficient[5, 6]. Frictional performance also depends on normal load, speed, temperature and humidity conditions. Behavior of the friction debris is sensitive to the environment and to the adsorption and desorption kinetics of the surrounding gases[7, 8]. C/C composites are reacting with environment and their oxidations remain a challenge.

The goal of this research is to understand the nature of volatile products released during aircraft braking at different energies and humidity levels and to relate these products to brake properties and frictional performance. This task was addressed by analyzing the "adsorption/desorption" capacity and chemical reactions of friction debris generated during simulated landings.

EXPERIMENTAL

Debris samples were obtained during simulated landing stops at different energies. Three types of C/C material were used in this research:
- Material A (2D randomly chopped pitch fiber charred resin, CVI),
- Material B (as material A, SiC modified)
- Material D (3D non woven PAN fiber, CVI).

Subscale dynamometer tests simulated friction conditions corresponding to braking performed at 25 and 100 % of normal landing energy (NLE) at three different humidity levels (2, 50 and 90 % relative humidity (RH)). A landing stop was always followed by three taxi stops (4.5 % of NLE) and this sequence was repeated 50 times. The C/C discs have the following dimensions: outer diameter = 3.75" (95.25 mm), inner diameter = 2.75" (69.85 mm). An environmental chamber was used to control humidity level during testing with accuracy ± 5 %.

Friction debris was collected directly in the environmental chamber and represents mostly the wear particulates that were released from the friction surfaces during testing. Samples were characterized using Thermogravimetric Analysis/Fourier Transform Infrared Spectroscopy, Temperature Programmed Reduction and surface area analysis.

TGA experiments were carried out using Cahn TGA-171 instrument. Approximately 300 mg of the sample was placed into alumina crucible, thermally stabilized in an argon atmosphere with flow rate equal to 40 cm^3/min at 25 °C for 16 minutes, then heated to 1000 °C at 5 °C per minute and held at this temperature for 1 hour.

FTIR data were collected using a Nicolet Nexus 670 FTIR instrument connected to TGA analyzer. Outgassing products were heated to 200 °C before entering FT-IR chamber. This tandem allowed to real-time analysis of gases that created during thermal decomposition of studied samples in inert atmosphere.

TPR experiments were performed using ChemBET 3000 TPR/TPD instrument (Quantachrome Instruments). About 50 mg of friction debris sample were placed into the quartz cell and outgassed for 3 hours at 300 °C under N_2 (30 cm^3/min). The sample was then cooled and transferred to the TPR furnace. Reduction was carried out using a mixture of hydrogen and nitrogen (5 % H_2 + 95 % N_2) over the temperature range 25-1000 °C at a heating rate of 5 °C/min. Hydrogen consumption was monitored using a thermal conductivity detector (TCD).

The specific surface area of friction debris particulates was measured using the Brunauer-Emmet-Teller (BET) method. Experiments were carried out using Nova 1200 BET analyzer (Quantachrome Instruments). Before determining the specific surface area, samples were outgassed in high vacuum at 150 °C for 15 hours. Specific surface area measurements were performed by nitrogen adsorption using the multipoint BET method taking six points from the linear region of the isotherm over the P/P_0 range of 0.01 to 0.35.

RESULTS AND DISCUSSION
Thermogravimetric Analysis

Figures 1, 2 and 3 illustrate TGA data (mass change and first derivative of mass change versus temperature) for friction debris samples generated from materials A, B and D, respectively.

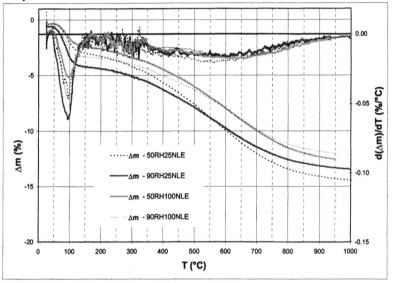

Fig.1. TGA analysis of friction debris of material A.

Because of difficulties with collecting of sufficient amount of material A the experiments for 2 % RH were not performed. Note all plots are color coded for clarity:
green = 2 % RH, 25 % NLE;   brown = 2 % RH, 100 % NLE;
purple = 50 % RH, 25 % NLE;   red = 50 % RH, 100 % NLE;
deep blue = 90 % RH, 25 % NLE;   pale blue = 90 % RH, 100 % NLE.

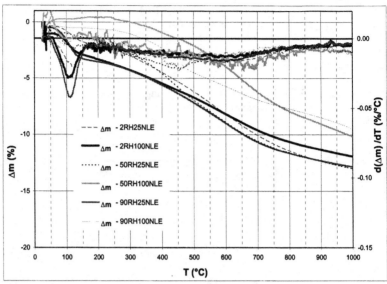

Fig.2. TGA analysis of friction debris of material B.

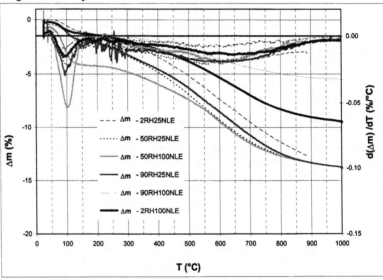

Fig.3. TGA analysis of friction debris of material D.

In all cases, the first derivative curves show a minimum of intensity at ~ 100 °C, corresponding to the maximum desorption rate of physisorbed water molecules. Subsequently, additional mass loss occurs across a broad temperature range extending up to ~ 1000 °C. In some cases, the first derivative data indicate that second and third minima are reached at approximately 425 °C and 600 °C, respectively. These temperatures correspond to the maxima of carbon dioxide and carbon monoxide release, as identified by FT-IR method. The observed mass changes of friction debris of all samples were in range from 12 to 14 % of their initial weights for material A, from 9 to 13 % for material B, and from 6 to 14 % for material D.

Generally, friction debris, generated from materials B and D at lower normal landing energy conditions (% NLE), is releasing relatively larger amounts of physisorbed water. Also amounts of released carbon monoxide and carbon dioxide were larger during thermal decomposition. In material A, where the most intensive physisorption occurs at 90 % RH, the smallest amount of decomposed species can be observed, and in the opposite direction – the smallest physisorption means the biggest chemical reactivity of material (50 % RH).

FTIR data obtained in conjunction with TGA analysis indicate that all of the mass loss occurring after removal of physisorbed water is related to release of CO and CO_2. Figure 4 illustrates three dimensional (temperature versus absorbance versus wavenumber) TGA-FTIR spectra for friction debris obtained from material D tested at 50 % RH and 100 % NLE. Similar data were obtained for all studied debris samples. Release of physisorbed water is observed in the temperature range from 50 to 200 °C, with maximum release at 100 °C. CO_2 release is observed between 225 and 650 °C and CO release is observed over the temperature range 300-900 °C.

None of these products were observed in TGA-FTIR analysis of bulk C/C brake material. Prior investigations have shown that "pure" carbons as well as different carbonaceous matter, are very stable up to 1000 °C in argon atmosphere[9, 10, 11].

It was concluded from this data that the release of carbon oxides is the result of decomposition of oxygen-containing functional groups produced by oxidation of the C/C material during braking. This conclusion is in agreement with other authors. Metzinger and Huttinger[12], for example, utilized mass spectrometry and observed that CO is formed between 350 and 800 °C. These authors suggested that this product is the result of decomposition of carboxylic anhydrides and carbonyl groups. According to their assumptions, CO_2 formation between 200 and 400 °C is attributed to the decomposition of carboxylic groups of different acidity, and the CO_2 formation above 400 °C is a result of decomposition of esters and anhydrides. The detected reaction products (Fig. 4) correlate with data published by Marchon et al.[13] who found a carbon monoxide as a prominent decomposition product of O_2, CO_2 and H_2O adsorption of polycrystalline graphite. The relative amount of these groups is proportional to the intensity of detected FTIR absorbance, adsorption capacity and reactivity of samples.

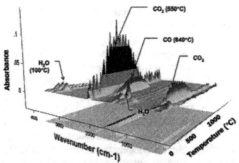

Fig.4. Characteristic 3-dimentional IR spectra of gases released during heating of friction debris of material D. Tested at 50 % RH and 25 % NLE (corresponds to the purple line plotted in Fig. 3).

Temperature Programmed Reduction

TPR data for friction debris from materials A, B and D are given in Fig. 5, 6 and 7, respectively. From these data it can be seen that two reduction peaks were detected over the temperature range investigated, for all debris samples of material A, B and D. The first peak was detected in the temperature range from 350 to 600 °C, and the second reduction peak was from 450 °C to 800 °C with maximum at 475 °C and 680 °C, respectively. All presented plots of material A, B and D have very similar shape. Sometimes, it is not easy to detect the first peak in plots shown at given magnification. Negative values above ~750-800 °C, indicate hydrogen release or nitrogen uptake, observed in all samples. At this time, any one of these mechanisms have not been proved or rejected. All presented plots below are color coded for clarity: the black line corresponds to typical TPR plot and the red one corresponds to its first derivative.

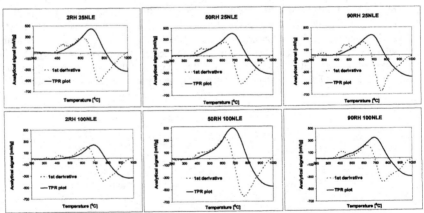

Fig.5. Temperature Programmed Reduction graphs for all studied samples of friction debris of material A used in friction experiments.

162

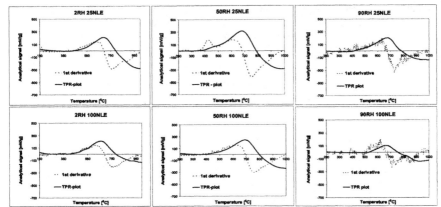

Fig.6. Temperature Programmed Reduction graphs for all studied samples of friction debris of material B used in friction experiments.

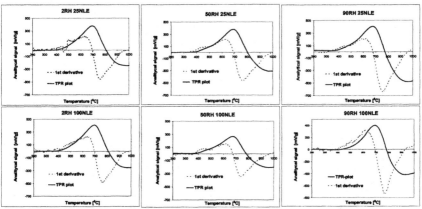

Fig.7. Temperature Programmed Reduction graphs for all studied samples of friction debris of material D used in experiments.

Specific Surface Area

The relationship between landing energy (NLE), relative humidity (RH) and specific surface area (SSA) of collected friction debris for material A, B and D are illustrated in Fig. 8, 9 and 10, respectively. SSA values obtained for the various debris samples ranged from 800 to 1300 m^2/g. No apparent direct correlation between SSA and RH or NLE is observed. It is noteworthy to mention, however, that the SSA of debris materials obtained with 2 and 50 % RH increases with increasing landing energy, whereas for debris obtained at 90 % RH, SSA values decrease with increasing landing energy.

This is in agreement with B.K. Yen[14] and J.D. Chen et al.[15] who report that friction and wear of graphite, non-graphitic carbon as well as carbon-carbon composites can be similarly influenced by water vapor and oxygen in air. However, they found only the qualitative relations between wear morphology and presence of water vapor and other condensable vapors in the environment.

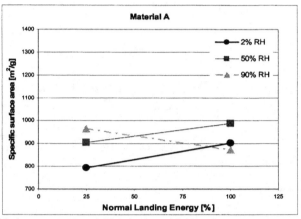

Fig 8. Relationship between NLE and SSA for material A.

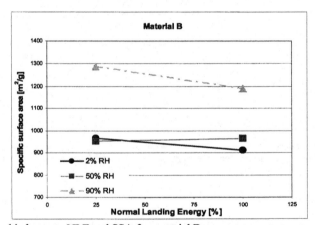

Fig 9. Relationship between NLE and SSA for material B.

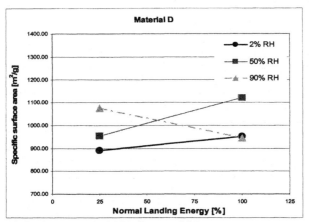

Fig 10. Relationship between NLE and SSA for material D.

CONCLUSIONS

Data obtained by TGA, TGA-FTIR and TPR analysis of friction debris produced from carbon-carbon composite brake materials indicate the presence of oxidized carbon species. These can be thermally decomposed in an inert atmosphere to carbon dioxide and carbon monoxide. Landing energy, RH and even the nature of the C/C used during braking have no significant influence on the nature of the products observed nor do they influence the temperature ranges over which these carbon oxides are released.

Normal landing energy and relative humidity do, however, influence the degree of oxidation of the C/C brake material during braking. For materials B and D, TGA data indicate that the degree of decomposition is positively correlated to landing energy. That means that for material A, however, the correlation is negative, which means that higher NLE gives as a result smaller amount of carbon-oxygen compounds.

No direct correlation between specific surface area of friction debris and landing energy and relative humidity has been observed for the materials studied. However, it was observed that in general, the SSA of produced friction debris increases with NLE for higher RH (50 % and 90 %), and decreases for smaller RH condition (2 %) used during subscale dynamometer tests.

ACKNOWLEDGMENT

This research was sponsored by NSF (EEC 9523372), State of Illinois, and consortium of 12 industries (www.frictioncenter.com).

REFERENCES

[1] M. Gouider, Y. Berthier, P. Jacquemard, B. Rousseau, S. Bonnamy, and H. Estrade-Szwarckopf, "Mass spectrometry during C/C composite friction: carbon oxidation associated with high friction coefficient and high wear rate," *Wear*, **256**, 1082-1087 (2004).

[2] K. J. Lee, J.H. Chern Lin, and C. P. Ju, "Electron microscopic study of worn PAN-PITCH based carbon-carbon composite," *Carbon*, **35**, 613-620 (1997).

[3] C. Blanco, and J. Bermejo, H. Marsh and R. Mendez, "Chemical and physical properties of carbon as related brake performance," *Wear*, **213**, 1-12 (1997).

[4] B. K Yen, and T. Ishihara, "An investigation of friction and wear mechanisms of carbon-carbon composites in nitrogen and air at elevated temperature," *Carbon*, **34**, 489-498 (1996).

[5] B.K. Satapathy, and J. Bijwe, "Performance of friction materials based on variation in nature of organic fibers," *Wear*, **257**, 573-584 (2004).

[6] L. M. Manocha, M. Patel, S. M. Manocha, C. Vix-Guterl, and P. Ehrburger, "Carbon/carbon composites with heat-treated pitches I. Effect of treatment in air on the physical characteristics of coal tar pitches and the carbon matrix derived therefrom," *Carbon*, **39**, 663-671 (2001).

[7] B. K. Yen, T. Ishihara, and I. Yamamoto, "Influence of environment and temperature on "dusting" wear transitions of carbon-carbon composites," *Journal of Materials Science*, **32**, 681-686 (1997).

[8] C. C. Li, and J. E. Sheehan, "Frictional and wear studies of graphite and a carbon composite in air and helium," *Inter. Conference, ASME, N.Y.*, 525-533 (1981).

[9] F. Cataldo, "A study on the thermal stability to 1000 °C of various carbon allotropes and carbonaceous matter both under nitrogen and in air," *Fullerenes, nanotubes and Carbon nanostructures*, **10**, 293-311 (2002).

[10] H. W. Chang, and R. M. Rusnak, "Contribution of oxidation to the wear of carbon-carbon composites," *Carbon*, **16**, 309-312 (1978).

[11] H. W. Chang, "Correlation of wear with oxidation of carbon-carbon composites," *Wear*, **80**, 7-14 (1982).

[12] T. Metzinger, and K. J. Huttinger, "Investigations on the cross-linking of binder pitch matrix of carbon bodies with molecular oxygen – Part I. Chemistry of reactions between pitch and oxygen," *Carbon*, **35**, 885-892 (1997).

[13] B. Marchon, J. Carrazza, H. Heinemann, and G. A. Somorjai, "TPD and XPS studies of O_2, CO_2 and H_2O adsorption on clean polycrystalline graphite," *Carbon*, **26**, 507-514 (1988).

[14] B. K. Yen, "Influence of water vapor and oxygen on the tribology of carbon materials with sp^2 valence configuration," *Wear*, **192**, 208-215 (1996).

[15] J. D. Chen, J. H. Chern Lin, and C. P. Ju, "Effect of humidity on the tribological behavior of carbon-carbon composites," *Wear*, **193**, 38-46 (1986).

FRICTION AND WEAR OF CARBON BRAKE MATERIALS

John A. Tanner and Matt Travis
The Boeing Company
PO Box 3707
Seattle, Washington 98124-2207

ABSTRACT

A literature review has been undertaken to establish what information on carbon friction and wear has been generated from past research activities. These past research activities have been summarized, and their results have been compared with subscale dynamometer test results. The

Water vapor in the atmosphere produces a direct lubricant effect on carbon. Observed transition temperatures within the range of 140°C to 200°C, associated with increases in friction and wear of carbon brake materials are attributed to water vapor desorption. Friction and wear transitions in the range of 500°C to 900°C may be associated with oxygen desorption.

INTRODUCTION

Friction and wear characteristics of carbon brakes are affected by a number of environmental parameters. Adsorption and desorption of water vapor, oxygen, and contaminates such as cleaning compounds or de-icing fluids can dramatically change the friction coefficient and wear mechanisms of carbon during normal braking operations. These changes can occur rapidly at discrete temperatures that are within the normal operating range of aircraft brakes. Wear debris accumulation on the friction interfaces further complicates this picture. As an airframe manufacturer, Boeing must understand how these environmental factors influence carbon brake materials if we are to reduce the risk of losses in brake performance that affect our airplanes and understand the cause of costly brake vibration problems. This is, indeed, a formidable task.

To gain some knowledge of these complex factors that affect carbon aircraft brakes, Boeing conducted a literature review in the late 1990's to establish what information on carbon friction and wear had been generated from past research activities, and to document those efforts to understand this complex interaction between the carbon brake and the environment in which it must operate. The purpose of this paper is to summarize those past research activities and to look for possible insights into the friction and wear mechanisms that we have observed to date in our own testing. The following paragraphs provide a chronology of prevailing wisdom on the friction and wear characteristics of carbon materials, highlight some details on various research investigations to better define these characteristics, and draw comparisons with our own experimental measurements to date with these earlier results.

REVIEW OF EARLY CARBON FRICTION AND WEAR RESEARCH

A Brief Historical Overview

Prior to the late 1960's, the use of carbon material in sliding contact applications was generally limited to motor brush commutators, seals, and bearings [1]. Contact pressures for these applications were low and their strength and toughness characteristics prevented their use for higher contact pressure applications such as aircraft brakes. In the 1960's, the introduction of carbon-carbon composites that incorporated pitch or polyacrylonitrile (PAN)-based carbon fibers removed these strength and toughness restrictions and led to the development of the carbon brake industry.

Bragg published one of the earliest studies of the friction and wear characteristics of carbon materials in 1928 [2]. In this study, Bragg used X-ray diffraction techniques to define a theory for the lubrication characteristics of carbon based on the intrinsic structure of graphite crystals. This theory involved shearing between adjacent basal planes, and was based, in part, on the relative spacing of carbon atoms within the hexagonal lattice that forms the graphite layers or basal planes and the vertical separation of the planes. The wide spacing between the graphite layers was assumed to be an indication of low inter-atomic forces, and thus, the root cause of good dry lubrication and excellent wear resistance of these materials [3, 4]. The Bragg supposition of carbon lubrication was the prevailing theory for about the next ten years.

In the late 1930's it was discovered that the lubrication and wear resistance characteristics of carbon materials were not intrinsic characteristics of the crystalline structure as suggested by Bragg, but instead were strongly influenced by the environment in which they were used. For example, carbon brushes used in the control circuits of variable pitch airplane propellers, which exhibited low or normal wear rates during low altitude flights, were shown to exhibit more rapid wear rates when aircraft began flying at altitudes of 20,000 to 40,000 feet [5]. The short-term fix to this problem was to develop a family of organic lubricants to be used with the electric brushes to reduce the high wear rates associated with high-altitude flight operations. The long-term impact has been a substantial body of research over the last seventy-five years to study the tribology of carbon friction surfaces.

Water-Vapor Adsorption and Desorption

The accelerated wear rates of carbon brushes associated with high-altitude flight was shown by Savage [6] to be the result of reduced water vapor in the upper atmosphere. His conclusion was that low wear and low friction of graphite and other forms of carbon is dependent upon the presence of adsorbed vapors on the friction surfaces.

In his study, Savage concluded that over a range of water-vapor pressures of 3 to 10 mm Hg only a monolayer of water molecules was formed on the surface since there did not appear to be any evidence of hydrodynamic lubrication effects. Figure 1 is a plot of atmospheric water-vapor pressure as a function of dewpoint temperatures [7]. The rectangular box in the figure encloses the range of Savage's results for full carbon lubrication with a monolayer of water molecules. The data indicate that this type of lubrication can be expected over a range of dewpoint temperatures from approximately −7°C to +12°C. Airline operators who consistently operate aircraft in climates with dewpoints lower than −7°C might experience an increase in carbon brake wear rates. Aircraft operations in climates with dewpoints higher than about 12°C might experience weak braking during cold taxi operations with carbon brakes due to excessive water-vapor surface contamination. This weak braking condition under cold taxi conditions is sometimes referred to as "morning sickness".

Data presented in figure 1 define the amount of water vapor in the atmosphere under 100 percent relative humidity conditions, but the figure can be used to establish the relationship between ambient temperature and dewpoint temperature for relative humidity conditions that are less than 100 percent. For example, at an ambient temperature of 21.1°C (70°F) and at 100 percent relative humidity, the water vapor pressure is about 18 mm Hg. If the ambient temperature remains unchanged at 21.1°C (70°F), but the relative humidity is reduced to 50 percent, then the water vapor pressure in the atmosphere is reduced by 50 percent to about 9 mm Hg, and the dewpoint temperature is reduced to 10°C (50°F). Now consider the case when the relative humidity is held constant at 50 percent, and the ambient temperature is increased from

168

21.1°C (70°F) to 25°C (77°F). For this case, the water vapor pressure is increased from 9 mm Hg to about 12 mm Hg or 50 percent of the water vapor pressure of 24 mm Hg for saturated air at to 25°C (77°F), and the dewpoint temperature is increased from 10°C (50°F) to about 14°C (57.2°F). Thus 50 percent relative humidity air at 25°C (77°F) contains 33 percent more water vapor than 50 percent relative humidity air at 21.1°C (70°F).

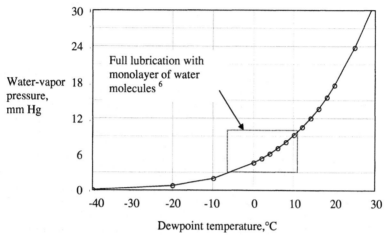

Figure 1: Atmospheric water vapor pressure plotted as a function of dewpoint temperature.

Lubricating Efficiency of Various Condensing Vapors

A number of condensing vapors other than water vapor are efficient in lubricating the friction interface of carbon materials. Since carbon brakes are exposed to a wide variety of contaminants during airline revenue operation, it is important to understand the potential role of these contaminants in brake performance. Such understanding may also provide explanations for anomalous brake performance that can arise during qualification testing, aircraft certification, and rejected take off (RTO) quality assurance compliance tests.

Savage and Schaefer [8] and Lancaster and Pritchard [9] studied the lubrication characteristics of a number of these different vapors including various hydrocarbons. Both studies found a strong tie between the chain length of the vapor molecule and the efficiency of its lubrication attributes.

Yen [4] presented results from a study of the friction characteristics of pins make of the four different carbon materials in contact with a machined cast iron disk under various atmospheric conditions. This series of experiments was a well thought out investigation of the friction characteristics of the various carbon materials associated with carbon aircraft brakes. The purpose of this investigation was to establish the influence of dry nitrogen; dry air (actually the oxygen partial pressure in air), and water vapor, in various combinations, on the friction characteristics of the carbon brake materials.

The friction characteristics of a graphite-cast iron interface with an area of approximately 75 mm^2 and a velocity of 5 cm/s under alternating wet and dry nitrogen atmospheres were tested in this study. The moisture content inside a closed 300-cm^3 chamber surrounding the test specimens was controlled from 100 parts per million (ppm) (76 microns Hg) to 100 percent relative humidity with a gas flow rate of 1 liter/s. Table I gives details on the water vapor and oxygen content of the various test environments in Yen's experiments.

Table I: Water vapor and oxygen contents in various test environments (reference 4)

Environment	Water vapor content	Oxygen content
Dry nitrogen	76 microns Hg	7.6 microns Hg
Wet nitrogen	9 mm Hg	7.6 microns Hg
Dry air	76 microns Hg	155 mm Hg
Wet air	9 mm Hg	155 mm Hg

Yen's experiment [4] indicates that graphite, nongraphitic carbon, and carbon-carbon composite material react in a similar manner when subjected to alternating dry nitrogen and dry air environments with higher friction coefficients associated with the dry nitrogen environment than the dry air environment. Data indicate that the three carbon materials exhibit friction coefficients in the range of 0.35 to 0.5, with a high frequency variation about the mean friction value, when in contact with a rotating cast iron disk in a nitrogen environment. The high frequency variation about the mean friction value is most prominent for the carbon-carbon composite in the dry nitrogen test environment, and this variation in friction coefficient is about ±0.05 about a mean of approximately 0.35. In a dry air environment, the three carbon materials exhibit friction coefficient values ranging from about 0.2 to 0.3 when in contact with a rotating cast iron disk. Both the graphite and nongraphitic carbon materials exhibit relatively uniform friction coefficients over sliding distance, associated with the dry air test environment, for the test conditions. The carbon-carbon composite material, on the other hand, exhibits gradually increasing friction coefficients with sliding distance in the dry air environment. These comparisons indicate that while the friction coefficients of graphite, nongraphitic carbon, and carbon-carbon composites may differ in detail, they do exhibit similar friction characteristics, in a general way, over the range of environmental conditions evaluated in the study. Unfortunately, the influence of additional contaminants on the amorphous carbon results, reported in the study [4] prevents a similar comparison of the friction characteristics of amorphous carbon with those of graphite, nongraphitic carbon, and carbon-carbon composite materials.

Those subtle differences in the friction responses of graphite, nongraphitic carbon, and carbon-carbon composites observed in Yen's experiment may be the result of thermal conductivity differences among the various carbon types tested and the cast iron test surface used in that investigation. This effect is examined more fully in the next section.

Desorption Transition Temperatures

The ability of the carbon surface to adsorb water vapor, oxygen or other vapors is dependent on the surface temperature. Savage first noticed this phenomenon [10] when he initiated testing of graphite on graphite. In his earlier research involving graphite on copper, he did not observe any speed effect on the lubricating effects of water vapor, up to the maximum speed of his testing. When the copper disk was replaced with a graphite disk, then the lubricating ability of the water vapor was seen to be a function of velocity. In reality this speed effect was actually a surface temperature effect due to differences in thermal conductivity shown in table II.

170

Table II: Typical thermal conductivity values for selected materials at room temperature.

Material	Thermal conductivity, W/m°C
Graphite	5
Carbon-carbon composite	
• Perpendicular to fiber direction	5-10
• Parallel to fiber direction	50-60
Cast iron	46
Copper	400

Lancaster.[11, 12] addressed the problem of determining the transition temperatures associated with desorption events for a number of carbon materials. Once it was established that carbon on carbon testing resulted in large transitions in friction and wear in his investigation, Lancaster used the thermal analysis of a sliding heat source on a semi-infinite plane, developed by Jaeger [13] to approximate the flash temperature in the friction interface of his carbon test specimens.

Lancaster observed transition temperatures in the range of 150°C-200°C that he associated with desorption of water vapor. Lancaster also noted higher transition temperatures in his experiments in the range of 500°C-900°C that might be associated with oxidation.

HEAT DYNAMIC MODEL OF FRICTION AND WEAR

The Heat Dynamic Model of Friction and Wear developed by Dr. Chichinadze and his colleagues can be summarized by the following equations[14, 15]:

$$\theta_{body} = \frac{\alpha_{hf} \cdot W \cdot b}{A_c \cdot \lambda \cdot t_T} \left[\left[\left[-\zeta \cdot \left(1 - \frac{\zeta}{2}\right) + \frac{1}{3} \right] - \sum_{av} \right] \cdot \tau_N + F_0 \cdot \tau_W \right]$$ (eqn. 1)

The term $\sum_{av}$ is defined as a series of exponential functions such that:

$$\sum_{av} = \left(\frac{2}{\pi^2}\right) \sum_{n=1}^{\infty} \left(\frac{1}{n^2}\right) \cos(n\pi\zeta) e^{-(n\pi)^2 F_0 \tau}$$ (eqn. 2)

$$\theta_{flash} = \frac{\sqrt{2}+1}{\sqrt{2}} \cdot \frac{d_r \cdot a^{0.5} \cdot \tau_N \cdot W}{A_r \cdot t_T \cdot \left[4\lambda \cdot a^{0.5} + \lambda \cdot (\pi \cdot d_r \cdot v_0 \cdot \tau_v)^{0.5}\right]}$$ (eqn. 3)

$$\theta_{surface} = \theta_{body}|_{(\zeta=0)} + \theta_{flash} + \theta_{ambient}$$ (eqn.4)

where:

α_{hf} = Heat flow shearing coefficient
W = Energy dissipated at friction interface
b = Disk thickness
A_c = Actual contact area
λ = Thermal conductivity
t_T = Stop time

ζ = Non-dimensional disk thickness
τ_N = Brake power normalized time function
τ_W = Frictional work normalized time function
F_0 = Fourier number
d_r = Mean diameter of real contact
a = Thermal diffusivity
A_r = Real area of contact
v_0 = Initial velocity
τ_v = Sliding velocity normalized time function
$\theta_{surface}$ = Temperature at friction interface
θ_{body} = Temperature of carbon disk
θ_{flash} = Temperature at surface asperities
$\theta_{ambient}$ = Ambient temperature

The first equation states that the body temperature of the carbon material, at any point through the thickness, is a function of both the rate at which energy is absorbed by the brake and the total work produced during the braking effort. This body temperature is directly proportional to the energy absorbed at the friction interface and inversely proportional to the thermal conductivity and contact area associated with the friction interface. Equation 3 defines the flash temperature associated with the surface asperities of the friction interface that define the real contact area as a function of the carbon thermal properties, sliding velocity at the friction interface, and surface roughness characteristics. The flash temperature peaks early in the braking effort then decreases rapidly as the braking effort continues. Equation 4 defines the surface temperature of the friction interface as the sum of the body temperature evaluated at the friction surface ($\zeta = 0$), the flash temperature of the surface asperities, and the ambient temperature. Equations 1 to 4 assume that the friction surface is heated uniformly across its area and that the heat flow is a one-dimensional flux through the thickness of the carbon disks.

BOEING EXPERIMENTAL RESULTS
Typical Results from Dynamometer Tests

Typical results from Boeing's subscale dynamometer testing of carbon brake materials are shown in figures 2 and 3. The figures are time history plots of temperature and friction coefficient obtained from a stator-rotor pair of ring specimens subjected to light braking at service energy levels (energy per unit mass of approximately 150 – 200 J/gm). Identified in each figure are two temperature measurements from thermocouples located 2 mm and 7 mm below the friction surface of the stator. Also identified in each figure is a prediction of the surface temperature, based upon a one-dimensional thermal analysis of Chichinadze [14, 15]. The surface temperature prediction includes the effect of flash temperature which is associated with the rapid heating of surface asperities at the initiation of contact. This flash temperature effect is characterized by a rapid, but momentary, increase in surface temperature at the onset of braking. In figures 2 and 3 this flash temperature peaks at about 760 °C and 300°C respectively for the conditions shown.

In figure 2 results are shown for Material A tested under service energy conditions. In this test the surface temperature initially peaks at about 760°C due to the flash temperature, quickly reduces to a value of about 140°C, and then rapidly increases again to a value of about 650°C within the first 10 seconds of testing. At the end of the test the surface temperature is approximately 475°C. At the onset of braking the friction coefficient is at about 0.28 and goes through a minimum of about 0.16 within the first second of testing. As the surface temperature begins its second rapid increase, the friction coefficient is shown to rapidly increase from its minimum value to a peak value of about 0.58. From this point the friction coefficient ratchets down to a local minimum value of approximately 0.27 about 12 seconds into the test and ratchets up to a secondary peak of about 0.4 which occurs 22 seconds into the test.

The minimum friction coefficient, 0.16, observed in figure 2 is consistent with water vapor adsorption lubrication [4, 6, 11, 16]. The abrupt increase in friction coefficient to about 0.58 is consistent with water vapor desorption [4, 11, 16], and the secondary peak in friction coefficient of 0.4 which occurred 22 seconds into the test is consistent with oxygen desorption [16, 17]. The irregular ratcheting in the friction coefficient trace over the time interval between 2 seconds and 22 seconds is probably the result of a number of factors. These factors might involve oxygen adsorption/desorption events, the accumulation of wear debris at the friction interface, non-uniform surface temperatures, and variations in the friction radius due to localized thermal expansion of the carbon material.

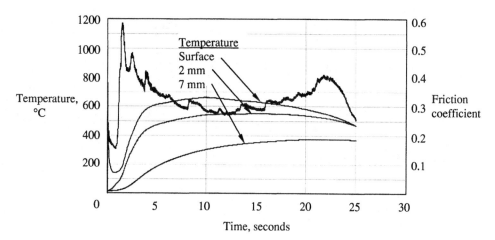

Figure 2: Time histories of temperature and friction coefficient obtained from Boeing dynamometer testing of material A. Service energy stop

In figure 3 results are shown for material B. This material is also tested at a service energy level (150 – 200 J/gm). In this test the surface temperature initially peaks at about 600°C due to the flash temperature, quickly reduces to a value of about 100°C, and then rapidly increases again to a value of about 550°C within the first 25 seconds of testing. At the end of the test the surface temperature is approximately 410°C.

At the onset of braking the friction coefficient is at about 0.28 and goes through a minimum of about 0.2 within the first 6 seconds of testing. As the surface temperature increases through 300°C, the friction coefficient is shown to rapidly increase from its minimum value to a peak value of about 0.5. From this point the friction coefficient ratchets down to a local minimum value of approximately 0.3 about 21 seconds into the test and then increases to a secondary peak of approximately 0.46 about 23 seconds into the test.

The minimum friction coefficient, 0.2, observed in figure 3 is again consistent with water vapor adsorption lubrication [4, 6, 11, 16]. The two peaks in the friction coefficient observed within the first 2 seconds of the test are consistent with water vapor desorption [4, 11, 16], and the secondary peak in friction coefficient of 0.46 which occurred about 23 seconds into the test is consistent with oxygen desorption [16, 17]. The peak friction coefficient observed about 7 seconds into the test occurred as the surface temperature was increasing through about 300°C. This friction peak may be associated with a water vapor desorption event, however, the surface temperature at this point is about 100°C higher than the range of temperatures generally associated with water vapor desorption events. Therefore, the possibility that this apparent desorption event could be associated with some other atmospheric gas or contaminant should not be ignored. The decrease in friction coefficient over the last 17 seconds of the test in figure 3 is probably the result of a combination of oxygen re-adsorption and accumulation of wear debris at the friction interface.

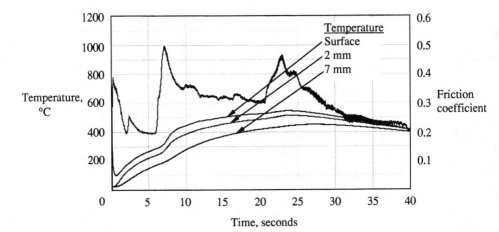

Figure 3: Time histories of temperature and friction coefficient obtained from Boeing dynamometer testing of material B. Service energy stop

A comparison of the data in figure 2 with the data in figure 3 reveals a number of similarities in the friction behavior of materials A and B and some subtle differences as well. Both materials appear to be affected by water vapor and oxygen adsorption/desorption events during the course of testing. However, the abrupt increase in friction coefficients associated with water desorption over a surface temperature range of 140°C to 200°C is more pronounced for material A than for material B. A major desorption event occurred for material B at a surface temperature of 300°C which may have been caused by some contaminant other than water vapor. The oxygen desorption event for material B, on the other hand, was more abrupt than the oxygen desorption event noted for material A. For material B the oxygen desorption event caused the friction coefficient to increase from about 0.3 to approximately 0.46 over a time interval of about 2 seconds. For material A the oxygen desorption event occurred over a time interval of about 10 seconds.

The thermal response of the two materials to the braking tests differs substantially. The peak surface temperature of material A was about 100°C hotter that the peak surface temperature for material B. This difference in peak surface temperature is attributed to two factors, differences in through-thickness thermal conductivity, 14.3 W/m°C at room temperature for material A and 43.9 W/m°C at room temperature for material B, and disk thickness, 22 mm for material A and 26 mm for material B.

Illustrating Carbon Brake Operational Characteristics with Subscale Dynamometer Tests

Figure 4 presents time history plots of temperature and friction coefficient for a subscale dynamometer test of material A subjected to a service energy stop. Occasionally, during the first braking test following an overnight or longer break in testing, the desorption event may be delayed for several seconds as indicated in figure 4. In airline service these delayed desorption events typically occur during the first taxi operations in the morning after the airplane has been parked overnight and have become known popularly as "morning sickness". For the specific test illustrated in figure 4 there had been a two-week break in testing. The data in figure 4 indicate that the friction coefficient during the first 7 seconds of testing is about 0.15. At this point in the test, as the surface temperature was increasing through about 300°C, material A apparently experienced a desorption event and the friction coefficient abruptly increased to about 0.42. There was also a corresponding abrupt increase in the temperature gradient at the time of the desorption event because the rate of energy input abruptly increased. The remainder of the braking test illustrated in figure 4 was typical of other service energy braking tests performed on material A as demonstrated in figure 2, for example. The abrupt increase in friction coefficient 7 seconds into the test may be associated with a water vapor desorption event, however, the surface temperature at this point is about 100°C higher than the range of temperatures generally associated with water vapor desorption events. Therefore, the possibility that this apparent "morning sickness" event could be associated with some other atmospheric gas or contaminant should not be overlooked.

Figure 5 highlights the results of an examination of the material B test specimens following a series of about 230 service energy stops. Figures 5 (a) and (b) are photographs of the stator and rotor, respectively. These photographs reveal a variation in surface brightness under room lighting conditions for both the stator and rotor. For both specimens the bright band extends across approximately a third of the contact area near the inner radius. Also visible in the photographs are a pair of dark bands generally associated with accumulated wear debris within the bright contact band of each specimen. Figures 5 (c) and (d) denote the surface profiles for the stator and rotor respectively. These profiles reveal the formation of a ridge on the stator and a corresponding groove on the rotor within their respective bright contact bands. The surface profile of the stator is inverted and superimposed on the rotor profile in figure 5 (e) to demonstrate that these two profiles are, in deed, mating surfaces. Taken together, the information in figure 5 appears to document the initiation of "record grooving" on the material B test specimens.

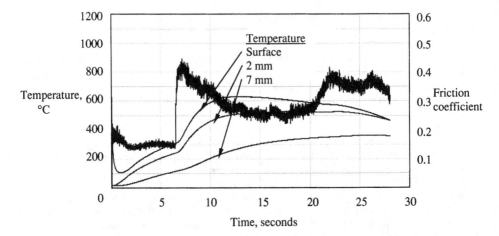

Figure 4: Time histories of temperature and friction coefficient obtained from Boeing dynamometer testing of material A. Service energy stop illustrating a "morning sickness" event.

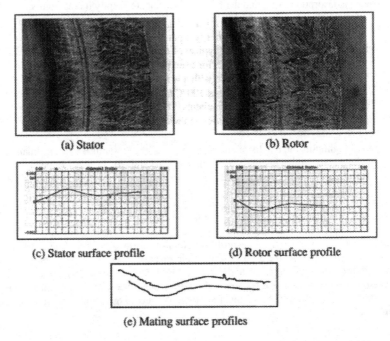

(a) Stator   (b) Rotor

(c) Stator surface profile   (d) Rotor surface profile

(e) Mating surface profiles

Figure 5: Initiation of "record grooving" during dynamometer testing of material B

Figure 6 is a photograph of a material B brake stator returned from service exhibiting extensive "record grooving". The causes of the formation of "record grooving" on carbon brake disks as illustrated in figures 5 and 6 are not well understood.

Figure 6: Photograph of a brake stator returned from service exhibiting "record grooving".

CONCLUSIONS

A literature review was undertaken to document the effort to understand the complex interaction between the carbon brake and the environment in which it must operate. These past research activities have been summarized, and their results have been compared with the operational experiences with carbon-carbon composite brakes. This review has provided some insights into the friction and wear mechanisms that have been observed to date in our testing.

Water vapor in the atmosphere produces a direct lubricant effect on carbon. The amount of water vapor in the environment and the surface temperature of the brake disk influence the degree of lubrication. Observed transition temperatures within the range of 140°C to 200°C, associated with increases in friction and wear of carbon brake materials may be attributed to water vapor desorption. Friction and wear transitions in the temperature range of 500°C to 900°C may be associated with oxygen desorption. Other contaminants such as alcohols, cleaning solvents, wear debris, and the debris from some machining operations can also significantly affect the friction and wear properties of carbon. Adsorption/desorption of water vapor, oxygen, and other atmospheric gases can cause abrupt changes in friction and wear of carbon materials, and may explain such operational brake characteristics as "morning sickness" and "grabby brakes". Graphite, nongraphitic carbon, and carbon-carbon composites are all susceptible to these influences.

REFERENCES

[1]Yen, B. K. and Ishihara, T: "The Surface Morphology and Structure of Carbon-Carbon Composites in High-Energy Sliding Contact". *Wear* **174** 111-117 (1994)

[2]Bragg, W. L.: *Introduction to Crystal Analysis*. G. Bell and Son, Ltd., London (1928)

[3]Midgley, J. W. and Teer, D. G.: "An Investigation of the Mechanism of the Friction and Wear of Carbon". *Transactions of the ASME* **85** 488-494 (1963)

[4]Yen, B. K.: "Influence of Water Vapor and Oxygen on the Tribology of Carbon Materials with SP2 Valence Configuration". *Wear* **192** 208-215 (1996)

[5]Ramadanoff, D. and Glass, S. W.: "High-Altitude Brush Problem". *Transactions of the AIEE* **63** 825-829 (1944)

[6]Savage, R. H.: "Carbon-Brush Contact Films". *General Electric Review* **48** 13-20 (1945)

[7]Shortley, G. and Williams, D.: *Elements of Physics* Third Edition. Prentice-Hall, Inc. (1963)

[8]Savage, R. H. and Schaefer, D. L.: "Vapor Lubrication of Graphite Sliding Contacts". *Journal of Applied Physics* **27** [2] 136-138 (1956)

[9]Lancaster, J. F. and Pritchard, J. R.: "The Influence of Environment and Pressure on the Transition to Dusting Wear of Graphite". *Journal of Physics D Applied Physics* **14** 747-762 (1981)

[10]Savage, R. H.: "Graphite Lubrication". *Journal of Applied Physics* **19** [1] 1-10 (1948)

[11]Lancaster, J. K.: "Transitions in the Friction and Wear of Carbons and Graphites Sliding Against Themselves". *ASLE Transactions* **18** [3] 187-201 (1975)

[12]Lancaster, J. K.: "The Friction and Wear of Nongraphitic Carbons". *ASLE Transactions* **20** [1] 43-54 (1975)

[13]Jaeger, J. C.: "Moving Sources of Heat and the Temperature at Sliding Contact". *Journal and Proceedings of the Royal Society of New South Wales* **76** [3] 203-224 (1942)

[14]Chichinadze, A. V. (Editor): "Polymers in Friction Assemblies of Machines and Devices – A Handbook" Allerton Press, Inc. (1984)

[15]Amberg, R. L. and Tanner, J. A.: "Thermal Response of Carbon Aircraft Brake Materials", presented at 4th International Symposium on the Tribology of Friction Materials, Yaroslavl, Russia. Published in Supplement to Proceedings (2000)

[16]Yen, B. K. and Ishihara, T.: "An Investigation of Friction and Wear Mechanisms of Carbon-Carbon Composites in Nitrogen and Air at Elevated Temperatures". *Carbon* **34** [4] 489-498 (1996)

[17]Peszynska-Bailczyk, K.; Filip, P.; Krkoska, M.; Anderson, K. and Pawliczek, A.: "Study of Adsorption/Desorption Phenomena on Friction Debris of Aircraft Brakes"; presented at American Ceramic Society 29th International Conference on Advanced Ceramics and Composites, Cocoa Beach, FL. To be published in Proceedings (2005)

PROCESSING AND FRICTION PROPERTIES OF 3D-C/C-SiC MODEL COMPOSITES WITH A MULTILAYERED C-SiC MATRIX ENGINEERED AT THE NANOMETER SCALE

A. Fillion, R. Naslain, R. Pailler, X. Bourrat, C. Robin-Brosse
Laboratoire des Composites Thermostructuraux, Université Bordeaux 1, 33600 Pessac, France

M. Brendlé
Institut de Chimie des Surfaces et Interfaces, CNRS
68057 Mulhouse, France

ABSTRACT

3D-C/C-SiC model composites are used to study the influence of the addition of small amounts of SiC on friction properties of C/C composites. Samples are prepared by pressure-pulsed CVI (P-CVI) from toluene and MTS/H_2 precursors. SiC is present in matrix as thin layers (1 to 15 layers, 100-170 nm thick, with $0.4 < V_{SiC} < 6.2\%$) at different distances from fiber surface. Friction properties, as assessed from pin-on-disk (of same nature) experiments in controlled atmosphere at a given temperature, are presented. For $V_{SiC} < 3\%$, the addition of SiC does not change markedly the friction properties whatever the localization of a SiC single layer with respect to fiber surface. Conversely, for $3 < V_{SiC} < 6\%$, the addition of SiC (as 3 to 15 layers) increases the coefficient of friction and the wear rate, both being linearly inter-related. The effect of SiC on the friction properties is tentatively discussed.

INTRODUCTION

C/C composites are extensively used in aircraft and racing car braking systems in place of conventional iron-based alloys on the basis of their excellent friction properties at high temperatures, lower density and longer lifetime. Conversely, they are oxidation-prone and their friction properties at low temperatures are affected by moisture (with occurrence of a low coefficient of friction (COF) regime related to adsorption of species from the environment, such as water, on the active sites of carbon). It has been recently suggested, with a view to extend these materials to other braking systems (including those for cars), to introduce some silicon carbide in the matrix of C/C composites in order to improve their friction properties in terms of COF and wear resistance [1-8]. The best results so far have been achieved with C/C-SiC (Si) composites fabricated through a hybrid process combining : (i) a polymer impregnation and pyrolysis (PIP) step to produce a porous C/C fiber substrate, (ii) a reactive liquid silicon impregnation (LSI) step to fill its open porosity with a SiC + Si mixture and (iii) a SiC-based coating. Steps (i) and (ii) yield a strong and tough material whereas step (iii) provides improved friction properties and oxidation resistance [1, 2, 9, 10]. Car braking systems combining disks of such C/C-SiC (Si) composites with sintered metal-based pads have been reported to exhibit attractive friction properties (high COF, low wear rate), the lifetime of the disks being that of the car. In this example, the friction properties are closely related to the use of a SiC-based coating, the core providing the required mechanical and thermal properties. Further, the composite is usually not sliding against itself but against a material of different nature displaying higher wear rates.

The aim of the present contribution was to assess what could be the effect of the addition of SiC, in small amount and controlled manner, on the friction properties of a 3D-C/C model composite *sliding against itself* in a controlled atmosphere and at a given temperature, on the basis of *pin-on-disk* experiments. It has a fundamental aspect since the material design is at least partly different from that of actual braking materials and the friction test conditions relatively different from those seen in service.

EXPERIMENTAL

The starting material was a 3D-multidirectional carbon fiber preform fabricated from unidirectional (1D) PAN-fiber plies stacked together with a 60° relative orientation, needled in the z-direction, oxidation-cured and heat-treated. The preform densification was achieved by pressure-pulsed chemical vapor infiltration (P-CVI) from toluene for the deposition of pyrocarbon and methyltrichlorosilane (MTS) diluted in hydrogen for SiC. As schematically shown in Fig. 1, SiC was introduced in the pyrocarbon matrix, at the nanometer scale and in a very controlled manner, as one or several (up to 15) thin layers located at various distances from the fiber surface, with an overall volume fraction ranging from 0.4 to 6.2%, that is in small amounts [11, 12]. Three 3D-C/C reference materials have also been used : (i) two composites also fabricated by P-CVI from toluene (RL-To) or propane (SL-Pro) but with no SiC addition and (ii) a composite produced by conventional I-CVI (I standing for isothermal/isobaric, SL for smooth laminar and RL for rough laminar) from natural gas (RL-ICVI). The as-prepared composites were characterized in terms of morphology and phase by scanning electron microscopy (SEM)

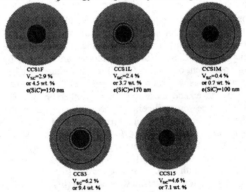

Fig. 1 : Distribution at the nm-scale of SiC within the pyrocarbon matrix in 3D-C/C-SiC model composites : in CCS3, e(SiC) values for the different SiC layers are the same as for CCS1F, CCS1L and CCS1M, in CCS15, they are too low to be accurately measured, adapted from ref. [12, 13].

or transmission electron microscopy (TEM), polarized light microscopy (to assess the pyrocarbon anisotropy through the so-called extinction angle Ae measurement), and X-ray diffraction (XRD) [13].

Friction properties were assessed through pin-on-disk experiments at a nominal temperature up to 400°C (but mostly 200°C) in controlled atmosphere (argon or dry air and, in some cases, wet air) with a sliding velocity ranging from 0.001 to 0.6 m s⁻¹ and an applied stress of 2 to 7 MPa (Fig. 2a) [13, 14]. Both the pins and the disk of same nature and composition, were machined from given composite cylinder as shown in Fig. 2b, their axes being the z-direction of the composite. The friction surfaces were ground with SiC-paper and pre-worn in the tribometer before friction parameter measurement, according to a procedure depicted elsewhere [13]. The coefficient of friction, μ, was calculated as the ratio between the tangential force F_t and the applied load F_n. The wear, w, was defined as the ratio between the mass loss Δm of the disk and the product of the sliding distance l_s time the applied stress σ_n.

RESULTS AND DISCUSSION

The morphology of the 3D-C/C-SiC composites, as observed by SEM, is illustrated in Fig. 3. As expected (see Fig. 1), silicon carbide is dispersed in the pyrocarbon matrix as thin layers, either continuous (when their thickness e(SiC) is high enough, namely of the order of 150 nm) or discontinuous (when it becomes lower than 10 nm), and located at different distance from fiber surface (this feature justifying the use of P-CVI to precisely control the architecture of the matrix) [12]. SiC is mainly present as the cubic β–modification (3C-polytype) with some stacking faults and close to stoichiometry. Conversely and as discussed in a detailed manner elsewhere [15], the nature of the pyrocarbon matrix strongly depends on the nature of the precursor and deposition conditions [13, 15]. In two of the reference materials, the matrix corresponds to conventional pyrocarbon microtextures, namely : (i) a rough laminar (RL) pyrocarbon (dense, $\rho = 2.13$ g cm⁻³ and strongly anisotropic, Ae = 22°) for the RL-ICVI composite and (ii) a smooth laminar (SL) pyrocarbon (less dense, $\rho = 1.95$ g cm⁻³, and less anisotropic, Ae = 12°) for the SL-Pro reference material produced by P-CVI from propane. By contrast, the pyrocarbon in the RL-To reference material as well as in all the 3D-C/C-SiC composites, is dense ($\rho = 2.11$ g cm⁻³) and strongly anisotropic (Ae = 22°) as the RL-pyrocarbon but it displays under polarized light a behavior similar to that of a SL-pyrocarbon. Despite its specific and new microtexture, it will be considered in the following as a RL-pyrocarbon on the basis of its strong anisotropy (see ref. 13 and 15 for a detailed discussion).

Generally speaking, C/C composites display two friction regimes : a high friction regime (with a high COF and wear) and a low friction regime (low COF and wear) with a transition between them depending on different parameters including the sliding velocity, the applied load, the temperature and the atmosphere [14, 16]. For the purpose of comparison between the 3D-C/C and 3D-C/C-SiC composites, the friction tests have been performed under conditions where all the materials were in the high friction regime, namely a temperature of 200°C, an applied normal stress of the order of 3 MPa, a sliding velocity of 0.1 to 0.3 m s⁻¹ and a dry atmosphere (argon or air). The effect of moisture will only be briefly presented in a second step.

181

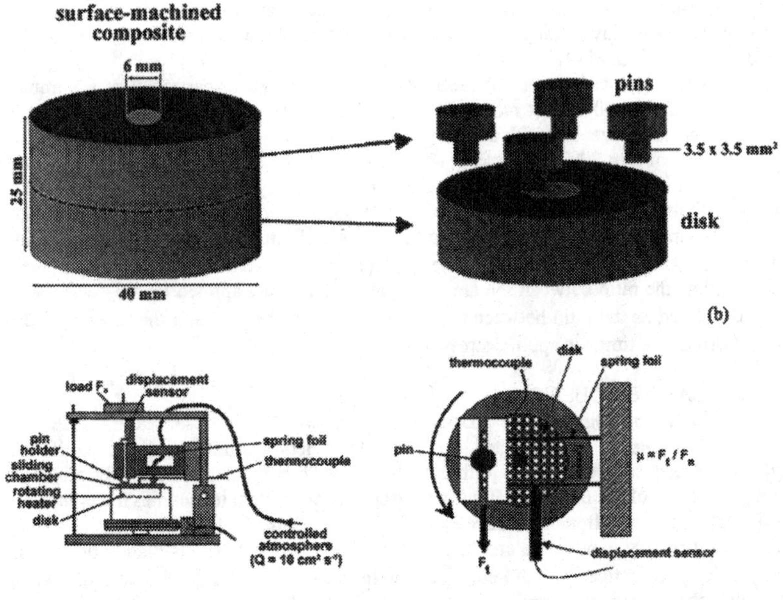

surface-machined composite

6 mm

25 mm

40 mm

pins

3.5 x 3.5 mm²

disk

(b)

Fig. 2 : Pin-on-disk friction tests : (a) pins and disk sample machining, (b) pin-on-disk tribometer, adapted from ref. [13].

The variations of the friction properties of the composites as a function of the SiC-content are shown in Fig. 4 for tests in dry air. It appears that both the coefficient of friction and wear do not change markedly (remaining actually close to those for the related RL-To reference material) for $V_{SiC} \leq 3$ vol. %. In other words, adding very small amounts of SiC as a thin single layer located at different distance of the fiber surface (see Fig. 1) has little effect on the friction behavior of the composites. Conversely, for $3 < V_{SiC} < 6$ vol. %, the addition of SiC as multilayers (3 and 15 layers, respectively) with an overall volume fraction which still remains relatively low, strongly increases both the coefficient of friction (x 2) and the wear of the disk (x 5). Again, the role played by the SiC dispersion mode (3 or 15 layers) is not obvious since the overall SiC content for the two composites is not exactly the same, namely 6.2 and 4.6 vol.% (a feature which suggests that V_{SiC} might be a more important parameter than the SiC dispersion mode, particularly in terms of debris formation in the contact). Further and as shown in Fig. 5, there is at least in a first approximation and for all the composites fabricated by P-CVI from toluene precursor, a linear relationship between the coefficient of friction and the wear of the disk beyond a μ-threshold (where the wear is almost nil) of the order of 0.2. In other words, increasing μ beyond this threshold (by adding more SiC to the pyrocarbon matrix) up to values as high as 0.5 also results in a significant increase in wear of the disk. Interestingly, the effect of SiC on the friction properties of the composites is more pronounced in dry air than in dry argon (where the values of μ and w are already relatively high for $V_{SiC} = 0$).

Surprisingly and as depicted elsewhere [13], the two 3D-C/C reference composites fabricated by P-CVI (RL-To and SL-Pro) with two different microtextures displayed an uncommon friction behavior when tested in *ambient atmosphere* (wet air RH = 38% at 20°C) under conditions where classical RL-CVI composites are in stable high friction regime, even when increasing the sliding velocity and the applied normal load [13]. Adding one single thin SiC-layer to the pyrocarbon matrix ($V_{sic} \leq 3\%$) only results in small change, the friction behavior in wet air remaining unstable. Conversely, the effect of SiC becomes significant when more SiC is introduced in the matrix (3 SiC layers with $V_{sic} = 6.2$ vol.%) : the μ–value was higher (0.5 instead of 0.2 – 0.3) and was achieved for a relatively low sliding velocity (0.03 m s^{-1}). Finally, when SiC was dispersed as 15 discontinuous layers ($V_{sic} = 4.6$ vol. %), the μ–value was lower (close to 0.3) and achieved for a higher sliding velocity (0.1 m s^{-1}).

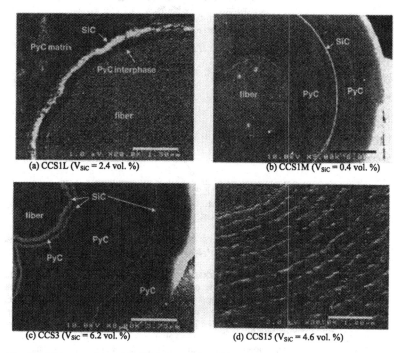

(a) CCS1L (V_{SiC} = 2.4 vol. %) (b) CCS1M (V_{SiC} = 0.4 vol. %)

(c) CCS3 (V_{SiC} = 6.2 vol. %) (d) CCS15 (V_{SiC} = 4.6 vol. %)

Fig. 3 : SEM-images of some 3D-C/C-SiC composites with : one SiC-layer, near the fiber surface (a) or within the pyrocarbon matrix (b), three SiC-layer (c) or 15 SiC-layers (d), adapted from ref. [13].

These friction instabilities correspond to the alternating transition between low and high friction regimes, and are apparently the result of an increased sensitivity of the

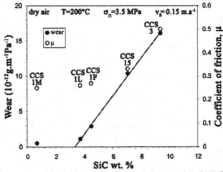

Fig. 4 : Variations of the friction properties of 3D-C/C-SiC composites as a function of SiC content (200°C, 3.5 MPa, 0.15 m s⁻¹, dry air), adapted from ref. [12] and [13].

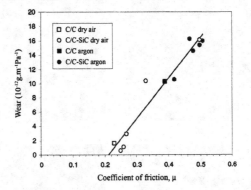

Fig. 5 : Linear relationship between wear and COF for 3D-C/C-SiC and related 3D-C/C fabricated by P-CVI from toluene (200°C, 3 MPa, 0.15 – 0.3 m s⁻¹, dry air or argon), adapted from ref. [12, 13].

critical conditions (characterizing the transition between the low friction regime to the high friction regime) toward a new parameter. Referring to Lancaster [16] the friction transition occurs beyond a critical surface coverage of adsorbed species from enviroment (like water, oxygen, etc ...). For a given adsorbant and a given adsorbate, this surface coverage is mainly controlled by the partial pressure and the temperature. However, while this hypothesis enables us to explain the influence of the nature of the adsorbants [17], it is unsufficient to explain the influence of the material itself. The influence of this latter may result from its chemical interaction with the environment within the contact. Indeed and as described elsewhere [18] a tribocontact may be considered as a triboreactor, where the mechanically created fresh wear debris display dangling bonds able to chemically react with the environment. Within the confined volume associated with the actual area of contact, this may drastically decrease the partial pressure of some adsorbants and thus favor the transition to the high friction regime. In other words, the increased production of wear debris induced by the introduction of small quantities of SiC (with abrasive properties) may well explain the above observed stability of friction. Moreover, the simultaneous increase in friction and wear as a function of the SiC content even suggests that both may derive from the same phenomenon : an increased proportion of dangling bonds within the contact. Conversely, it seems difficult to obtain high values of the coefficient of friction without high wear rates. This is apparently in contrast with the low wear rates reported by previous authors for C/C-SiC (Si) composites. We are tempted to believe that this is only valid for the disk (coated with a SiC-based deposit). The whole system, characterized by the use of pads of different material, probably still displays larger wear rates [1, 2, 9, 10].

CONCLUSION

From the results presented above, the following conclusions can be tentatively drawn : (i) P-CVI is a suitable technique to introduce, in a very controlled manner, SiC in the pyrocarbon matrix of C/C composites, (ii) SiC, even in a small amount, increases the coefficient of friction (up to 0.5) but decreases the wear resistance in the so-called high friction regime, under dry argon or air atmospheres, during pin-on-disk tests at 200°C, (iii) in wet air, the addition of SiC is effective for establishing a stable high friction regime and (iv) the effect of SiC might be related to an increase in the debris formation rate.

References

[1]W. Krenkel, "Design of ceramic brake pads and disks", *Ceram. Eng. Sci. Proc.*, [23] 319-330 (2002).

[2]W. Krenkel, B. Heidenreich and R. Renz "C/C-SiC composites for advanced friction systems", *Adv. Eng. Mater.*, 4, [7] 427-436 (2002).

[3]R. Gadow, "Current status and future prospects of CMC brake components and their manufacturing technologies", *Ceram. Eng. Sci. Proc.*, 21 [3] 15-29 (2000).

[4]J.Y. Paris, L. Vincent and J. Denape, "High-speed tribological behaviour of a carbon/silicon carbide composite", *Composites Sci. Technol.*, 61 (2001) 417-423.

[5]C.P. Ju, C.K. Wang, H.Y. Cheng and J.H. Chern Lin, "Process and wear behavior of monolithic SiC and short carbon fiber-SiC matrix composite", *J. Mater. Sci.*, 35 (2000) 4477-4484.

[6]V.K. Srivastava, "Wear behaviour of C/C-SiC composites sliding against high-Cr steel discs", *Z. Metallkd.*, 94 [4] 458-462 (2003).

[7]H.N. Ko, S. Hanzawa and N. Hashimoto, "Friction characteristics of C/C composite impregnated with silicon", *Key Eng. Mater.*, 247 (2003) 301-304.

[8]C.D. Cho, B. Lee, S.U. Lee and H.J. Cho, "Computer simulation on mechanical evaluation of ceramic matrix composite automobile brake disks", *J. Ceram. Soc. Japan*, Suppl. 112-1 Pac Rim Sp. Issue, 112 [5] S 423 - S 427 (2004).

[9]W. Krenkel, "C/C-SiC composites for hot structures and advanced friction systems, *Ceram. Eng. Sci. Proc.*, 24 [4] 583-592 (2003).

[10]W. Krenkel and R. Renz, "C/C-SiC components for high performance applications", in "*Proc. ECCM-8*, Naples 3-6, 1998" (I. Crivelli-Visconti, ed.), Vol. 4, pp. 23-29, Woodhead Publ., Cambridge, 1998.

[11]R. Naslain, R. Pailler, X. Bourrat and G. Vignoles, "Processing of ceramic matrix composites by pulsed-CVI and related techniques", *Key Engineering Mater.*, 159-160 (1999) 359-366.

[12]R. Naslain, R. Pailler, X. Bourrat, J.M. Goyheneche, A. Guette, F. Lamouroux, S. Bertrand and A. Fillion, "Engineering non-oxide CMCs at the nanometer scale by pressure-pulsed CVI", in High Temperature Ceramic Matrix Composites 5 (M. Singh et al., eds.), pp. 55-62, *Amer. Ceram. Soc.*, Westerville (OH), USA, 2004.

[13]A. Fillion, "C/C and C/C-SiC composites for tribological applications", *PhD Thesis*, n° 2168, Univ. Bordeaux 1, Jan. 13, 2000 (in French).

[14]R. Gilmore, "Friction of carbon-carbon composites and the role played by surface chemistry", *PhD Thesis* n° 94-MULH-0330, Haute Alsace Univ., Mulhouse, 1994 (in French).

[15]X. Bourrat, A. Fillion, R. Naslain, G. Chollon, M. Brendlé, "Regenerative laminar pyrocarbon", *Carbon* **40** (2002) 2931-2945.
[16]J.K. Lancaster, *ASLE Trans.*, 18 (1975), pp. 187-201.
[17]J.K. Lancaster, J.R. Pritchard, *J. of Appl. Phys.*, D, **14**, 4 (1981), pp. 747-762.
[18]M. Brendlé, P. Stempflé, *Wear* **254** (2003) pp. 818-826.

CARBON FIBER-REINFORCED BORON CARBIDE FRICTION MATERIALS

Robert J. Shinavski
Kuo-Chiang Wang
Hyper-Therm HTC, Inc.
18411 Gothard Street, Units A, B & C
Huntington Beach, CA 92648

Peter Filip
Tod Policandriotes
Center for Advanced Friction Studies
Southern Illinois University
Carbondale, IL 62901

ABSTRACT

Ceramic matrix composites are being examined as an alternative to carbon/carbon (C/C) for high performance aircraft brake friction applications. In particular, carbon fiber-reinforced boron carbide (C/B₄C) is of interest for next generation heat sinks due to its higher volumetric heat capacity. Composites were fabricated by chemical vapor infiltration processing of boron carbide into carbon fiber preforms. Thermal and mechanical properties of the resulting composites were then characterized. The effect of elevated temperature exposure, such as would be experienced during use as an aircraft brake friction material, was examined with regards to the stability of the composite microstructure and the effect of the heat treatments on the composite properties.

Tribological properties were evaluated on sub-scale specimens at scaled energy conditions similar to an F-16 aircraft brake. Very low wear rates were measured at low rates of energy dissipation. Significantly higher wear rates resulted at higher rates of energy dissipation. Finite element modeling supports a thermal origin of the observed behavior.

INTRODUCTION

Boron carbide-based ceramic composites are being investigated as next generation friction materials for high performance aircraft braking applications due to their refractory nature and high volumetric heat capacity. An improved heat capacity can result in greater heat dissipation or the design of a smaller brake package. In addition, improvements are desired in minimum friction coefficient and wear rate. An improved wear rate is particularly important as this can result in a longer service life and subsequently a reduced life-cycle cost, which would mitigate the higher predicted cost of a boron carbide matrix composite as compared to carbon/carbon.

The high performance aircraft brake environment is unique in that the material must meet not only mechanical and thermal specifications, but must also demonstrate a low wear rate and a stable friction coefficient over a wide range of energy dissipation rates. The material must also have a low density and a high use temperature. Temperatures of greater than 1600°C can occur in C/C friction materials during a rejected take-off, and thus any candidate material must also be capable of sustaining such extreme temperatures.

Boron carbide ceramics have an acceptably high melting temperature and a low density in addition to its high volumetric heat capacity. Boron carbide is also widely reputed for its extreme hardness and excellent wear resistance in contact with other materials. However, limited studies of boron carbide-boron carbide friction couples were found in the literature.[1-2] The study of Larsson et al.[1] demonstrated that a range of friction coefficients from 0.1-0.9 can be obtained, and that widely disparate wear rates can result that are dependent on humidity and temperature. Low wear rates were associated with tribochemical film formation that was more likely to occur at higher humidity levels and higher sliding speeds. The present work differs substantially in that the energy dissipated by the frictional contact is substantially higher, and that a fiber-reinforced boron carbide matrix composite is being examined. Evaluation of a C/B$_4$C composite including mechanical, thermal and tribological properties with respect to high energy friction applications, such as in a high performance aircraft brake, is examined.

MATERIALS

A carbon-fiber-reinforced boron carbide composite material was produced by chemical vapor infiltration for this study. The fiber preform was composed of needled polyacrylonitrile (PAN)-based carbon fibers with a 24% fiber volume obtained from Goodrich Corporation. Carbon fiber was selected based on its combination of high temperature strength retention and low cost. The preform utilized was similar to those currently used in aircraft brake friction materials. The preform selection also allowed direct comparison with previously studied silicon carbide matrix ceramic friction materials.[3]

The first step in material fabrication was the deposition of a 0.5 μm pyrolytic carbon fiber coating deposited by the thermal decomposition of hydrocarbon gas. The pyrolytic carbon fiber coating prevents chemical bonding of the boron carbide to the carbon fiber surface, while possessing a sufficiently compliant structure to arrest or deflect matrix cracks from causing fiber failure. These properties result in a significantly tougher and strain tolerant composite than a monolithic ceramic.

The boron carbide matrix was then deposited in the preform by chemical vapor infiltration (CVI). Boron carbide was formed by the reduction of boron trichloride in the presence of a hydrogen-hydrocarbon gas mixture at elevated temperature and reduced pressure. The CVI process was continued until the densfication rate of the composite ceased to increase when surface porosity was fully densifed with boron carbide; thereby removing access to the remaining internal porosity. Figure 1 shows a metallographic cross-section of the carbon fiber-reinforced boron carbide. Image analysis indicates a residual porosity of 14.7%. Helium pycnometric measurements of 13.9% confirm this porosity level. A bulk density of 1.81 g/cm^3 was measured. This bulk density suggests a matrix density of 2.1-2.2 g/cm^3; significantly less than the 2.51 g/cm^3 for crystalline boron carbide. Utilizing transmission electron microscopy, electron energy loss spectroscopy and energy dispersive spectroscopy, the matrix was shown to be of stoichiometric composition, within the accuracy of the measurements, and amorphous in structure[4].

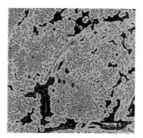

Figure 1. Metallographic cross-section of C/B₄C.

PROCEDURES

After fabrication of C/B₄C, mechanical and thermal properties were measured before and after elevated temperature exposures. Heat treatments of 1400°C-2000°C were conducted under vacuum with a dwell time of one hour. The mechanical and thermal properties were then re-evaluated to determine the influence of the thermal conditioning. Wear testing was conducted on a Link dynamometer with a single rotor and stator. The frictional contact area of the rotor and stator were 11.75 cm OD and 9.21 cm ID. Both the rotor and the stator were 12 mm thick. Details of the dynamometer test conditions utilized to simulate an F-16 aircraft brake have been summarized elsewhere.[5] Wear testing consisted of fifty simulated landings for each energy examined. Three simulated taxi stops accompanied each landing. Wear rate was measured after each set of fifty landings and associated taxis.

RESULTS

Significant non-linear stress-displacement behavior was observed in flexural testing of the materials (Figure 2). A fracture toughness significantly greater than a monolithic material was also obtained. Mechanical properties measured on the as-fabricated C/B₄C are summarized in Table I.

The heat capacity of the C/B₄C was found to be 0.846 J/g/K at ambient conditions. Calculation of the heat sink energy density by integration of the heat capacity up to 1600°C shows that 5.5 kJ/cm3 of energy can be absorbed by the heat sink. This calculated value exceeds the predicted value for C/B₄C and represents a significant improvement in heat sink capacity.

Table I. Mechanical properties of as-fabricated C/B₄C.

Flexural Strength	(through-thickness)	152±34 MPa
	(transverse)	147±12 MPa
Compressive Strength		292±78 MPa
In-Plane Shear Strength	(double-notched)	97±14 MPa
Fracture Toughness	(single-edge notched bend)	7.3±0.9 MPa m$^{\frac{1}{2}}$

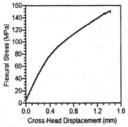

Figure 2. Significant non-linearity exists in the flexural properties of C/B₄C as shown by the through-thickness stress-displacement.

The laser flash method was utilized to measure the thermal diffusivity of the C/B₄C. The diffusivity decreases with increasing temperature, however an increase is measured on cooling as a result of the heating to 1400°C (Figure 3). This suggests that a change occurred within the material that increased its thermal diffusivity as a result of the thermal treatment. X-ray diffraction (Figure 4) shows strong crystalline peaks for boron carbide that were not present in the as-fabricated material. The thermal conductivity can be calculated to be 3.8 W/m/K at ambient, and increases to 4.4 W/m/K after being heated to 1400°C.

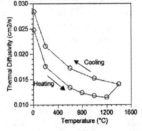

Figure 3. Thermal diffusivity decreases with increasing temperature and shows the effect of boron carbide crystallization that occurs between 1200-1400°C

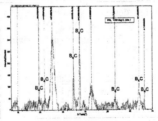

Figure 4. X-ray diffraction shows the presence of crystalline boron carbide after heating to 1400°C. The other diffraction peaks correspond to graphitic carbon.

Further investigation into thermal and mechanical property changes induced by heat treatments as high as 2000°C was also investigated. Figure 5 summarizes the mechanical properties after elevated temperature exposure. Flexural properties of the composites do not

190

change significantly due to the crystallization of the boron carbide matrix at 1400°C. However, the matrix dominated shear strength was observed to decrease; presumably due to shrinkage of the matrix during crystallization. At temperatures of 1600°C and greater, mechanical properties were significantly reduced by nominally 50%. A near complete loss of mechanical integrity occurred after heating to 2000°C to the extent that many of the mechanical tests could not be conducted due to failure of the test specimens during handling after heat treatment. Metallographic examination showed extensive reaction of the carbon fiber reinforcement and the boron carbide matrix (Figure 6). Such reaction was not observed on metallographic cross-sections of the C/B$_4$C heated to lower temperatures. A likely cause of the interaction between the carbon and boron carbide is interdiffusion of the constituents and the solubility of excess boron and carbon in boron carbide. Thermal diffusivity was also measured after the elevated temperature exposures, and thermal conductivities were calculated for ambient conditions (Figure 7). The ambient thermal conductivity increases steadily with increasing heat treatment temperature, which suggests that in addition to crystallization that either grain growth or that limited interdiffusion of the carbon and boron carbide may occur at temperatures less than 2000°C.

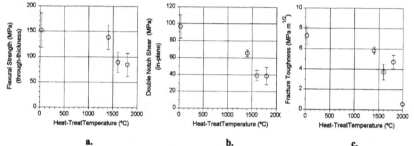

a.   b.   c.

Figure 5. Room temperature mechanical properties after heat treatment: a. flexural strength, b. in-plane shear strength, c. fracture toughness.

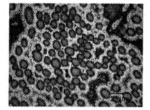

Figure 6. Microstructure of C/B$_4$C after heating to 2000°C shows significant reaction of the fibers and matrix.

High energy density dynamometer testing was performed to measure the friction and wear properties under the service conditions of an aircraft brake. The average friction coefficient of the C/B$_4$C for landing stops was consistently between 0.45-0.50 with the exception of the conditions simulating low energy taxi stops and the rejected take-off (RTO), where the average coefficients were 0.64 and 0.42, respectively (Figure 8).

191

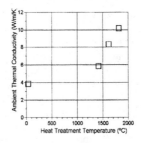

Figure 7. Ambient thermal conductivity increases as a result of heat treatment.

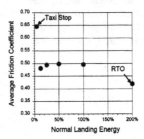

Figure 8. Average friction coefficient as a function of landing energy.

However, the friction coefficient was found to change significantly during an engagement. The friction coefficient was low when significant energy (high inertia and high rotational velocity) was in the system, and increased significantly when the rotational velocity was lower. This behavior is most dramatically shown by examining the evolution of the friction coefficient during the RTO stop as shown in Figure 9. The remaining kinetic energy or rotational velocity was noted to be more significant than the energy that had been dissipated in determining the friction coefficient (i.e both high and low energy stops showed the same high friction coefficient near the end of the stop). This high friction coefficient may be a test artifact of the minimal pressure required at the low energy levels to result in a constant deceleration rate. The friction coefficient was found to decrease with increasing applied pressure.

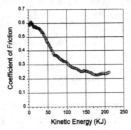

Figure 9. RTO friction coefficient changes significantly during the engagement.

The wear rate was found to be significantly less than C/C tested at similar conditions[5] at energy levels up to 50% normal landing energy. However at higher energy levels, the wear rate increased dramatically to an unacceptably high level. Figure 10 presents a comparison of C/C and C/B₄C wear rates that show the transition of C/B₄C from low wear to high wear conditions when the energy increases from the 50% to 100% energy levels.

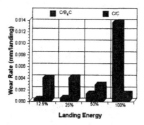

Figure 10. Wear rate of C/B₄C determined from sub-scale dynamometer testing compared with C/C[5].

DISCUSSION

C/B₄C was demonstrated to have a significantly higher volumetric heat capacity than C/C, and exceptional wear properties at low energy levels. The low friction coefficient that accompanied the low wear rates found by Larsson was not observed. As-fabricated mechanical properties also provide the required strength and toughness necessary for utilization as an aircraft brake friction material. However, reaction of the carbon fiber and B₄C was found to be significant at temperatures >1800°C, but mechanical and thermal properties indicate that the reaction may begin at 1600°C. For utilization of B₄C as a matrix with high heat capacity, the carbon fiber must be protected from reacting with the B₄C. Additionally an unacceptably high wear rate was measured at higher energy levels.

Finite element modeling was utilized to provide insight into potential thermal origins of the high wear rate. The modeling assumed adiabatic conditions of the rotor and stator. The modeling indicated that the steel attachment fixtures did not significantly change the rotor and stator temperatures from a fully adiabatic case given the time frames and thermal conductivity of the material. Figure 11 shows a comparison of the maximum temperatures from the modeling and measured temperatures at 6.3 mm from the surface. The agreement between the experimental data and the modeling validates the assumptions and the input data utilized for the

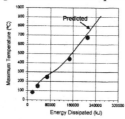

Figure 11. Comparison of finite element model and experimental data for predicting maximum temperature at 6.3 mm from the friction surface.

193

model. Figure 12 shows the maximum thermal gradient that results midway through the simulated rejected take-off, where a maximum surface temperature of 1361°C is predicted and the thermal gradient is in excess of 150°C/mm near the surface. At 100% normal energy conditions the maximum surface temperature of the stator is predicted to be a more modest 900°C. Therefore overheating of the surface cannot explain the accelerated wear observed. However given the thermal gradient, the dimensional change of the composite may be a factor. The high wear region was characterized by very "noisy" stops when the applied pressure was low near the end of the stop suggesting that the friction surfaces may not remain parallel.

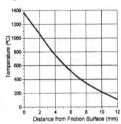

Figure 12. An extreme thermal gradient is predicted due to the low thermal diffusivity of the material.

CONCLUSIONS

Carbon-fiber reinforced boron carbide is a promising candidate for high energy friction applications due to its high heat capacity. However, the thermal diffusivity must be significantly improved such that the enhanced heat capacity can efficiently be utilized and that extremely high surface temperatures and/or extreme temperature gradients do not result. Interdiffusion of the carbon and boron carbide at elevated temperatures was identified as a potential limitation, but finite element modeling suggests that such temperatures do not occur under overload conditions.

ACKNOWLEDGMENT

This Work was funded by the Air Force Research Laboratory Materials and Manufacturing Directorate under contract F33615-01-C-5209.

REFERENCES

[1] P. Larsson, N. Axén, and S. Hogmark, "Tribofilm Formation on Boron Carbide in Sliding Wear," *Wear*, **236**, 73-80, 1999.

[2] E. Rabinowicz, *Friction and Wear of Materials*, 2nd edition, John Wiley & Sons, 1995.

[3] S.Vaidyararaman, M Purdy, T. Walker, and S. Horst, "C/SiC Material Evaluation for Aircraft Brake Applications," *High Temperature CMC Conference*, Paper No. 119, 2001.

[4] R. J. Shinavski and M. K. Cinibulk, "Boron Carbide Matrix Composites for Advanced Aircraft Brakes," *Proc. of the 27th Annual Conf. on Composites, Materials, & Structures*, 2003.

[5] T. Policandriotes, T. Walker and P. Filip, "Subscale Aircraft Dynamometer Testing on C/SiC," *Proc. of the 28th Annual Conf. on Ceramics, Metals, and C/C Composites, Materials & Structures*, 2004.

THERMAL SHOCK IMPACT ON C/C AND SI MELT INFILTRATED C/C MATERIALS (SIMI)

Dale E. Wittmer and Peter Filip
Southern Illinois University
1230 Lincoln Drive-MS6603
Carbondale, IL 62901

ABSTRACT

Wear of carbon-carbon materials for use in brake systems is a product of many dimensions. In operation, brakes see very high temperatures and are subject to high thermal shocking loads. This investigation deals with a baseline a commercial pitch fiber/charred resin matrix/CVI carbon/carbon (C/C) composite and the same composite infiltrated with silicon by melt infiltration (SiMI). The objectives of this work were to explain the improved frictional performance and fracture resistance of the SiMI materials, following the thermal shocking experiments, and compare their behavior to that of the commercial C/C composite.

Friction properties and wear were determined following sub-scale dynamometer testing of disc on disc, where the testing conditions were dry (5 to 18% relative humidity) or wet (>50% relative humidity). Different energy levels were used as a proportion of normal landing energy (NLE), with three taxi and one landing sequence per cycle. A total of 50 cycles were run on each disc pair.

The results showed that the SiMI materials had higher and more stable coefficient of friction (μ) than the C/C composite; however there was more oscillation in μ and more noise related to the SiMI materials compared with the C/C. In addition, the SiMI process eliminated the sensitivity to moisture.

INTRODUCTION

Frictional carbon/carbon based composite materials are repeatedly subjected to sudden temperature increases as a result of numerous braking operations. The kinetic energy of a vehicle (airplane) is predominantly converted into heat during braking [1]. While carbon fiber reinforced carbon matrix composites (C/C) represent a very suitable material for environments with extreme temperature loads [2], the performance of modified silicon melt infiltrated C/C materials (SiMI) was questioned. This is in part due to differences in thermal expansion between the constituents of the composite: carbon (fiber and matrix, coefficient of thermal expansion being approximately 0.5×10^{-6}/K) and Si-containing products (Si, SiC, SiO_2, SiCO, coefficient of thermal expansion ranging between 3.8 to 0.8×10^{-6}/K). It was hypothesized that differences in thermal expansion between the carbon and Si-containing materials would lead to the formation of additional cracks, resulting in a negative impact on the mechanical and frictional properties of these SiMI composites. Nevertheless, the frictional performance (stability and level of the coefficient of friction, sensitivity to moisture, and wear rate) as detected in the subscale dynamometer tests was reported to be superior for materials containing Si [3]. The objectives of this work were to explain the improved fracture resistance of the SiMI materials, following the thermal shocking experiments, and compare their behavior to that of the commercial C/C composite.

EXPERIMENTAL PROCEDURES

Discs for subscale dynamometer tests were machined from commercial C/C material. Pitch fiber (2 dimensional randomly chopped), charred resin and chemical vapor infiltration (CVI) were the components of the C/C material. Discs with an outer diameter of 95 mm were used in the subscale dynamometer test, at different simulated normal landing energy (NLE) conditions. The energy conditions ranged from that of taxi (4.5% of NLE) to rejected take off (RTO which is 200% of NLE), at dry (5 to 18%) or wet (50 % relative humidity) conditions. The identical discs made of the same C/C material were infiltrated with silicon at temperatures higher than the melting point of Si (1401°C) to make the SiMI materials. Identical friction tests were performed on the SiMI materials. Two samples randomly cut from different C/C discs and two randomly selected SiMI samples were used in the thermal shock experiment (Table I).

Table I. Identification of samples.

Sample notification	formulation
1	C/C
2	C/C
3	SiMI
4	SiMI

After subscale dynamometer testing, three segments were cut out of each disc and subjected to thermal up-shock. The thermal up-shock treatments were performed in a Centorr-Vacuum Industries belt furnace under a flowing nitrogen atmosphere. Temperature increases of approximately 1000°C/minute were applied to the samples in several repeated cycles. Three maximum temperatures were selected for thermal shock treatment: a) 1200°C, when Si does not melt, b) 1450°C, above melting point of Si, and c) 1800°C, which represents a calculated surface temperature for extreme landing conditions. Four thermal shocks cycles were applied to samples at 1200°C and 1450°C, and two thermal shocks were applied to samples at 1800°C. In order to achieve the up-shock conditions desired, the belt furnace conditions were: the belt speed was 11.43 mm/min and cycle duration was 24 min at 1800°C, and the belt speed was 16.51 mm/min and cycle duration was 17 min at both 1200 and 1450°C.

After each run, the mass of each of the samples was monitored using a Mettler-Toledo microbalance. The surfaces of each sample were inspected using a Hitachi D2460 SEM, equipped with energy dispersive X-ray microanalysis (EDX). Samples were also subjected to standard X-ray diffraction (40 KV, 30 ma, $Cu_{K\alpha}$ radiation, Ni filter, 1.8°-2θ/min scan rate).

RESULTS AND DISCUSSION
Friction Studies
Coefficient of Friction

The average values of the coefficient of friction as detected in of the dynamometer tests are given in Fig. 1. These results show that the SiMI materials had higher and more stable coefficient of friction (μ) than the C/C composite; however there was more oscillation in μ and more noise related to the SiMI materials compared with the C/C. In addition, the SiMI process eliminated the wear sensitivity to moisture that has been observed previously.

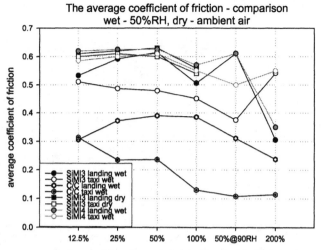

The average coefficient of friction - comparison
wet - 50%RH, dry - ambient air

testing conditions (amount of normal landing kinetic energy)

Figure 1. Coefficient of friction test results for C/C and SiMI composites.

Friction Surface

Figure 2 shows a comparison of the friction surfaces of the C/C composites compared with the SiMI composites, following friction testing. The C/C is obviously worn with the CVI matrix fractured and pulled out of the composite. There is also significant damage to the carbon fiber. In contrast, the SiMI composite is very smooth with little wear damage observed.

Up-Shock Treatment

Mass changes during shock treatment

Table II lists the detected mass changes of samples following each individual up-shock treatment. Since a protective nitrogen atmosphere was used in the experiment, the mass of C/C and SiMI samples remained practically unchanged during all up-shock treatments performed at different temperatures. This indicated that no significant oxidation of carbon occurred, however the SiMI samples developed a green SiC coating on their surface following up-shock, indicating a reaction between the remaining free Si and the free carbon in the composites. This is also why there was no measurable change in weight observed.

X-ray diffraction

While the X-ray diffraction experiment did not show any significant changes in the microstructure of C/C materials treated at different conditions, it was possible to see changes in the diffraction spectra of samples containing Si. The SiMI samples, inspected before exposure to thermal shocking, exhibited the presence of crystalline carbon, silicon, and silicon carbide. XRD taken after the up-shock treatment indicated the absence of metallic silicon, indicating that the free silicon in the composites had reacted with the free carbon to form SiC. These results agreed with the observed green SiC tint on the samples surfaces following the up-shock treatments.

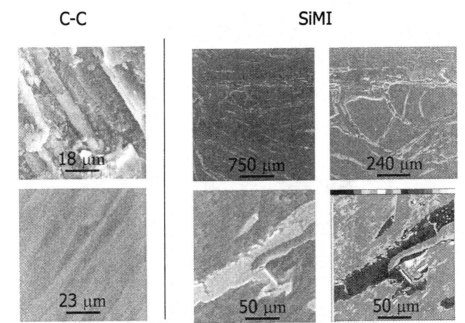

C-C	SiMI

Figure 2. SEM micrographs of C/C and SiMI composites following friction testing.

Table II. Detected mass of samples after thermal up-shock treatment.

Temperature and sample	ORIGINAL MASS [g]	MASS AFTER THERMAL UP-SHOCK [g]			
		Cycle 1	Cycle 2	Cycle 3	Cycle 4
1800°C					
1	8.963	8.963	8.957		
2	17.747	17.742	17.73		
3	14.386	14.388	14.376		
4	11.777	11.782	11.777		
1450°C					
1	8.557	8.555	8.555	8.553	8.553
2	18.954	18.955	18.962	18.955	18.954
3	14.332	14.333	14.335	14.338	11.337
4	11.429	11.432	11.432	11.435	11.435
1200°C					
1	7.403	7.403	7.399	7.405	7.403
2	16.092	16.083	16.084	16.09	16.083
3	14.131	14.124	14.122	14.125	14.121
4	8.525	8.520	8.523	8.524	8.520

198

SEM Studies

A detailed SEM analysis addressed the character and frequency of microcracks, present on the samples surfaces. Figure 3 shows the typical surface morphology of the samples (C/C in Figs. 3(a) and (b)) and SiMI Figs. 3(c) and (d)) before the up-shock treatment was applied. The C/C materials are typified by the presence of numerous microcracks, which are typically located in the vicinity of charred resin pockets, as easily seen in Fig. 3(a).

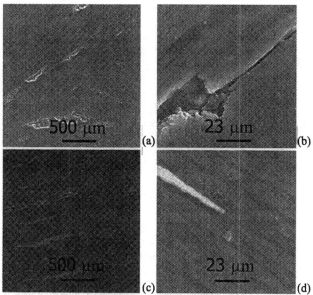

Figure 3. Surface of samples before thermal up-shock was applied. (a) and (b) C/C, (c) and (d) SiMI. The bright areas in (c) and (d) correspond to Si and its reaction products with C and O.

Figure 3(b) is a detailed view of the C/C material surface, indicating that charred resin (right bottom) is separated from the fiber bundle (upper left) by a deep microcrack. It is well known that the formation of these microcracks is related to differences in thermal expansion between the charred resin and carbon fiber during the heating and cooling processes applied during heat treatment [1].

In contrast with the C/C samples, the microcracks in the SiMI materials were typically filled by Si and products of Si reactions with C and O (SiC, SiO_2, and SiCO). An example is shown in Figure 3(c) and a detailed view is given in Figure 3(d). It is easily seen from the detailed view (Figure 3(d)) that the microcracks are very well filled and sealed by the silicon and products of its reaction. However, it is necessary to note that the general level of filling of the original microcracks would depend on the amount of silicon available.

SEM studies revealed that the metallic silicon present in the microstructure of original SiMI materials (before shock treatment) evaporated during the thermal shock treatment and subsequently condensed on the surfaces of samples and reacting with the surface free carbon in the process. This process is quite remarkable at 1800°C and progresses to a lesser extent at lower temperatures. Figure 4(a) shows a detailed view from the surface of the sample after two up-shock cycles were applied at 1800°C. Energy dispersive microanalysis (Figure 4(b)), in combination with XRD results, indicates that bright particulates visible on the surface of the sample, shown in Figure 4(a), correspond to SiC. Apparently, SiC must have been formed after evaporation and subsequent condensation of silicon on the surface of the composite. Obviously the silicon condensing on the C/C surface reacted with C and formed SiC crystals. As mentioned previously, green hexagonal SiC was detectable by the naked eye, covering the surface of the SiMI samples. Figure 2(a) also shows a microcrack in the vicinity of the Si infiltrated region.

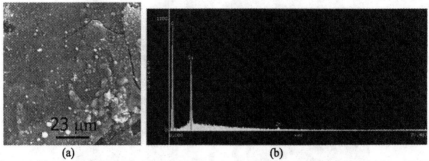

| (a) | (b) |

Figure 4. The surface of SiMI sample subjected to two up-shock cycles at 1800°C (a) and EDX spectrum taken from a bright particle (b). Bright particles represent SiC.

A careful quantitative analysis of the microstructure revealed that the C/C materials contained approximately 11.6 vol. % voids (microcracks, pores), whereas the SiMi samples contained only 5.6 vol. % voids. Clearly, the silicon infiltration reduced the presence of voids and microcracks on the surface of samples before the up-shock treatment was applied.

Thermal shocks applied to C/C materials did not change the structure significantly. Figures 5(a) through 5(c) show the presence of microcracks in the C/C samples subjected to thermal up-shocks at 1200°C, 1450°C, and 1800°C, respectively. Characteristic separation between fiber and matrix is easily visible. In addition, it seems that the charred resin matrix might have undergone additional cracking when shock treatment at 1450° and 1800°C were applied (see Figs. 5(b) and (c), respectively). Again, the bright particles, visible on the surface of C/C materials shown in Fig. 5, represent SiC generated during the shock treatment. These samples were located in the vicinity of SiMi samples and Si released from SiMI was also deposited on C/C samples.

The characteristic surface of SiMI materials subjected to thermal shocks at 1200°C, 1450°C, and 1800°C are shown in Figs. 6(a), (b) and (c). When they are infiltrated with Si, the reaction products are formed in these voids and metallic silicon remains in the microstructure (Figs. 6(a) and (b)).[3] Figure 6(b) provides a view of the filled (upper) and unfilled (lower part) cracks. The separation between resin matrix pockets and fiber bundles, which is typical for C/C composites, is missing in the SiMI composites when sufficient amounts of Si are available. This behavior was very observable in all three micrographs, which are shown in Fig. 6.

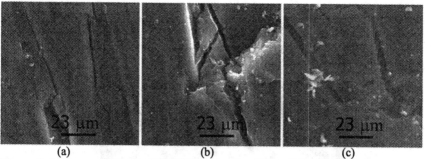

| (a) | (b) | (c) |

Figure 5. Microcracks detected in C/C materials when subjected to up-shock treatment at temperatures of: (a) 1200°C, (b) 1450°C and (c) 1800°C.

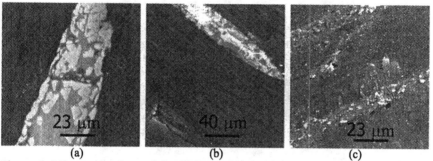

| (a) | (b) | (c) |

Figure 6. Microcracks detected in SiMI materials when subjected to shock treatment at temperatures of: (a) 1200°C, (b) 1450°C and (c) 1800°C.

In contrast with the C/C materials, the SiMI samples exhibited a lower frequency of microcracks at every level of shock temperature. Si infiltrated the composites and very often sealed the original microcracks which were present in C/C material. It is easily visible during the up-shock cycles applied at lower temperatures (1200°C and 1450°C), where the differences in the coefficient of thermal expansion between C and SiC, SiO_2, or SiCO do not play a significant role. Heating to 1200°C and 1450°C did not cause any additional crack formation. At the highest applied temperature of 1800°C, two types of newly formed microcracks were detected (Fig. 6c). Perpendicular microcracks with respect to a longitudinal axis of the infiltrated Si formed preferentially, however, parallel microcracks were also observed.

Apparently, these microcracks originated in the Si infiltrated composites. It is very probable that the sharp edges of SiC located in Si metallic matrix, which are very visible in Fig. 6(a), as well as differences in thermal expansion of Si and SiC, represent the major contributors to crack formation initiated in Si the infiltrated regions. Apparently, at lower temperatures, when Si is present in metallic form, the microcracks do not propagate to the C/C surrounding material. A Si metallic matrix can easily compensate the thermal stresses by plastic flow. The crack formation and growth is significantly limited at these conditions. At 1800°C, however, the metallic Si evaporates (see Fig. 6(c)) and the larger thermal stresses are not easily compensated.

Figure 7 shows distributions of C, Si, and O in the area already shown in Fig. 4. This represents a detail from Fig. 6(c). It can be seen that Si was smeared over the surface and the microcrack propagates between Si-rich (probably SiC) and carbon-rich (carbon) regions. It is obvious that the original hypothesis assuming the role of differences in the coefficients of thermal expansion is based on a realistic supposition.

Figure 7. EDX mapping indicating distribution of C (upper left), Si (upper right), and O (lower left) over the area shown in the microstructure (lower right).

It is also obvious that the "perpendicular" microcracks were not present in C/C materials without Si infiltration. The "parallel" microcracks observed in SiMI materials are rare and significantly smaller when compared to microcracks detected in C/C samples after shock treatment.

In general, the microcracks formed in SiMI are less frequent and smaller when compared to microcracks detected in C/C materials. Since the frequency and size of microcracks is lower in SiMI when compared to C/C samples, it is possible to conclude that Si melt infiltration reduces the amount of voids and thermally induced microcracks within the investigated temperature range. The distribution of oxygen is preferentially limited to areas where Si is present. Since the formation of SiO_2 type glass and microcrystalline matter is well known for its anti-oxidation protection [4], it is obvious that the Si infiltration not only improves the resistance towards crack formation but also may inhibit the oxidation of carbon.

CONCLUSIONS

SiMI materials had a higher and more stable coefficient of friction (μ) than the C/C composite; however there was more oscillation in μ and more noise related to the SiMI materials compared with the C/C. In addition, the SiMI process eliminated the wear sensitivity to moisture.

Silicon infiltration reduced the presence of microcracks on the surface of carbon/carbon discs. A significantly smaller size and frequency of microcracks was detected in SiMI samples when compared to C/C samples. The silicon present in the structure evaporated during the heating of brake discs, condensates on the disc surfaces and forms silicon carbides and oxides (heating applied in protective nitrogen atmosphere). This leads to increased oxidation protection.

REFERENCES

[1] C. Blanco, J. Bermejo, H. Marsh, and R. Menendez, Chemical and Physical Properties of Carbon as Related to Brake Performance, in: Carbon and Carbonaceous Composite Materials – Structure-Property Relationship, Ed. By K Palmer et al., 1996, World Scientific, Singapore, pp. 331 - 363.

[2] I. L. Stimson and R. Fisher, Design and Engineering of Carbon Brakes: New Fibers and Their Composites, Phil. Trans. R. Soc. London, A294 (1980) pp. 583 - 590.

[3] P. Filip and A. Pawliczek, Influence of Varying Energy and Humidity Conditions on Frictional Performance of C/C and C/C/SiC Composite Materials, CAFS Quarterly Report Vol. 7, No. 2, April 2003, pp. 27 – 125.

[4] W. Kowbel, J. C. Withers, and P. O. Ransone, CVD and CVR Silicon-Based Functionally Gradient Coatings on C-C Composites, Carbon, Vol.33, No. 4, (1995), 415 -426.

Reliability of Ceramic and Composite Components

POST ENGINE TEST CHARACTERIZATION OF SELF SEALING CERAMIC MATRIX COMPOSITES FOR NOZZLE SEALS IN GAS TURBINE ENGINES

E. Bouillon, C. Louchet, and P. Spriet
SNECMA Propulsion Solide
33187 Le Haillan, FRANCE

G. Ojard, D. Feindel, C. Logan, and K. Rogers
Pratt & Whitney
400 Main Street
East Hartford, CT 06108

T. Arnold
SNECMA Moteurs, Villaroche
77550 Moissy-Cramayel, FRANCE

ABSTRACT

The advancement of self-sealing ceramic matrix composites offers durability improvements in hot section components of gas turbine engines. These durability improvements come with no need for internal cooling and with reduced weight. Building on past material efforts, ceramic matrix composites based on either a carbon fiber or a SiC fiber with a sequenced self-sealing matrix have been developed for gas turbine applications. The specific application being pursued on this effort is an F100-PW-229 nozzle seal. Ground engine testing has been done to full design life for some seals. One material system has been tested to 150% of design life. The ground testing has demonstrated a significant durability improvement from the baseline metal design. Residual properties are being determined for both systems by extracting tensile and microstructural coupons from the ceramic matrix composite seal. This was done as a function of design life. Nondestructive interrogation was used as a guide in setting cutting diagrams. The results from this effort will be presented.

INTRODUCTION

There are multiple considerations that need to be balanced as ceramic matrix composites (CMC) are considered for advanced gas turbine engine applications. Not only must the CMC be able to perform as a direct replacement, additional benefits need to be realized to make the whole replacement and demonstration effort worthwhile. A specific application that has been seeing a lot of recent efforts as well as multiple CMC systems is nozzle applications for gas turbine engines for jet fighters[1-3]. This is due to the improved high temperature capability that a CMC offers as well as durability improvements. There is also the added benefit that the CMC components do not need cooling air for them to operate in these conditions. In addition, there are weight benefits since the CMC has a fourth of the density of the metal being considered for replacement.

There is interest in looking at a new class of CMC beyond those referenced above[4-6]. Through generational material developmental work by SNECMA Propulsion Solide, a novel sequenced CVI matrix that is self-sealing (protects the fiber matrix interface) has been developed for gas turbine engine applications[6]. For the application of interest, this is coupled with a multilayer woven reinforcement that inhibits delamination formation[5,6]. A novel matrix technology has been developed that combines carbides deposited by CVI process with specific sequences of Si,C,B which forms protective oxides at a wide range of temperatures[6].

MATERIAL

This effort is focused on evaluating two ceramic matrix composites as nozzle seals for the F100-PW-229 fighter engine that powers both the F-16 and F-15 fighters. The first material considered was the CERASEP® A410, which is a Hi-Nicalon SiC fiber in the sequenced self-sealing matrix. In addition, SEPCARBINOX® A500 was looked at and this varied from the A410 material as a carbon fiber replaced the SiC fiber as a cost reduction.

Coupon Fabrication

Plates were woven as an 8HS weave with a multilayer reinforcement. The fiber was either carbon or Hi-Nicalon™ (SiC). The number of layers was chosen to obtain a composite thickness of 4 mm before the start of the Chemical Vapor Infiltration (CVI) process. A CVI cycle is performed to obtain the desired shape, while in a mold, and appropriate fiber volume fraction prior to further infiltration. The final phase after de-molding, consists of performing CVI cycles for densification and sealing protection. If any machining to coupon shape is to be done, it will occur between these CVI cycles.

Nozzle Seal Fabrication

Multiple nozzle seals were fabricated for testing. The process was the same as for the coupons listed above (same maximum thickness) but different cross sections of seals were made: constant and variable thickness as shown in Figure 1. These were initial attempts to manufacture both the A410 and A500 as nozzle seals. The following seals were made: of A410 (SiC/sequenced Si,C,B), 2 of constant cross section and 2 of variable cross section (See Fig. 1); of A500 (C/sequenced Si,C,B), 2 of constant cross section were made. After the last CVI cycle, metallic hardware was installed on the seals. This hardware was required for the installation of the seal into the divergent nozzle.

a) constant cross-section   b) variable cross-section
Figure 1. Cross-sections of F100-PW-229 Nozzle Seals Fabricated for Testing

ANALYSIS

Before testing could begin (coupon or engine), reviews were undertaken to assure that the expected stresses that would be seen in the CMC part were consistent with the material database and material capability reported previously[6,7]. The thermal and mechanical analysis has been carried out taking into account the worst-case flight point corresponding to maximum after-burner conditions. The authors have reported these efforts previously[4,5,8]. All of the analysis has shown that the material capability would be greater than the stresses imposed during engine testing.

TESTING AND RESULTS

Coupon Testing

Coupon testing was mainly focused on fatigue behavior to probe the material. The fatigue cycle was a 2-hour hold cycle as established by the Enabling Propulsion Material program where the maximum load is achieved within 5 seconds and then held for 2 hours before being unloaded and reloaded within 10 seconds[9]. This test is aggressive as the average load is the hold stress and the load/unload cycle is intended to break any oxide or glass that has formed in cracks that may be protecting the material at temperature. For the CERASEP® A410 CMC material all the testing was done at 1204°C. This is significantly above the operating temperature that the F100-PW-229 nozzle seal would see in operation. Some of the testing was done in air and limited testing was done in a 90% steam environment at 1204°C. The stress lifetime results for this effort are shown in Figure 2.

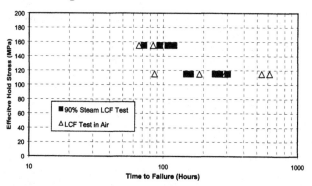

Figure 2. Plot of Stress versus Failure Time for A410 Fatigue Tests (Air and Steam)

For the SEPCARBINOX® A500 material, the fatigue testing was only done in air and the test temperatures were changed to 600°C and 1000°C as this was more representative of the conditions that would be seen for the nozzle seal application being pursued. Testing was done both by Pratt & Whitney (PW) as well as Snecma Propulsion Solide (SPS). The results for this testing are shown in Figure 3.

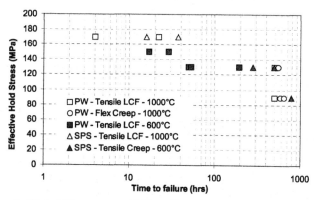

Figure 3. Plot of Stress versus Failure Time for A500 Fatigue Tests (Air)

Ground Engine Testing

Each CMC divergent seal consists of a CMC detail that is both a liner and structural member. Seal size is 520.7 mm in length with a tapered width that ends at the trailing edge width of 119.4 mm. The thickness is slightly larger than the current metallic seal and is over 4 mm as described earlier. Special shaping at the forward edge limits the effect of the thickness increase that could shift the cooling film provided at the nozzle throat. To provide attachment to two adjacent divergent flaps and the upstream convergent seal, Inconel 718 details were designed. The attachment hardware is locally bolted to the CMC liner with metallic bolts that penetrate to the gas path and held tight with self-locking nuts on the cold side. (The metallic features on the cold side of the seal are shown in Figure 4.) The bolt heads remain flush with the gas path surface to avoid disturbance of the cooling film layer. Weight savings of 2.9 lbs. per nozzle are achieved with these material systems as alluded to earlier.

The CMC divergent nozzle seals were ground tested in two different F100 engines by Pratt & Whitney at the P&W Florida facility and also at Arnold Engineering Development Center (AEDC). 87 % of the accumulated time was performed on an F100-PW-229 engine, while the remaining 13 % was performed on an F100-PW-220 engine. The divergent seal is 100 % interchangeable between the –100, -200, -220, and the –229 engine configurations of the F100 engine. Since the –229 has the highest thrust capability and hence the highest temperature and pressure profiles of the F100 series, the divergent seals exhibit the most severe conditions during that portion of testing. This created the best scenario for evaluating hardware improvements for increased durability. The various engine nozzle locations used during the test were chosen to always have an A410CT and an A500 seal in a hot streak location.

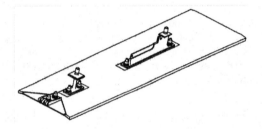

Figure 4. View of CMC Seal with metal hardware

Based on operational data from F-15E and F-16 aircraft, Mission and Duty Cycle requirements are calculated to provide design life standards. To achieve representative results in a reasonable timeframe, Accelerated Mission Tests (AMT) were developed. Hardware life is measured by equivalent cyclic damage or Total Accumulated Cycles (TAC). The test cycles are accelerated by eliminating the "non-damaging" time from the mission profiles, such as cruise and idle times. Acceleration factors of 2.5 to 4 are typical increases in operational time to AMT time.

Since the seals were not all installed at the same time, the engine time per part varies. Table I summarizes the total engine experience achieved per seal. F100 Parts are tracked based on the Total Accumulated Cycles (TAC) as well as engine time (hours) as described above. The hot hours (afterburner) are a subset of the total engine hours. Full life for the metal seals is 2150 TACs.

Table I. Engine Testing Summary

Seal Type	TACs	Engine Hours	Hot Hours
Completed Testing – Full Life			
A410CT (1)	4851	1294.9	97.6
A410VT (2)	4623	1341.8	94.8
A500CT (1)	4609	1307.2	94.4
Completed Testing – High Time			
A410CT (1)	6582	1750.3	117.4
A500CT (1)	5611	1485.3	102.0
(#) = number of seals			

TAC = Total Accumulated Cycles = Measurement of equivalent cyclic damage (throttle movements).

During engine testing, there were various times where the parts were removed and made available for detailed nondestructive and photographic documentation. The nondestructive evaluation chosen for this effort was pulse echo and through transmitted thermography. During the course of testing there was only one significant indication in the high time A410 constant thickness seal. Photo-documentation also showed durability improvements. This is shown in Figure 5. As can be seen, there is limited seal coat loss even though this part saw 6582 TACS. This is 3 times the standard metal seal life. For the A410 variable thickness and the A500 constant thickness seals in Table I, equivalent or less seal coat loss was seen indicating the durability of the material in this application. The only seal that had significant seal coat loss was and A410 constant thickness seal that had an upstream metal part event occur that disrupted the cooling film flow leading to increased surface stresses. Even though there was seal coat loss, there was no formation of delaminations[4,8].

Figure 5. A410 Constant Thickness Seal showing limited Seal Coat Loss

Sectioning was done on one A410 constant thickness seal (4851 TACS) and one A500 constant thickness seal (4609 TACS). In addition, tensile bars were extracted from the seals to perform room temperature tensile tests. These later tests were done to note if there was any debit in the room temperature tensile tests after engine testing as compared to pristine values. For the A410 testing, there was no debit in the properties noted[5]. The A500 constant thickness seal results showed a slight decrease in capability from the baseline. (In this case, the baseline was samples extracted from a witness seal.) The average ultimate tensile strength from the seal was found to be 192 MPa where the unexposed average was 204 MPa (percentage decrease of 6%). The strain to failure showed a similar percentage decrease from a seal value of 0.76% strain to failure whereas the witness seal had a strain to failure capability of 0.81%. The modulus also

211

decreased from a witness value of 82 GPa to an average value of 71 MPa (a 15% decrease). Even though there was limited property degradation seen in the tensile tests results, fiber pullout was still present (See Figure 6).

Figure 6. SEM image from tensile bar extracted from A500 Seal (4609 TACS)

DISCUSSION

The coupon testing had significant results for both material systems. The A410 results shown in Figure 2 indicating that there is negligible differentiation between the results in air and steam. The lifetimes do not differentiate the two testing environments as seen for other high performance CMC systems[10]. The A500 results achieved significant lifetimes at the stress levels tested when compared to other C fiber based systems[11]. Both of these results are achieved by the formation of glass at temperature that protects the fiber interface. The glass formation can be seen in post-test photos of the material. This is shown in Figure 7 for the two material systems. For the A500 system, the glass formation is clearly seen in the cracks hence indication that it is performing the protection required to achieve significant life in fatigue testing. For the A410 testing in steam, glass beads are present indicating that glass formation is occurring. Even though glass beads are present indicating it is weeping out of the material, glass formation was still protecting the interface since there was not a decrease in lifetime in the steam tests.

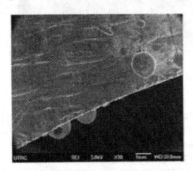 

a) A410 (90% Steam LCF Test)   b) A500 (1000°C LCF Air Test)
Figure 7. Glass formation in the two material systems being investigated

The seal testing is just as informative. All the seals exceeded the metal seal life of 2150 TACS. The seals exceeded this goal from 2x to 3x. Most importantly, these seals have achieved these results with no indication of significant debit in material capability. The A410 constant thickness seal showed no debit in stress, strain or modulus. The A500 constant thickness seal is almost as impressive showing a drop of 6% in stress and strain capability. This drop was for a total of 10 tensile coupons extracted from the seal. What is more important is to look at the test results averaged based on location in the seal as shown in Figure 8. As can be seen, the greatest change in strength is shown in the aft location of the seal.

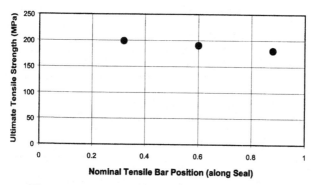

Figure 8. Tensile strength versus location along seal

These results become clearer when the cross section work is taken into account. Cross sections at various points along the seal are shown in Figure 9. Near the aft end of the seal, the seal coat has been compromised and oxidative attack has occurred in the top ply of the carbon fibers. While this decreased the strength capability it was not a large attack since the self-sealing matrix protected the material at the tow level.

a) cross section near center position of seal

b) cross section near aft position of seal
Figure 9. Cross sections from A500 constant thickness seal

CONCLUSIONS

The coupon testing had shown that the sequenced matrix of Si,C and B can lead to significant improvements over a standard matrix scheme. This was seen in the fact that the steam testing was not as aggressive as would be thought based on other material testing. The A500 testing (C fiber) was an even clearer example with the fatigue testing in air showing 2 orders of magnitude increase in lifetime over other C fiber SiC matrix CMC systems.

The material performance was also seen in the ground engine testing capability of the CMC seals tested. All of the seals tested exceeded the lifetime of the metal seal that it is meant to replace. As long as there was no upstream events, all of the seals saw only limited surface degradation (seal coat loss). Non-destructive interrogation at various points in the seal life did not indicate any concerns or delamination formation.

Residual testing of both an A410 and A500 constant thickness seals saw significant residual capabilities. The A410 seal saw no degradation. The A500 seal saw limited degradation and this was limited mainly to the aft end of the seal. Even though the seal coat was lost, the protective nature of the self-sealing matrix around each tow, protected the fiber from greater oxidative attack.

All of these results, coupon and ground engine testing, have allowed this effort to progress to flight testing that will occur in 2005. This is a significant accomplishment.

REFERENCES

[1] L. Zawada, G, Richardson and P. Spriet, "CMCS for Aerospace Turbine Exhaust Nozzles", to be published in the proceedings of the HTCMCV, Seattle, WA, September, 2004.

[2] R. Kestler and M. Purdy, "SiCf/C for Aircraft Exhaust" presented at ASM International's 14th Advanced Aerosapce Materials And Processes Conference, June 9-13, 2004, Dayton, OH.

[3] Staehler, J.M. and Zawada, L.P., "Performance of Four Ceramic-Matrix Composite Divergent Flap Inserts Following Ground Testing on an F110 Turbofan Engine", J. Am. Ceram. Soc. 83 [7] 1727-38 (2000)

[4] Bouillon, E.P., Ojard, G.C., Habarou, G., Spriet, P.C., Arnold T., Feindel, D.T., Logan, C., Rogers, K., Doppes, G., Miller, R., Grabowski, Z. and Stetson, D.P., "Engine Test Experience and Characterization Of Self Sealing Ceramic Matrix Composites For Nozzle Applications in Gas Turbine Engines", ASME Turbo Expo 2003, Atlanta, Georgia, June 16-19, 2003, ASME Paper 2003-38967.

[5] Bouillon, E.P., Ojard, G.C., Habarou, G., Spriet, P.C., Arnold T., Feindel, D.T., Logan, C., Rogers, K., and Stetson, D.P., "Engine Test and Post Engine Test Characterization Of Self Sealing Ceramic Matrix Composites For Nozzle Applications in Gas Turbine Engines", ASME Turbo Expo 2004, Vienna, Austria, June 14-17, 2004, ASME Paper 2004-53976.

[6] Bouillon, E.P., Lamouroux, F., Baroumes, L., Cavalier, J.C., Spriet, P.C. and Habarou, G., 2002, "An Improved Long Life Duration CMC for Jet Aircraft Engine Applications", ASME paper No. GT-2002-30625.

[7] Bouillon, E.P., Abbé, F., Goujard, S., Pestourie, E., Habarou, G., 2000, "Mechanical and Thermal Properties of a Self-Sealing Matrix Composite and Determination of the Life Time Duration", Ceram. Eng. and Sci. Proc., 21(3) pp. 459-467.

[8] Bouillon, E.P., Ojard, G.C., Habarou, G., Spriet, P.C., Lecordix, J.L., Feindel, D.T., Linsey, G.D. and Stetson, D.P., "Characterization and Nozzle Test Experience of a Self Sealing Ceramic Matrix Composite for Gas Turbine Applications", ASME Turbo Expo 2002, Amsterdam, Netherlands, June 3-7, 2002, ASME Paper 96-GT-284.

[9] Brewer, D., 1999, "HSR/EPM Combustion Materials Development Program", Materials Science & Engineering, 261(1-2), pp. 284-291.

[10] A. Calamino and M. Verrilli, "Mechanical performance improvements of an In-situ Boron Nitride-coated MI SiC/SiC Composite", to be published in the proceedings of the 26th Annual Conference on Composites, Materials and Structures, Cocoa Beach, FL, Jan. 28-Feb 1, 2002.

[11] Verrilli, M., Kantzos, P. and Telesman, J., 2000, "Characterization of Damage Accumulation in a Carbon Fiber-Reinforced Silicon Carbide Ceramic Matrix Composite (C/SiC) Subjected to Mechanical Loadings at Intermediate Temperature", ASTM STP 1392, Mechanical, Thermal and Environmental Testing and Performance of Ceramic Composites and Components, M.G. Jenkins, E. Lara-Curzio and S.T. Gonczy, eds., American Society for Testing And Materials, West Conshohocken, PA, pp. 245-261.

DIMENSION STABILITY ANALYSIS OF NITE SIC/SIC COMPOSITE USING ION BOMBARDMENTS FOR THE INVESTIGATION OF RELIABILITY AS FUSION MATERIALS

H. Kishimoto, T. Hinoki, K. Ozawa, K-H. Park, S. Kondo and A. Kohyama
Institute of Advanced Energy, Kyoto University
Gokasho, Uji, Kyoto 611-0011, Japan

ABSTRACT

A new process for fabricating of SiC/SiC composite, the Nano-powder Infiltration and Transient (NITE) method produces dense, high strength SiC materials with potential applications in fission or fusion reactors. Irradiation effects reduce the performance of materials due to accumulation of point defects. In this research, SiC/SiC composite fabricated by the NITE process was irradiated with Si-ions. Post-irradiation investigation was performed by TEM. The ion-irradiation experiments were performed using a 1.7 MeV tandem accelerator at DuET facility in Institute of Advanced Energy, Kyoto University. Irradiation temperature and dose were up to 1200 °C and 10 dpa. The SiC matrix was well crystallized and contained a small amount of secondary phases. The EDX analysis specified that the secondary phase was aluminum-rich. The SiC grains and the secondly phase exhibited enough microstructural and dimensional stability after the Si-ion.

INTRODUCTION

Silicon Carbide (SiC) has been expected to be used as a structural material for the application of fusion and fission energy sources because of the attractive properties at elevated temperature. Fusion and fission reactors of coming generation such as Gas-cooled Fast Reactor (GFR), which operate near or over 1000 °C, require excellent high temperature materials. SiC and SiC fiber-reinforced SiC matrix (SiC/SiC) composites are promising candidates for the reactor systems. An important factor on the structural materials in fusion and fission reactors is the radiation damages. The damage accumulation behaviors and the irradiation effects have been researched, and the results shows that stoichiometric monolithic SiC has enough resistance ion irradiation damage accumulation up to 1400 °C. [1]

The processing technique of SiC/SiC composites is one of the most important issues. It needs to satisfy several conditions such as quality of materials, cost and convenience for mass production. Additionally, the neutron bombardments in fusion and fission reactors complicate the matters because of radiation effects. Previous researches about the dimensional stabilities of SiC under irradiation environments showed that amorphized SiC was crystallized by neutron bombardments, and it brought about the large volume shrinkage over 10%. [2] Because the shrinkage of fibers causes the reduction of mechanical properties and dimensional stabilities, SiC as structural materials for fusion and fission application need to be crystallized perfectly.[2] Tyranno-SA (UBE Industry, Ltd., Yamaguchi, Japan) SiC fibers are stoichiometric and high crystallinity.[3] The Tyranno-SA fibers are expected to have sufficient irradiation stability and, were used in the NITE SiC/SiC composites.

The technique for fabrication of matrix was also a big issue. There were many kinds of techniques such as Reaction Sintering (RS), Polymer Infiltration and Pyrolysis (PIP) and Chemical Vapor Infiltration (CVI).[4, 5, 6] RS method did not satisfy the conditions of stoichiometric composition and low impurity, and CVI method could receive high purity SiC matrix but it was not enough dense. The CVI method was too expensive and not convenient for mass production. PIP method is difficult to fabricate dense composites. An innovative processing for fabrication of SiC/SiC composites was developed called Nano-Infiltration and Transient Eutectic Phase (NITE) Process is the promising answer for the needs of innovation.[7] NITE process is able to attain high-quality SiC/SiC composites having following characteristics (1) dense and robust (2) fairly high thermal conductivity (3) chemically stable (4) potentially low production cost.[8] These characters are very appropriate for the blanket materials in fusion reactor.

In this research, the chemical analysis of microstructure using FE-TEM on NITE SiC/SiC composites and study of microstructural and dimensional stabilities of the composite under irradiation environments using the ion irradiation method are shown.

EXPERIMENTAL

Used reinforcement was Tyranno[TM]-SA grade-3 polycrystalline fiber. Tyranno-SA fibers were coated with pyrolytic carbon (PyC) by the chemical vapor deposition method. The thickness of PyC coating was about 0.5 μm. NITE SiC/SiC composites were fabricated at Ube Co. Ltd using β-SiC nano-powder (Marketech International Inc., USA). The grade of this NITE SiC/SiC composite is named Pirotgrade-2.[9] The particle size of β-SiC nano-powder was below 30 nm. T he powders and fibers were fabricated at 1800 °C, 20 MPa using hot pressing. Al_2O_3-Y_2O_3-SiO_2 ternary system additives were used for the fabrication of composites. The composites were cut to square shapes and their surfaces were polished by a diamond particle polishing. The dimension of specimens was 4×2×2mm. The surface for irradiation was selected to be normal to the fiber direction.

1.7 MeV tandem (Model 4117HC, HVEE) accelerator at Dual-Beam Material Irradiation Facility for Energy Technology Research Facility (DuET) in Institute of Advanced Energy, Kyoto University was used for this research. Used ion was 5.1 MeV Si ion, and the irradiation temperature was from 800 °C to 1200 °C. The fluence was nominally up to 1.4×10^{21} Si/m^2, this corresponded to 10 displacement per atom (dpa). The dose rate was 6×10^{-4} dpa/s. A focused ion beam (FIB) processing was used for the preparation of TEM specimens from the irradiated SiC/SiC composites. The thinned foils were cut off from the materials and lifted up using a micro pick-up system. The foils were put on the carbon film supported copper grids. The microstructural investigation and chemical composition analysis was carried out with a JEOL JEM-2200FS high-resolution analysis TEM, EDX, EELS. The irradiation induced swelling of monolisic NITE SiC was also investigated by the step height method. The dimension of monolisic NITE SiC plates was 3mm square and 0.2 mm thickness. The swelling of irradiated region produced steps on the surface. The step heights between unirradiated and irradiated surfaces were measured by interferometric profilometry (Micromap 128, Micromap Co. Tucson, Arizona, USA). The detail of investigation was described in a previous report. [10]

RESULTS

Figure 1 shows secondary electron image (SEI) and a backscattering electron image (BEI) of NITE SiC/SiC composite after Si-ion irradiation experiment at 1200 °C and 10 dpa. These images are a. The pictures show the specimen surface near border of irradiated and irradiated regions. In the SEI image, the irradiated region makes a better contrast because of slight thermal etching of carbon interphases. In the irradiated region, significant modifications such as fracture of matrix, fiber shrinkage, debonding of interphases and fibers were not detected. The BEI image shows the distribution of composition. Bright contrast regions are the additives left in matrix. Since the EDX investigation showed the composition of these regions were mainly aluminum or yttrium. These additives did not move during irradiation experiments from 800 °C to 1200 °C, they looked keeping stabilities under the condition in this research. Pure SiC swells under the irradiation environment. Figure 2 shows the temperature dependence of swelling on SiC under irradiation environments. High-purity CVD-SiC irradiated by Si-ion was used as

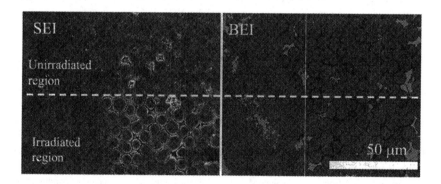

Figure 1   Secondary electron image (SEI) and back-scattering electron (BEI) image of NITE SiC/SiC composite after Si-ion irradiation at 10 dpa and 1200 °C

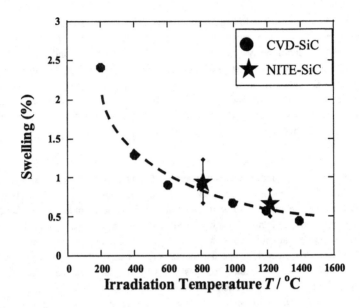

Figure 2   Temperature dependence of Si-ion irradiation induced swelling of CVD-SiC and monolithic NITE SiC

reference materials to compare with monolithic NITE SiC.[1] The swelling of CVD-SiC decreases with irradiation temperature. The swelling of monolithic NITE-SiC were plotted on Figure 2, and they were on the trend of CVD-SiC.[1] Because irradiation induced swelling of monolithic NITE-SiC is almost the same as the CVD-SiC, the interphase fractures are not expected to be caused by dimensional stabilities in NITE SiC/SiC composites.

From the view of macro-scale dimension, NITE SiC showed enough stability under the irradiation environment, but it is also important to investigate micro- and nano-scale microstructural behaviors. The fracture toughness of SiC/SiC composite depends on the progress of cracks in matrix and the pull out of fibers from matrix, thus the microstructural construction significantly affects to the mechanical strength of composites. Figure 3 shows a TEM image of NITE SiC/SiC composite irradiated at 1200 °C. The microstructure of the NITE SiC matrix looks dense and well crystallized. The secondary phase was not clear except for carbon which was supposed to peel off from the interphase during NITE processing. The boundary of irradiated and unirradiated region is the center of the image, and the radiation damages are observed in the

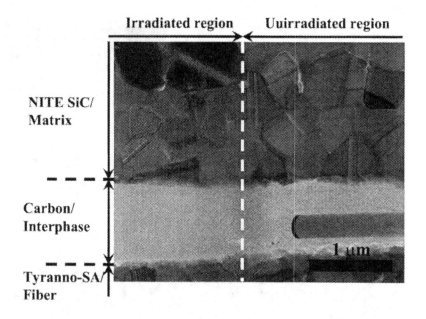

NITE SiC/
Matrix

Carbon/
Interphase

Tyranno-SA/
Fiber

1 µm

Figure 3   TEM image of NITE SiC/SiC composite irradiated at 1200 °, 10 dpa

irradiated region. The carbon interphase slightly swells by the irradiation, but SiC grains in matrix and fiber were not modified. The microstructure of NITE SiC in this research was very similar to the Tyranno-SA fiber. EDX analysis using STEM showed the difference of them clearly. Figure 4 shows the EDX image of composition on the NITE SiC/SiC composite. The chemical compositions of silicon, carbon and yttrium were uniform, but aluminum rich phases were detected in the NITE SiC matrix. The phases mainly exist on the interface between matrix and carbon interphase. Thin aluminum phases were also detected on the grain boundaries in the NITE SiC matrix. These aluminum rich phases were not modified significantly by the Si-ion irradiation and kept stability in this research.

DISCUSSION

In the history of development of SiC/SiC composites, the processing had been key issues because of the difficulty to attain the dense matrix of stoichiometric SiC with low costs. NITE processing is actually an innovation in the field of fusion and fission materials. The technique opened the road to new reactor systems, thus the development of technique of NITE process is

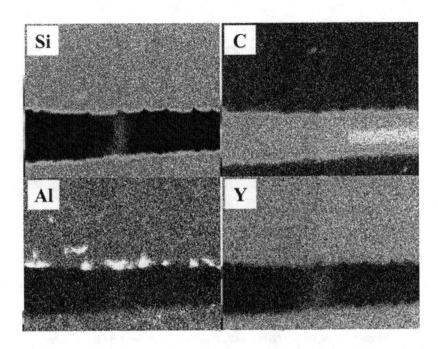

Figure 4   STEM-EDX analysis of Figure 3.   Distribution of silicon, carbon, aluminum
and yttrium on NITE SiC/SiC composite irradiated at 1200 °C, 10 dpa

rapidly progressing. Because of the high-speed of development, the analysis of NITE SiC/SiC composite could not follow the development of processing. Especially there is almost no research for the investigation of performance as fusion and fission materials. This research is the start-up of it, so the investigation of as-received composites was necessary at first. The SiC grains were crystallized very well, and there was not amorphized silica or other amorphized phase produced during the NITE processing. This microstructure is the most important point as fusion and fission materials because such amorphized phases may cause unstable dimensional changes under the irradiation environments. The irradiation effects for additives may be problem, thus it is necessary to investigate more. In this research, additives also kept stable against the irradiation. The irradiation condition were not enough hard for the investigation as fusion materials, some heavier irradiation experiments at higher temperature are scheduled. The construction of additives left in the NITE SiC matrix could not be determined in the present research, it is on progress now. The additives were also left on the grain boundaries. The construction of them should be paid attention because the mechanical properties are affected. As SiC is a brittle material, and additives leave surrounding the SiC grains, the improvement of additives may make the strength of NITE SiC/SiC composites better.

CONCLUSION

The microstructural research and chemical analysis of the NITE SiC/SiC composite using Si-ion irradiation and TEM investigation were performed. The irradiation experiments up to 1200 °C and 10 dpa were performed DuET facility at Institute of Advanced Energy, Kyoto University. The NITE SiC/SiC composites had dense, stoichometric SiC matrix. TEM investigation showed that SiC grains were crystallized, and the NITE SiC matrix did not include the amorphous-like phases which bring about unstable behaviors of microstructure under the irradiation environment. STEM-EDX investigation showed the aluminum-rich phase exists on the interface between carbon interphase and NITE SiC matrix. The very thin layer of aluminum-rich phase also distributed on the grain boundaries of NITE SiC matrix. These phases showed microstructural stability under the irradiation environment up to 1200 °C and 10 dpa.

ACKNOWLEDGEMENT

The ion-irradiation experiments were carried out with the assistance of H. Ogiwara, O. Hashitomi, S. Ikeda and Dr K. Jimbo, Institute of Advanced Energy, Kyoto University.

REFERENCES

[1] H. Kishimoto, Y. Katoh, K.H. Park, S. Kondo and A. Kohyama, "Dual-Beam Irradiation Effects in SiC," *Advanced SiC/SiC Ceramic Composites: Developments and Applications in Energy Systems, Ceramic Transactions* **144**, American Ceramic Society 343-352 (2002).

[2] L.L. Snead, M.C. Osbone, R.A. Lowden, J. Strizak, R.J. Shinavski, K.L. More, W.S. Eartherly, J. Bailey, A.M. Williams, "Low Dose Irradiation Performance of SiC Interphase SiC/SiC Composites," *Journal of Nuclear Materials*, **253** 20-30 (1998).

[3] T. Ishikawa, Y. Kohtoku, K. Kumagawa, T.Yamamura, T. Nagakawa, "High-Strength Alkali-Resistant Sintered SiC Fiber Stable to 2200°C," *Nature*, **391** 773-775 (1998).

[4] L.L. Snead, R.H. Jones, A. Kohyama and P. Fenici, "Status of Silicon Carbide Composites for Fusion," *Journal of Nuclear Materials*, 233-237 26-36 (1996).

[5] D.P. Stino, A. J. Caputo, R. A. Lowden, "Synthesis of Fiber-Reinforced SiC Composites by Cgamical Vapor Infiltration," *American Ceramic Society, Bulletin*, **65** 347-350 (1986).

[6] M. Kotani, A. Kohyama, K. Okamura, K. Inoue, "Fabrication of High Performance SiC/SiC Composite by Polymer Impregnation and Pyrolysis Method," *Ceramic Engineering and Science Proceeding*, **21** 339-364 (2000).

[7] A. Kohyama et. Al., *IAEA-CN-94, 19th Fusion Energy Conference*. Lyon, France, 14-19 October 2002, FTP1/02 (2002)

[8] A. Kohyama, T .Hinoki, J-S Park, "Advanced SiC/Composite Materils for Advanced

Quantum Energy System," *Proceeding of The Joint International Conference on* *"Sustainable Energy and Environment (SEE)* " Hua Hin, Thailand 1-3 December 2004, 1-5 (2004).

[9] A. Kohyama, "Effort on Large Scale Production of NITE-SiC/SiC with Tyranno-SA," *Proceeding of 29th International Conference on Advanced Ceramics and Composites*, Cocoabeach, FL, USA, 23-28 January (2005)

[10] H. Kishimoto, Y. Katoh, A. Kohyama and M. Ando, "The influence of temperature, fluence, dose rate and helium production on defect accumulation and swelling in silicon carbide," *Effects of Radiation on Materials*, ASTM STP **1405**, American Society for Testing and Materials, 775-785 (2001).

FRACTURE STRENGTH SIMULATION OF SiC MICROTENSILE SPECIMENS – ACCOUNTING FOR STOCHASTIC VARIABLES

Noel N. Nemeth
NASA Glenn Research Center
21000 Brookpark Rd.
Cleveland, OH, 44135

Glenn M. Beheim
NASA Glenn Research Center
21000 Brookpark Rd.
Cleveland, OH, 44135

Osama M. Jadaan
University of Wisconsin – Platteville
Platteville, WI, 53818

William N. Sharpe
The Johns Hopkins University
3400 North Charles St.
Baltimore, MD, 21218

George D. Quinn
NIST
100 Bureau Dr.
Gaithersburg, MD, 20899

Laura J. Evans
NASA Glenn Research Center
21000 Brookpark Rd.
Cleveland, OH, 44135

Mark A. Trapp
Carnegie Mellon University
5000 Forbes Ave.
Pittsburgh, PA, 15213

ABSTRACT

This paper summarizes work performed to predict the room-temperature strength of SiC micro-scale tensile specimens with introduced stress concentration using NASA Glenn Research Center's Ceramics Analysis and Reliability Evaluation of Structures Life prediction program (CARES/Life). A polycrystalline and a single crystal SiC material were tested. CARES/Life tended to over predict strength of the polycrystalline material but had relatively better success predicting the strength response for the single crystal material. Large columnar grains and rough side-walls for the polycrystalline material likely reduced the applicability of continuum analysis and therefore the accuracy of the CARES/Life analysis. The single crystal material did not have (or had less of) these complications. Significant specimen-to-specimen dimensional variation existed in the specimens. To simulate the additional effect of the dimensional variability on the probabilistic strength response for the single crystal specimens the ANSYS Probabilistic Design System (PDS) was used with CARES/Life. The ANSYS/PDS – CARES/Life simulation was not rigorous. Isotropic material and fracture behavior was assumed in the analysis. However the feasibility of using these programs to account for multiple stochastic variables on the strength response of a structure was demonstrated.

INTRODUCTION

MicroElectroMechanical Systems (MEMS) that are being developed for power generation and propulsion (PowerMEMS) pose difficult design challenges with respect to strength and durability. For example miniature turbomachinery may rotate in excess of 1 million rpm and experience stresses of several hundred MPa while operating in a hot combustion environment[1]. Silicon carbide (SiC) would be an excellent material choice for these harsh environment applications because of its ability to maintain strength, resist creep and resist oxidation at gas turbine operating temperatures. We at NASA's Glenn Research Center (GRC) are working to develop SiC micro-fabrication technology as well as characterization and appropriate life prediction design methodology. This work is performed under NASA's Advanced Micromachining Technology for SiC Microengines project within the Alternate Energies Foundation Technologies (AEFT) program.

In this paper we summarize the work performed to date to predict the room-temperature strength of SiC micro-tensile specimens with introduced stress concentration using the Weibull distribution. This was done using GRC's Ceramics Analysis and Reliability Evaluation of Structures Life prediction program (CARES/Life)[2,3]. We also demonstrate the feasibility of using CARES/Life with the ANSYS Probabilistic Design System (PDS)[4] to simulate or predict the strength response of brittle material components while simultaneously accounting for the effect of the variability of geometrical features on the strength response.

SPECIMEN DESIGN, AND MATERIAL

Our testing program had three purposes. First, to demonstrate fabrication of simple structures – microtensile specimens in this case - that have high aspect ratios (vertical dimension or etch depth divided by lateral feature size) with sufficient strength, surface finish, and dimensional tolerance suitable for PowerMEMS. Highly directional deep reactive ion etching (DRIE) processes were used for this purpose[5]. Second, to correlate process improvements with strength response. And third, to test how well the Weibull probabilistic distribution – on which CARES/Life is based – works for miniature-sized components. This was done with specimen geometries with various levels of stress concentration. This tested a fundamental premise of Weibull theory – that strength increases as the area (or volume) under highest stress decreases.

Two materials – a polycrystalline SiC and a single crystal SiC – were tested. The polycrystalline specimens were tested from two separate fabrication runs during 2003 (designated by our bookkeeping as batches 3 and 4) and the single crystal specimens were tested from one fabrication run during 2004 (which we designated as batch 6). The polycrystalline specimens were made with a less mature etching recipe than those of the single crystal material hence showing a consistently poorer quality. This confounded the analysis results to an unknown degree. The quality of the single crystal specimens was better although dimensional variability was still significant. In the subsequent analysis results are presented first for the polycrystalline material followed by the single crystal material. Only room-temperature results are shown. High temperature testing at 1000°C was performed on the single crystal specimens but analytical results were not available for this paper. Fabrication details are not described here since the purpose of this paper was to report on Weibull analysis of SiC structures at the size scale appropriate for designing PowerMEMS turbomachinery.

The polycrystalline silicon carbide was manufactured by Rohm and Haas and consisted of cubic β-phase with grain sizes of 5-10 μm, as stated in the manufacturer's material specification[6]. The single crystal material was a single hexagonal crystal n-type 6H-SiC (resistivity 0.065 Ohm-cm) research grade material manufactured by Cree[7] with a 3.5° off-axis orientation. The specimens were oriented parallel to the flat where the flat was the $\{10\bar{1}0\}$ plane and with the flat face parallel to the $<11\bar{2}0>$ direction. All specimens were micromachined by DRIE using an inductively coupled plasma etcher (STS Multiplex ICP). However, the etch process for each material was different. The etch recipe used for the single crystal material resulted in improved surface finish. The etch mask was electroplated nickel, approximately 8 μm thick.

The key to a tensile test method is a specimen design that enables effective gripping. Screw or pin-in-hole grip ends are obviously impractical at this size scale. Wedge-shaped specimen ends that fit into inserts in the grips is a design that was developed earlier for testing steel and nickel microspecimens[8,9]. A traditional tensile specimen has a straight gage section, but there was concern that a straight specimen would break at the stress concentrations where it faired into the ends. This indeed turned out to be the case roughly half the time when tests were attempted on polycrystalline straight-gage-section specimens. We do not report on that data here. A specimen with a gentle curvature and smaller size avoids this problem while producing a uniform stress in the middle of the gage section. To generate a localized stress concentration various notch and hole specimen geometries were designed.

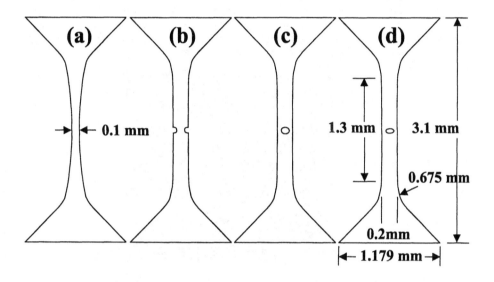

Figure 1. Basic schematic of the dogbone specimens (not to scale); (a) curved, (b) notched, (c) circular hole, (d) elliptical hole. The gage section is 1.3 mm in length.

Figure 1 is a schematic of the basic geometries used for dogbone microtensile specimens. These specimens are actually rather large to be considered true MEMS, however, they are of appropriate size for PowerMEMS turbomachinery applications. The flared-end is the gripping section of the specimen and the gage section length is 1.3 mm with a nominal cross section of 0.2mm by 0.125mm. From left to right in the figure these are (a) curved, (b) double-notched, (c) circular hole, and (d) elliptical hole. These yielded stress concentrations relative to the cross-section of (a) 1.01, (b) 2.3, relative to the minimum width, and (c) 3.8, and (d) 6.8, relative to the gross width, respectively based on results from finite element analysis. The curved specimen was ½ the nominal width or 0.1mm at its narrowest point at the center. The notch radius of the

225

notched specimen was nominally 0.025 mm. The central hole nominal diameter was ½ the width or 0.1 mm in diameter. The central ellipse was nominally 0.1 mm across its major axis (perpendicular to the load axis) and 0.05mm across its minor axis for a ratio of 2:1 between the major axis and the minor axis. The curved specimen geometry was our base-line from which we predicted the response of the other geometries. The notched specimen (fig. 1 (b)) was only used for the polycrystalline specimens – it represented an earlier design. The central hole specimens (circular and elliptical) were only used with the single crystal material – they represented a later design. It was felt that the central hole design would be less susceptible to bending stresses if they were present.

The specimens were etched at GRC and tested at Johns Hopkins University (JHU). The JHU team individually measured the dimensions of each specimen prior to fracture. The DRIE process used at GRC did not produce completely vertical side-walls, but rather introduced a gradual taper as material is etched away resulting in specimens with a trapezoidal cross-section. A major goal of this project was to produce specimens with nearly vertical side walls and steady progress had been made as fabrication improved. Interestingly, the slope of the taper reversed from the fabrication runs made with the polycrystalline material and the run made with the single crystal material. Figure 2 shows an example of a cross-section for a single crystal specimen with a circular hole.

Figure 2 – Cross-section of a single crystal SiC test specimen with circular hole etched to a depth of 140 μm, as observed using SEM. Nickel mask is at the top.

The DRIE process tended to produce specimens with rough side walls compared to the other surfaces. This is illustrated in figure 3. The fabrication process used for the polycrystalline material had rougher side-walls (shown in figure 3 (a)) than the improved process used for the single crystal material (shown in figure 3 (b)). Note in figure 3 (a) that the etch grooves worsened further away from the nickel mask side and that the notch feature appeared to be less distinct away from the nickel mask side.

Tables I – V show the average specimen cross-sectional dimensions at the notch, hole, or narrowest width in the case of the curved specimen for the two materials used. The top surface is the nickel mask side and the bottom surface is opposite the nickel mask side. The notch width in Table II is the distance between the opposing notch roots - it therefore represents the minimum specimen width on that surface. Note in Table V that the elliptical hole dimensions show significant scatter – particularly for the bottom minor axis.

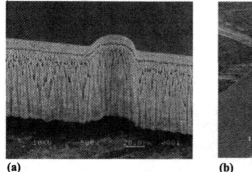

(a)   **(b)**

Figure 3. SEM of etched side-walls; (a) notched polycrystalline specimen (fabricated in 2003 – batch 3) and; (b) circular-hole geometry single crystal specimen (fabricated in 2004 – batch 6).

Table I. Polycrystalline SiC curved specimen average cross-sectional dimensions (at narrowest point); n = number of specimens, t = thickness. All dimensions in microns.

Batch #	n	t	Top Width	Bottom Width
Batch 3	22	136 ± 5	104 ± 2	81 ± 6
Batch 4	23	143 ± 3	107 ± 2	93 ± 2

Table II. Polycrystalline SiC notched specimen average cross-sectional dimensions; n = number of specimens, t = thickness. All dimensions in microns.

Batch #	n	t	Top Width	Top Notch Width	Bottom Width	Bottom Notch Width
Batch 3	26	138 ± 3	202 ± 2	156 ± 7	176 ± 6	146 ± 8
Batch 4	25	144 ± 4	207 ± 2	160 ± 4	193 ± 3	156 ± 5

Table III. Single crystal SiC curved specimen average cross-sectional dimensions (at narrowest point); for n = 19 specimens, t = thickness. All dimensions in microns.

Batch #	t	Top Width	Bottom Width
Batch 6	127 ± 2	111 ± 3	125 ± 7

Table IV. Single crystal SiC circular-hole specimen average cross-sectional dimensions; for n = 19 specimens, t = thickness. All dimensions in microns.

Batch #	t	Top Width	Top Diameter	Bottom Width	Bottom Diameter
Batch 6	126 ± 2	210 ± 3	91 ± 2	223 ± 7	77 ± 5

Table V. Single crystal elliptical-hole specimen average cross-sectional dimensions; for n = 18 specimens, t = thickness. All dimensions in microns.

Batch #	t	Top Width	Top Maj. Axis	Top Min. Axis	Bottom Width	Bottom Maj. Axis	Bottom Min. Axis
Batch 6	127 ± 1	210 ± 4	91 ± 4	42 ± 3	225 ± 7	73 ± 5	26 ± 8

EXPERIMENTAL PROCEDURE

Figure 4 shows a polycrystalline silicon carbide specimen in a set of aluminum grips. Not shown in the figure are tapped holes for 0-80 screws, which fasten washers that cover the ends of the specimen. They do not grip the specimen, but capture the broken ends which would pop out on fracture. (A narrow strip of cellophane tape served the same purpose in later tests). One grip is fixed, and the other moves through a linear air bearing to reduce friction; it is attached with a thin steel wire to a load cell mounted on a motorized translation stage.

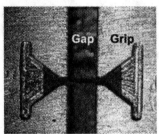

Figure 4. A SiC specimen in aluminum
grips. The specimen is 3.1 mm long.

A capacitance-based gage measured the displacement of an aluminum 'flag' attached to the movable grip so that a force-displacement plot could be obtained. Displacement measurement was not necessary since only the strength was needed, but the plots were made to see if there was any unusual behavior (such as slipping as the specimen seats in the grips) during a test. The motorized stage was run at a constant rate for all tests - each taking a minute or so. Some details of the setup are given in reference 10.

FRACTOGRAPHIC RESULTS FOR POLYCRYSTALLINE SPECIMENS

Extensive fractographic analysis was done on several specimens of each shape from the first processing run of the polycrystalline material – designated batch 3. Conventional scanning electron microscopy was effective for the fracture surface examination. The detailed examination lead to the following observations for the polycrystalline material:

- The primary fracture sources were grooves in the side walls of the specimens caused by the etching procedure. These grooves tended to spread out and get deeper away from the nickel mask and probably acted as stress concentrators.
- Large grains were often (but not always) at the origin. When this occurred, the large grain cleavage was approximately at right angles to the applied stress. Presumably the large grains were susceptible to a sharp crack popping in from the etch groove, due to cleavage along a preferred crystallographic plane in the SiC.
- There are a lot of etch grooves along the gage section of a specimen. Some probably connected to a region of normal SiC microstructure with fine grains. Other grooves had

large grains located next to them, but these may have had orientations resistant to fracture. Other grooves had large grains connected to them, and if oriented the right way, a cleavage crack could propagate into the grain.

In short the critical flaws were in many cases a combination of the sidewall grooves and a large grain. Figure 5 shows an example of a fracture surface of a notched specimen resulting from a long columnar grain at the bottom of an etch groove.

Figure 5. Two views of the fracture surface of a notched specimen with local fracture strength of 0.80 GPa. The initiation site is a crack in a large, long columnar grain at the bottom of an etch groove. The twist hackle in the cleaved grains is evident in the right view and shows the fracture path was from left to right.

A simple fracture mechanics analysis was made to estimate the size of flaws to be found. The nearly uniform stress field of the curved specimens was combined with a K_{Ic} estimate of 2.5 MPa√m [11]. The shape factors for the long surface cracks was Y = 1.99 and Y = 1.3 for the surface semicircular cracks. This predicted that flaws should have been between 7 μm and 30 μm deep. Note also that the predicted and observed flaws were large relative to the cross section size as well as to the grain size. Some of the large grains were as large as 150 μm in length and grains of 10 –50 μm size were quite common.

Fractographic results for the single crystal material were not available, however it was expected that sidewall grooves would also be the source of the strength controlling flaws since those surfaces still showed significant roughness. Also, as demonstrated in reference 12 the strength response of small brittle structures is influenced by surface finish. It is therefore reasonable to assume that the strength response of sub-millimeter or micron-sized structures is controlled by surface residing flaws that result from the micromachining process.

WEIBULL STATISTICS AND CARES/LIFE ANALYSIS

The CARES/*Life* software describes the probabilistic nature of brittle material strength using the Weibull cumulative distribution function. For uniaxially stressed components the 2-parameter Weibull distribution for surface residing flaws describes the component fast-fracture failure probability, P_f, as

$$P_f = 1 - \exp\left[-\frac{1}{\sigma_0{}^m} \int_A \sigma\ (x,y)^m\ dA \right] \tag{1}$$

where A is the surface area, $\sigma(x,y)$ is the uniaxial stress at a point location on that surface, and m and σ_0 are the shape and scale parameters of the Weibull distribution, respectively. The shape parameter (or Weibull modulus) is a unitless measure of the dispersion of strength while the scale parameter is the characteristic strength (at $P_f=0.6321$) of a unit area of material in uniaxial tension and has units of stress·area$^{1/m}$. An analogous equation based on component volume, V, can be shown for flaws that reside within the component.

Determination of the Weibull parameters comes from rupture experiments of specimens (ideally 30 or more) in simple tension or flexure. Regression techniques such as least squares and maximum likelihood have been developed that can determine these parameters from a simplified form of Equation (1);

$$
\begin{aligned}
P_f &= 1 - \exp\left[-\int_A \left(\frac{\sigma\ (x,y)}{\sigma_f} \right)^m dA \left(\frac{\sigma_f}{\sigma_o} \right)^m \right] \\
&= 1 - \exp\left[-A_e \left(\frac{\sigma_f}{\sigma_o} \right)^m \right] = 1 - \exp\left[-\left(\frac{\sigma_f}{\sigma_\theta} \right)^m \right]
\end{aligned}
\tag{2}
$$

where σ_f is the peak stress in the specimen, σ_θ is the specimen characteristic strength, and A_e is known as the effective area. For component reliability that is a function of the volume, an analogous set of equations can be developed where V_e is the effective volume.

The Weibull size effect is a direct consequence of Equation (2) and predicts that the average strength of a large sized component is lower than that for a smaller sized component for identical loading and geometry. The magnitude of the size effect is a function of the effective area (or volume) and the Weibull modulus. For two different component geometry/loading combinations the size effect strength ratio is obtained by equating the probabilities of failure for the two different components resulting in

$$\left(\frac{\sigma_{f,2}}{\sigma_{f,1}} \right) = \left(\frac{A_{e,1}}{A_{e,2}} \right)^{\frac{1}{m}} \tag{3}$$

In this work we study size-effect by using stress concentration. The notched and central-hole specimens were designed to have a predicted size effect relative to the curved specimen that was large enough to easily stand out from the inherent scatter in the data. We used the maximum likelihood estimated Weibull modulus m from the curved specimen data to predict the size effect for the specimens with stress concentration. The PIA fracture criterion (see reference 2 for a description) was used to calculate the effective areas A_e using the m obtained from the curved specimen data. The contribution from secondary and tertiary principal stresses was assumed to be negligible as well as any strength degradation that could have occurred from slow crack growth. Confirmation of a size effect and the ability to predict it is a direct test of the applicability of the Weibull distribution to model the strength response of these materials at this size scale.

SIZE EFFECT STUDY OF THE POLYCRYSTALLINE SiC

For the polysrystalline SiC the CARES/Life analysis was performed with deterministic finite element analysis using average specimen dimensions. Specimen fracture stresses were calculated from measured fracture loads, the stress concentration factor (determined from finite element analysis), and the specimen minimum cross-sectional area. For the i'th specimen then

$$(\sigma_f)_i = \frac{(fracture\ load)_i}{(cross-sectional\ area)_i}(stress\ concentration\ factor) \qquad (4)$$

From the specimen fracture stresses the Weibull parameters can be estimated. This is shown in Tables VI and VII for the respective batch 3 and batch 4 processing runs. Because fractography showed that critical flaws primarily originated along the side-walls CARES/Life was used to calculate side-wall effective effective area, A_{esw}, considering only the gage section of the specimens. This is also shown in Tables VI and VII for their respective Weibull parameters.

Table VI. Batch 3 polycrystaline SiC strength and maximum likelihood estimated Weibull parameters; n=number of specimens.

Specimen	n	Avg. strength (GPa)	m	σ_θ (GPa)	A_{esw} mm^2
Curved	22	0.47 ± 0.16	3.2	0.53	0.184
Notched	26	0.68 ± 0.19	3.8	0.75	0.0143

Table VII. Batch 4 polycrystaline SiC strength and maximum likelihood estimated Weibull parameters; n=number of specimens.

Specimen	n	Avg. strength (GPa)	m	σ_θ (GPa)	A_{esw} mm^2
Curved	23	0.47 ± 0.15	3.4	0.53	0.197
Notched	25	0.79 ± 0.28	3.1	0.88	0.0157

To predict the failure probability of specimen geometry 2 using the characteristic strength $\sigma_{\theta,1}$ of the baseline specimen geometry 1 we used

$$P_f = 1 - \exp\left[-\frac{A_{esw,2}}{A_{esw,1}} \left(\frac{\sigma_{f,2}}{\sigma_{\theta,1}} \right)^m \right] \qquad (5)$$

where in this case geometry 1 is for the curved specimen and geometry 2 is for the notched specimen and side-wall effective area A_{esw} is used. For this equation A_{esw} is recalculated for the notched specimen using the Weibull modulus m estimated from the curved specimen data. The predicted response for the notched specimens compared to the experimental data is shown in figures 6 and 7 for batches 3 and 4 respectively. For both batches a size effect can be seen in the

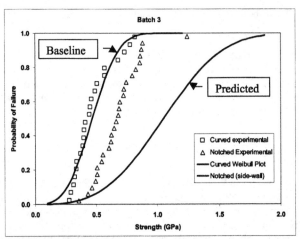

Figure 6. Batch 3 experimental fracture strengths and predicted strengths with the Weibull distribution for the notched specimens.

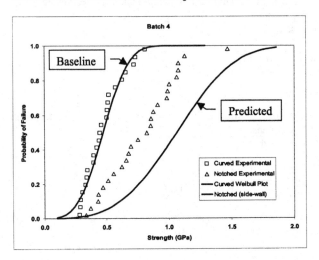

Figure7. Batch 4 experimental fracture strengths and predicted strengths with the Weibull distribution for the notched specimens.

experimental data, however the analysis based on the Weibull distribution tended to over predict this effect – particularly for batch 3. Batch 3 and batch 4 were fabricated using the same process recipe, however, because some dimensions significantly differed from one another as shown in Tables I and II it was decided not to combine the data from both batches. Batch 4 seemed to show less dimensional variation and better verticality of the side walls, however it can only be speculated that this played a role in the improved prediction of batch 4 relative to batch 3. The degree to which this material deviated from continuum behavior could well explain much of the discrepancy. These effects included large grains and etching grooves relative to specimen dimensions. Nor was the effect of stress gradients across the length of critical flaws considered in the Weibull theory.

SIZE EFFECT STUDY OF THE SINGLE CRYSTAL SiC WITH ANSYS/PDS

This analysis demonstrates using CARES/Life with the ANSYS Probabilistic Design System to account for stochastic variables – in this case the variability of specimen dimensions – on the specimen strength response. A rigorous analysis was not performed here and therefore results should only be viewed as approximate.

The ANSYS Probabilistic Design System is an analysis tool which works with the ANSYS finite-element analysis program. The ANSYS/PDS allows the effects of probabilistic loads, component geometry, and material properties to be considered in the analysis. It offers probabilistic analysis methods such as Monte Carlo simulation or Response Surface Method. Previously it has been shown that CARES/Life can be embedded within the ANSYS/PDS using ANSYS macros so that the effects of stochastic variables of component geometry, loading and material properties on the predicted strength and life can be assessed[13]. For the analysis of the single crystal SiC (batch 6) specimens we have attempted to use CARES/Life with ANSYS/PDS to simulate the effect of dimensional variation on predicted failure probability. This was done because dimensional variation of the elliptical hole specimen was significant (see Table V). Isotropic material and fracture behavior were assumed since CARES/Life does not have anisotropic reliability models for single crystals. All specimens had the same orientation parallel to the primary flat of the wafer and hence the same orientation relative to the crystallographic cleavage plane(s). The effect of multiaxial stresses on the cleavage plane(s) was not considered. As was done previously with the polycrystalline material the strength response of the circular-hole and elliptical-hole specimens was predicted using the curved specimen data as a baseline. Results are shown versus first principal stress.

Table VIII shows the estimated Weibull parameters from the experimental data. The peak stress σ_f for each specimen was determined with ANSYS using the individual measurements for each specimen cross section. Figure 8 shows the quarter-symmetry mesh and principal stress solution for the gage section of the elliptical-hole and circular-hole specimens using average specimen dimensions. A uniform pressure was applied at the specimen end to simulate the load. The estimated Weibull modulus m in Table VIII appears to be lower for the circular-hole and elliptical-hole specimens. However, there was one particularly high value in the elliptical-hole data that had the effect of lowering the Weibull modulus from 3.9 (with the data point removed) down to 2.0 (with the data point included).

Table VIII. Batch 6 single crystal SiC strength and maximum likelihood estimated Weibull parameters; n=number of specimens.

Specimen	n	Avg. strength (GPa)	m	σ_θ (GPa)
Curved	19	0.66 ± 0.12	5.9	0.71
Circular hole	19	1.26 ± 0.34	3.8	1.38
Elliptical hole	18	1.53 ± 0.84	2.0	1.74

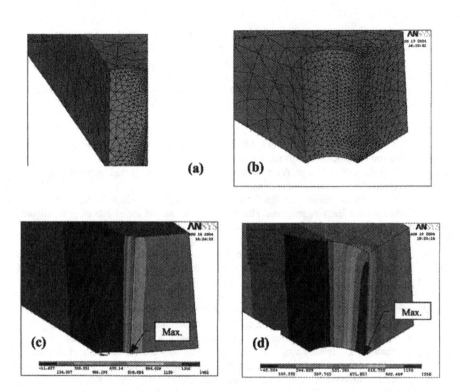

Figure 8. Quarter-symmetry mesh and principal stress solution for the elliptical-hole and circular-hole specimen using averaged specimen dimensions. Only the specimen gage section was modeled: (a) Elliptical-hole mesh. (b) Circular-hole mesh. (c) Elliptical-hole principal stress solution. (d) Circular-hole principal stress solution.

To simulate fracture strength data we took advantage of Equation (5) and Monte-Carlo simulation. For the PDS simulations a normal distribution was used to describe the variation in the dimensions shown in Tables IV, and V. For the nickel mask side (the top side) the average dimensions and respective standard deviations were used. For the side opposite the nickel mask (the bottom side) the ratio of the taper from the nickel mask side to the opposite side multiplied by the nickel mask side dimension was used. For a given trial within a simulation, a random number generator was used to choose a P_f, then using the ANSYS model generated from the trial, A_{esw} was calculated using the baseline Weibull modulus, CARES/Life and a predefined deterministic load. Finally σ_f for the trial was calculated from Equation (5) using the baseline data and the randomly generated P_f value. The result was a collection of n fracture strengths that could be ordered from lowest to highest value; $\sigma_{f,1}$, $\sigma_{f,2}$, ... $\sigma_{f,i}$, ..., $\sigma_{f,n}$. These values were then ranked similar to experimental fracture data using the formula

$$P_{f,i} = \frac{i - 0.5}{n} \qquad (6)$$

Using the procedure outlined above, simulations for 200 specimens of the circular-hole and elliptical-hole specimen types were run. The curved-specimen was used as the baseline. These results are shown in Table IX and figure 9. The simulations were generated using a Weibull modulus m of 6.5, which was the value obtained from the fracture forces for the curved specimen and σ_θ of 0.71 GPa. Table IX shows the elliptical hole-specimen had a lower estimated Weibull modulus compared to the curved specimen simulated fracture data. This was because the elliptical-hole specimen was affected to a greater degree by dimensional variability. This fact was confirmed when the effective side-wall area A_{esw} versus failure probability P_f was compared for the elliptical-hole specimen. The A_{esw} value tended to be larger at low P_f and smaller at high P_f.

Table IX – Weibull parameters regressed from the PDS simulations using maximum likelihood

Specimen	# Simulated specimens,n	m	σ_θ (GPa)
Circular hole	200	5.9	1.24
Elliptical hole	200	4.1	1.87

Figure 9 shows the CARES/Life & ANSYS/PDS simulation results compared to the experimental data. Also shown are CARES/Life predictions using average specimen dimensions similar to the analysis done for the polycrystalline material as discussed in the previous section. The experimental data did show an appreciable size effect – although less than was expected for the elliptical-hole data. The curved specimen data was used as the baseline and it can be seen that the Weibull parameters chosen fit the experimental data very well – as it should have (magnified views not shown also confirmed the good fit). Very good correlation was achieved for the circular-hole specimen. For the elliptical-hole specimen the simulation tended to over predict the strength. Nonetheless these results were encouraging. It is also interesting to see the close correlation between the simulation and the averaged dimension results for the circular-hole specimen, indicating the effect of dimensional variability on these specimens was small. The deviation was larger between the simulation and averaged-dimension predictions for the elliptical-hole specimen – showing the increased scatter in strength predicted from the simulation. Interestingly the simulation also indicated somewhat lower fracture stresses than the averaged-dimension prediction, correlating better to the experimental data. Further detailed investigation would be needed to understand this trend.

CONCLUSION

The ability of the Weibull distribution to be used to predict the strength response of room-temperature SiC polycrystalline and single crystal microtensile specimens with various stress concentrations was investigated using the NASA developed CARES/Life program. Analysis was performed assuming the strength controlling flaws resided on those material surfaces that were exposed to the etching process. CARES/Life tended to over predict the strength of the polycrystalline material but had better success predicting the strength response for the single crystal material. Large columnar grains and rough side-walls for the polycrystalline material likely reduced the applicability of continuum analysis and therefore the accuracy of the CARES/Life analysis. The single crystal material did not have (or had less of) these complications because it lacked discontinuous phases and because of an improved fabrication process that produced smoother side-walls.

The microtensile specimens had significant specimen-to-specimen variations in dimensions. We demonstrated with the single crystal SiC specimens that ANSYS-PDS and CARES/Life could be used together to account for this variation. This analysis was not rigorous mainly because isotropic material and fracture behavior was assumed. However, the simulation results showed good correlation for the circular-hole specimen geometry and some over prediction of the strength of the elliptical-hole specimens. These results tend to support using the Weibull distribution for design and analysis purposes of SiC MEMS – and probably for other MEMS brittle material systems as well – as long as the material can be modeled as a continuum and size-effect extrapolation from a baseline is kept modest.

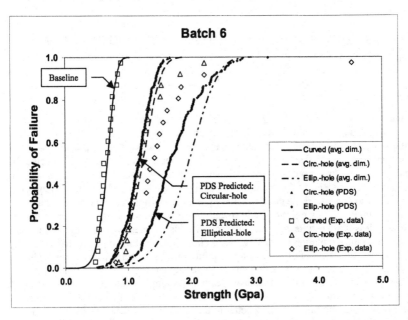

Figure 9. CARES/Life & ANSYS/PDS Monte-Carlo simulation of 200 predicted fracture stresses for the circular-hole and elliptical-hole specimen types. Simulation results are compared to experimental data and predictions from CARES/Life with averaged specimen dimensions.

ACKNOWLEDGEMENTS:
We would like to acknowledge the Principal Investigator Dr. Glenn Beheim and his team for the fabrication element of this work, Prof. William Sharpe Jr. and his team at Johns Hopkins

University for their testing element, Dr. Stefan Reh of ANSYS Inc. for his assistance with PDS, and Eric Baker of Connecticut Reserve Technologies for CARES/Life and programming assistance. This work was funded under the NASA Alternate Energies Foundation Technologies (AEFT) program and the Ultra Efficient Engines Technology (UEET) program.

REFERENCES:

1. A. Epstein, "Millimeter-Scale, MEMS Gas Turbine Engines," Proceedings ASME Turbo Expo 2003 Power for Land, Sea, and Air; June 16-19, 2003, Atlanta, Georgia, USA. Paper No. GT-2003-38866.

2. N. Nemeth, L. Powers, L. Janosik, and J. Gyekenyesi, "CARES/Life Ceramics Analysis and Reliability Evaluation of Structures Life Prediction Program." NASA/TM-2003-106316, (2003).

3. N. Nemeth, O. Jadaan, T. Palfi, and E. Baker, "Predicting the Reliability of Ceramics Under Transient Loads and Temperatures With CARES/Life." Probabilistic Aspects of Life Prediction, ASTM-STP 1450, W. Johnson & B. Hillberry, Eds., ASTM International, (2004). Also; J. ASTM Int., Vol. 1, No. 8, Paper ID JAI11578, (2004).

4. S. Reh, P. Lethbridge, D. Ostergaard, "Quality-Based Design with Probabilistic Methods." ANSYS Solutions, Vol. 2, No. 2, (2000). http://www.ansys.com/assets/tech-papers/mems-solutions-4.pdf

5. G. Beheim, "Deep Reactive Ion Etching for Bulk Micromachining of Silicon Carbide." The MEMS Handbook, M. Gad-el-Hak, ed., CRC Press, (2002).

6. WWW.cvdmaterials.com/silicon.htm, Rohm and Haas, (2004).

7. WWW.cree.com/ftp/pub/sicctlg_read_new.pdf, "Silicon Carbide Substrates," MAT-CATALOG.00D, Cree Materials, 4600 Silicon Drive, Durham, NC, 27703, (1998-2005).

8. W. Sharpe, D. Danley, and D. LaVan, "Microspecimen Tensile Tests of A533-B Steel," Small Specimen Test Techniques, ASTM STP 1329, W. Corwin, S. Rosinski, and E. Van Walle, Eds., Am. Soc. for Testing and Materials, 497-512, (1998).

9. W. Sharpe, D. LaVan, and R. Edwards, ''Mechanical Properties of LIGA-Deposited Nickel for MEMS Transducers'', Proceedings Transducers '97, Chicago, IL, 607-10, (1997).

10. W. Sharpe, O. Jadaan, N. Nemeth, and G. Beheim, "Strength of Polycrystalline Silicon Carbide Microspecimens at Room and High Temperature", Proceedings of the 2004 SEM Annual Conference, Costa Mesa, CA, Session 84, Paper No. 405, (2004).

11. G. Quinn, and J. Salem, "Effect of Lateral Cracks of Fracture Toughness Determined by the Surface Crack in Flexure Method," J. Am. Ceram. Soc., **85**, 873–80, (2002).

12. K. Chen, A. Ayon, and S. Spearing, "Controlling and Testing the Fracture Strength of Silicon on the Mesoscale." J. Am. Ceram. Soc., **83**, 1476-80 (2000).

13. S. Reh, T. Palfi, and N. Nemeth, "Probabilistic Analysis techniques Applied to Lifetime Reliability Estimation of Ceramics". Paper No.APS-II-49 Glass. JANNAF 39th CS/27th APS/21st PSHS/3rd MSS Joint Subcommittee Meeting, Colorado Springs, Colorado, December 1-5, (2003).

DESIGN AND RELIABILITY OF CERAMICS:
DO MODELERS, DESIGNERS, AND FRACTOGRAPHERS SEE THE SAME WORLD?

George David Quinn
National Institute of Standards and Technology
100 Bureau Drive
Gaithersburg, MD, 20899-8529

ABSTRACT

Techniques for design and reliability analyses of structural ceramic components are now well established. Laboratory test coupon data are used in design codes to predict the fast fracture and time dependent reliability of monolithic or dispersed-phase ceramic composite components. Fractographic analysis of fractured components shows that there can be surprises, however. A few case studies are reviewed to show that components often fail from an unanticipated cause. The reliability models make a number of assumptions that, if violated, put the predictions at risk.

INTRODUCTION

The general design concepts for reliability analysis of monolithic structural ceramics have been by now well established. These are not recapitulated in detail here and the reader is referred to any one of a number of good reviews. [1,2,3,4,5,6,7,8,9] Referring to Figure 1, the ceramic component is envisioned as having volume, surface and/or edge type strength limiting flaws. Which of these cause fracture depends upon the stress state in the part. So for example, a component loaded in uniform tension is more apt to trigger failure from the largest flaw than if the component were loaded in bending. Weibull analyses have long been used successfully to model the probability of fracture or survival [1]. Complex structures may be analyzed by finite element models and the stress distributions fed to other programs that integrate the risk of rupture over all elements in the body. Time-dependent reliability is customarily analyzed by modeling subcritical flaw growth that can weaken the component, until one flaw reaches criticality leading to component fracture. If the component is exposed to elevated temperature, then temperature dependent properties and crack growth behavior must be factored in.

Although this is the conventional analysis for mechanical reliability of structural ceramics exposed to tensile stresses, it should be kept in mind that other failure mechanisms such as wear, erosion, creep fracture, or compression fracture can also cause "failure" or loss of function.

Despite the widespread acceptance of this general approach, there are considerable variations in the details, and indeed, the devil may be in the details. What is the best Weibull analysis to use? Should a two parameter or three parameter distribution be assumed? What is the best way to estimate the distribution parameters? Are they biased? How much laboratory data should be collected? What type of lab data? Is stress rupture data better than

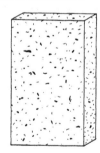

Figure 1 A ceramic component with a myriad of flaws.

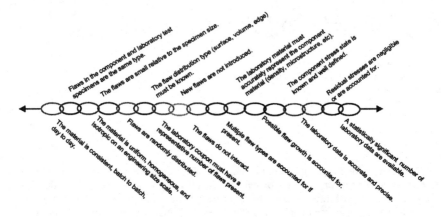

Figure 2 The weakest links in a design and reliability model may be the model assumptions.

dynamic fatigue or crack velocity data? Is there a threshold stress or stress intensity below which crack growth will not occur? What is the best multiaxial stress failure criterion? Should one use test data based on indentation flaws or genuine material flaws? What are the uncertainties and errors? How far, if at all, may one extrapolate?

One finds that the record is rather mixed for hundreds of ceramic reliability and design analyses. There have been some spectacular successes and many failures. This prompted Morrell in 1989 to question the value of any laboratory test coupon data for reliability analysis, and to suggest that component testing is a sounder approach.[10] A review by Quinn and Morrell [11] in 1991 investigated this matter in more detail, principally in the context of whether flexural strength test data had value for design. They concluded that the general design concepts for reliability analysis were sound, but that the analyses make a series of critical assumptions about the material, the flaws, and the nature of the loading. Only if all the assumptions were valid, could one expect the reliability models to be accurate. Otherwise, the models could make very erroneous predictions. Indeed, the whole affair may be likened to weakest link theory as shown in Figure 2, except that is this instance the links are the model assumptions. The final conclusion in the review was that laboratory scale flexural strength data could be used for design, but one had to be lucky. By this it was meant that more often than not nature violates one or more of the assumptions. This sobering conclusion should not be limited to flexure strength data either: any reliability analysis based on any laboratory coupon data has the same risks. Quinn and Morrell recommended that test coupons of different sizes be tested to search for multiple flaw types and to validate the size-scaling model. Ideally, specimens should be cut from the component to compare to the separate laboratory coupon data sets. They summarized: "The reliability of any design extrapolation is likely to be best when the test bar most closely related to the component in terms of size, stress state, and defect distribution."

One might wonder that if there were so many caveats, whether any design with brittle ceramics has ever succeeded. The record shows that indeed there are a number of successful cases, often because the investigators paid close attention to the assumptions in Figure 2 [11].

This paper gives four case studies wherein fractographic analysis revealed critical information that should be included in a reliability model. The primary point of this paper is that judicious use of fractographic analysis of laboratory strength coupons and destructive prototype testing can dramatically improve reliability models and databases.

CASE 1 The Ford Gas Turbine Rotor

This was an ambitious but well thought out small project conducted at the Ford Motor Company's Scientific Research Laboratory[a] in the late 1970's to mid 1980's.[12,13] It was part of a much larger endeavor to incorporate advanced ceramics into automotive gas turbine engines. Considerable effort was expended on improving ceramic materials, developing reliability codes, generating data bases, fabricating parts, and running them in test rigs. Full scale engine testing was extremely expensive and risky, so it was decided that a small project would be performed to verify the ceramic design and reliability codes by using a realistic model component shown in Figure 3. This simulated gas turbine rotor was mounted in a hot spin rig shown in Figure 4 and rotated at 50,000 rpm while heated by hot gasses to a peak rim temperature of 1260 °C (2300 °F).

Ten model rotors were fabricated and tested. The component was designed to have a high probability of survival on loading (> 95 %), but to probably fail within 25 h. The design analysis was based on the premise that failure would occur due to slow crack growth of preexisting flaws that were distributed throughout the volume. It was expected that fracture would occur from the thin web portion of the rotor where the stress and temperature were both high. The intent was to compare the actual rotor lifetimes to predictions from the reliability codes, refine the models as needed, and identify the best type of laboratory test data.

The model rotor was made of a state of the art hot-pressed silicon nitride[b] (NC 132) that was carefully machined to final dimensions. The particular grade was one of the most thoroughly analyzed structural ceramics of all time and was eventually used as the world's first reference material for the property fracture toughness.[14] Eventually enough data was available that a comprehensive fracture mechanism map was constructed,[15] but that was after the

Figure 3 Ford silicon nitride model gas turbine rotor #1323 which survived 25 h intact. The model rotor has a 95 mm diameter.

---

[a] Certain commercial materials or equipment are identified in this paper to specify adequately the experimental procedure. Such identification does not imply endorsement by the National Institute of Standards and Technology nor does it imply that these materials or equipment are necessarily the best for the purpose.
[b] Grade NC 132, Norton Co., Worcester Ma.

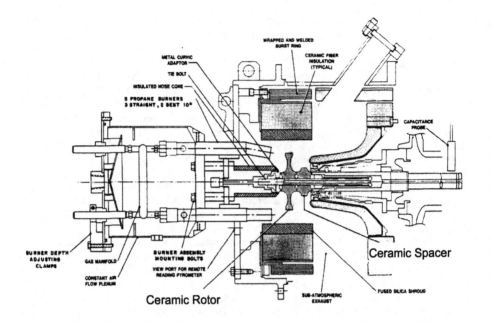

Figure 4 The Ford hot spin test rig (Figure from Ref. 12).

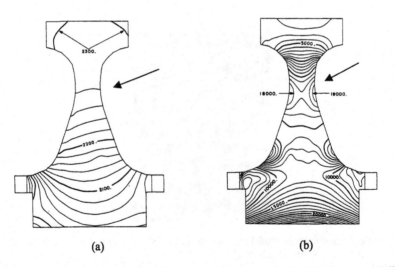

(a)   (b)

Figure 5 Isotherms (°F) (a), and maximum principal stresses (psi) (b) in the rotor. Failure was expected to occur in the thin web (large arrows) (Figures from Ref. 12). The maximum stress in the web was oriented primarily in a radial direction.

conclusion of the Ford rotor test project. A substantial amount of laboratory test coupon data were available including flexural and tensile strengths, elastic properties, and slow crack growth parameters. All were available as a function of temperature. Slow crack growth data were available from three types of tests: variable stressing rate strength tests (dynamic fatigue), crack velocity - fracture mechanics tests (double torsion), and stress rupture (flexural and tensile). The model rotor did not have turbine vanes, but an increased rim mass to simulate their effect. Figure 5 shows the temperature and principal stresses from the heat transfer and finite element models, respectively. The maximum principal steady state stress was 131 MPa (19,000 psi).

Figure 6 shows the reliability as a function of time. An estimate of the error in the calculated reliabilities was made using estimated temperature errors of $\pm$ 11 °C (20 °F). At time zero, the reliability was simply the fast fracture reliability that was evaluated by integrating the normal stress around the unit sphere in each element and evaluating the Weibull volume integral for each element. The three types of slow crack growth data gave very divergent predictions. Figure 6b shows the outcomes for the six rotors that failed at times from 0.2 h to 18.6 h. Three other rotors survived 24 h and a fourth at 13 h and were treated as censored outcomes (not shown). The predictions made with the stress rupture data set gave the best correlation to the actual lifetimes and the authors concluded that stress rupture data is the best for predicting reliability versus time.[12] This outcome was gratifying to this author since his data were used for the predictions.[16]

So in this case, the designer and modeler saw a ceramic component designed to fail from slow crack growth of preexisting volume flaws located in the rotor web. What did the fractographer see?

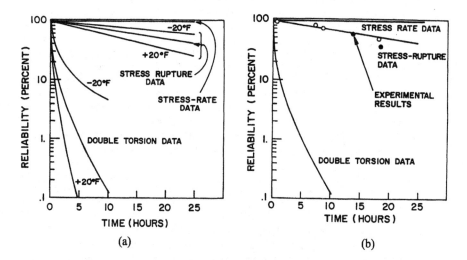

Figure 6 Reliability versus time. (a) shows the predictions and estimated uncertainty limits for expected temperature variations, and (b) shows the actual outcomes for six rotors that failed while running hot at 50, 000 rpm (Figures from Refs. 12, 13) The two solid points are the rotors that were fractographically analyzed by the author.

The fractographic analysis revealed that fracture did not occur quite as anticipated. The test rig was constructed in such a manner that most broken fragments were retrievable. The longest (18.6 h) and third longest (13.8 h) rotors were sent to the author for fractographic evaluation. It was felt that these would have large slow crack growth zones that would be easy to find. One virtue of NC 132 is that it was a very fine-grained fully dense material with minimal second phase. Fracture markings were quite clear and the local direction of crack propagation could be assessed in every piece. Rotor # 1324 (18.6 h) was reconstructed piece by piece over the course of one week culminating in the assembly shown in Figure 7. Many fragments from the rim were missing, but it did not matter. The figure also shows that the spacer had also fractured and its pieces were commingled with the rotor thereby hampering the assembly. Nevertheless, the central portion and web portion of the rotor was almost completely reassembled. After completion, the rotor was taken apart again and a map of crack propagation directions was made.

The interpretation was clear. Primary fracture commenced at the bore or the curvic coupling teeth and cracks branched and ran out to the rim as shown in the Figure 7 schematic. Every single fragment was examined in a futile search for a web origin and/or slow crack growth zone, but none were found. All primary fractures started from the bore or coupling area. There were three origins from the inner part of the rotor. Origin "A" was in the bore at the exact edge of the bore, which although chamfered nonetheless had surface grinding cracks (Figure 8). It could not be determined whether slow crack growth had enlarged the grinding cracks. The other two origins were located on the other side of the bore and were from the curvic coupling teeth machined into the rotor. These were part of the attachment scheme. One origin was a transverse machining crack and the other was an impact – contact crack.

Thus, fracture started on one side of the rotor, thereby opening up the disk structure whereupon unbalanced forces triggered a rupture on the opposite side. Origin A had a well-defined small fracture mirror allowing an stress estimate of a 630 MPa (92 ksi) which is much greater than the model estimated bore stress of ≈172 MPa (25 ksi) from Figure 5. This, plus the violent branching from site A, suggest it was a secondary fracture. Primary fracture occurred on the other side of the bore, causing the rotor to go completely out of balance leading to the overstress fracture at point A. Origin h was a large contact crack flaw with a very large mirror. The local stress was estimated to be 500 MPa (72 ksi) from the flaw size and approximately 580 MPa (84 ksi) from the fracture mirror. Origin g was determined to have fractured after site h. Figure 5 shows there were modest bore stresses (≈138 MPa, 20 ksi) and temperatures (1150 °C, 2200 °F), but the stress does not match the level assessed by fractographic analysis.

One other possibility is suggested by an observation from the fractographic examination. When the rotor and spacer were assembled in the hot test rig, it was customary to separate them with a thin platinum foil to prevent direct ceramic-to-ceramic contact. The fractographic examination showed uneven platinum wear traces and even some bare spots. Could it be that the foil deformed or crept with time such that ceramic-to-ceramic contact eventually occurred? If so, then the rotor stress distribution may have become unbalanced or local stress concentrations or contact stresses may have contributed to cause initial fracture at site h.

This exercise was a good example of one of the author's laws of fractography: "The first one is the hardest." While at first glance reassembly of an exploded rotor may appear to be a formidable task, it was merely time consuming. The second rotor, #1314, took less time to analyze. It also had fracture initiation sites at the bore or at the teeth. Years later, the author became aware of comparable work done on model silicon nitride rotors at Daimler-Benz.[17] They also reconstructed burst silicon nitride rotors and used fractography to find that grinding cracks in the bores sometimes were fracture origins.

Some lessons learned from our fractographic analyses were:

a. Fracture occurred from a different cause than expected and modeled.
b. Stress rupture data may have been the best for reliability estimation, but the correlation of failure times was fortuitous for the two rotors examined.
c. Volume flaws were not found in the two fractured rotors. Surface machining cracks were found and thus, Weibull area scaling should also have been included in the reliability model.

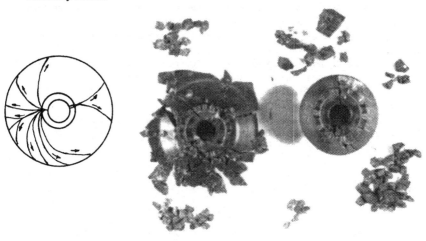

Figure 7 Reconstructed model rotor and spacer. Fracture initiated either at location A on the left side of the bore of the rotor or at two locations g or h on the opposite side of the bore. The schematic on the left shows the radiating pattern on cracks in the rotor.

Figure 8 Fracture surfaces showing a fracture mirror and the origin site A, which is a surface crack from grinding. The right figure highlights the fracture mirror with arrows.

CASE 2   The Ceramic Gun Barrel

The U. S. Army has off and on over the past twenty-five years investigated ceramic liners to improve gun barrel life and reduce mass. The low density, high compression strength, refractoriness, and erosion and wear resistance of ceramics could be used to advantage. Figures 9 and 10 show drawings for a 50-caliber machine gun breech with a ceramic liner from a project conducted for the U. S. Army in the 1980's.[18] One design placed the ceramic into compression residual stresses by shrink fitting a steel sleeve around the ceramic. The steel sleeve was heated and placed over the cool ceramic tube. As the assembly cooled, the steel contracted and put the ceramic into radial and hoop compression. The dimensions and temperature differentials were chosen so that the residual axial compression stresses were up to 170 MPa (25 ksi), the radial stresses up to 345 MPa (50 ksi), and the hoop tensile stresses up to 1034 MPa (150 ksi). These were sufficiently high to keep all portions of the ceramic tube in compression when firing stresses were superimposed. The environment is severe and dynamic, but if the ceramic were always under compression, perhaps it would not fracture. Several ceramics were tried, but most testing was on a sintered α-SiC.

So in this case, the designer saw a part that was fully constrained by compression residual shrink fit stresses. After testing, what did the fractographer see?

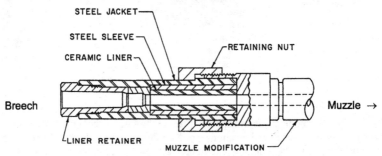

Figure 9 Schematic of the breech end of the 50 caliber (12.7 mm) gun barrel (Adapted from Ref. 18).

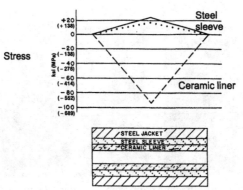

Figure 10 Axial stress distribution in the ceramic inner liner after shrink fitting. (Adapted from Ref. 18).

246

Some assemblies survived as many as a thousand single shot firings, confounding some skeptics who felt that the assembly would not survive one shot.[18] One assembly that did develop ceramic cracking after a few hundred shots had circumferential fractures as shown in Figure 11a. The steel sleeve and jacket were machined away to allow extraction of the ceramic as shown in Figure 11b. The fracture planes were perpendicular to the axial direction, suggesting that expansion-burst stresses were not the cause of failure, since the later would have created radial cracking. Fractographic analyses showed that every fragment fractured from one or more contact cracks that were periodically spaced on the *outer rim* of the ceramic.

So if the shrink fitting created compression stresses, where did the tensile stresses come from? The plane of the fractured surfaces and also of the initial semielliptical contact cracks informs us the tensile stresses were axial. Figure 10 shows the axial residual compression stresses did taper off towards the tube ends. The most likely sources of tensile stresses are dynamic stress waves generated during firing. Even if these are initially compressive, they can change phase and become tensile as the stress wave reflects off end faces. Furthermore, the sound velocity and impendence of the silicon carbide and the steel sleeve are not matched and stress waves will propagate at different rates in the axial direction.

The contact stress cracks often were periodic around the rim. Although the parts were machined to tight tolerances, it is likely that slight variations in the mating surfaces led to uneven fit, and hence stress concentration sites that triggered the contact cracks.

So in this case, the ceramic was designed to always be in compression, but fractography showed otherwise. Some design and modeling improvements were suggested. The tolerances and surface specifications for the mating parts could be changed. The elastic properties of the ceramic and the confining sleeve could be matched better. More sophisticated stress models could examine the transient stress states in the assembly.

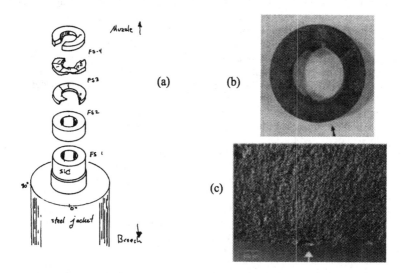

Figure 11 Fractured α-SiC gun barrel (a) is a schematic. (b) shows one fracture surface of one of the ring shaped fragments. The bore diameter is 12.7 mm. The arrow shows a contact crack origin on the outer surface where it contacted the steel sleeve. A close-up is shown in (c).

247

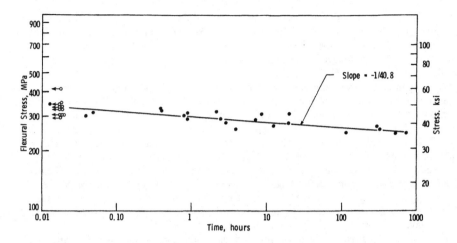

Figure 12 Stress rupture curve at 1200 °C in air for sintered α-SiC [19-21] Every time dependent fracture was from a pore or porous zone, and only if it was surface connected.

CASE 3   Conflicting Stress Rupture Data for Sintered α Silicon Carbide

This case is not about a specific component, but about the risk of treating all flaws as if they behave the same. As part of a program on characterization of structural ceramics for heat engine ceramics in the 1980's, the author conducted extensive stress rupture testing of] including pressureless sintered α-SiC.[c] [19,20,21 At elevated temperatures, there was a distinct trend of time dependent failure for bend bars tested in air such as shown in Figure 12. Trend like this are usually interpreted as evidence of slow crack growth. Tests of two other batches made in different years confirmed that time dependent failure always occurred at 1200 °C and above. On the other hand, another team [22] had data that matched the ≥ 1300 °C data, but contradicted the 1200 °C data. They did not observe any time dependent fractures at 1200 °C. Why did the two groups have such contrasting outcomes at 1200 °C?

Fractographic analysis provided the answer.[19-21] At 1300 °C and above, easily detected intergranular slow crack growth from volume-distributed flaws such as large grains, pores, or agglomerates caused fracture such as shown in Figure 13a. Some of these volume origins were located in the bulk. Both teams detected this behavior.

A different story emerged at 1200 °C. Every one of the time-dependent fracture origins in references 19-21 was a pore or porous region and every one was connected to the outer surface as shown in Figure 13b. These flaws not have slow crack growth markings. Evidently the pore type flaws were susceptible to very localized stress corrosion cracking (from oxidation) that sharpened or locally extended tiny microcracks on the pore periphery. Not much change in flaw size or severity was necessary for these flaws to go critical, since they broke at stresses very close to the fast fracture strength as shown in Figure 12.

The other testing team did not detect this stress corrosion mechanism since they used Knoop artificial flaws in all of their bend specimens.[22] Artificial flaws often are effective tools for studying fracture, but in this instance they produced misleading results. The Knoop

---

[c] Hexoloy SA, Carborundum, Niagara Falls, NY.

248

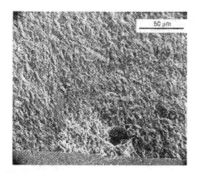

Figure 13. Fracture origins in stress rupture specimens. (a) shows a classic slow crack growth zone in a specimen that broke at 1300 °C. (b) shows a surface connected porous zone in a specimen that broke at 863 hours at 1200 °C.

flaws did grow by slow crack growth at 1300 °C and above. The Knoop flaws did not grow by slow crack growth or by stress corrosion at 1200 °C. One might ask why the specimens with Knoop flaws did not fracture from the pores and porous regions that were undoubtedly present. The answer is that the Knoop flaws reduced the strength of the specimens to less than 200 MPa, a stress well below that necessary to activate the pore flaw stress corrosion mechanism.

In summary, three flaws types controlled the room temperature strength as well as acted as sources of time dependent fracture at 1300 °C: pores and porous regions, large grains, and agglomerates. Only pores and porous zones were vulnerable to time dependent stress corrosion cracking at 1200 °C, and *only* if they were connected to the surface. A reliability model that assumed no time dependent failure at 1200 °C in air would have been faulty. This case demonstrates that it is often best to let a material reveal what type of flaws it is apt to fail from.

CASE 4 Strength Data for Miniature Silicon Carbide Tensile Specimens

Testing methodologies must keep pace with emerging technologies for miniature devices and structures for microelectromechanical systems (MEMS) and even smaller devices. Sharpe et a. [23] recently investigated the strength of miniature silicon carbide specimens with cross section sizes of ≈ 200 μm or less prepared by chemical vapor deposition (CVD) followed by deep reactive ion etching to final shape. Specimens with three types of gage section were used as shown in Figure 14. The curved gage section specimens were prepared to eliminate a stress concentration at the end tab - gage section junction, and consequently had narrower gage sections. The curved specimens had a good distribution of breakages in the gage section, unlike the straight specimens which sometimes broke at the end of the gage section. Consequently the curved gage section specimens were used as the baseline data set. Weibull statistics were applied to scale the strengths and to determine whether surface or volume flaw scaling gave better correlation.

Fractographic analysis was difficult but productive. The coarse microstructure created a very rough fracture surface that masked common fracture markings. Large SiC grains affected the crack propagation across the fracture surface, causing significant crack redirection and severe roughness as the propagating crack sought out preferred cleavage planes. The most helpful features for pinpointing the origin were twist hackle lines on cleaved grains and occasional large

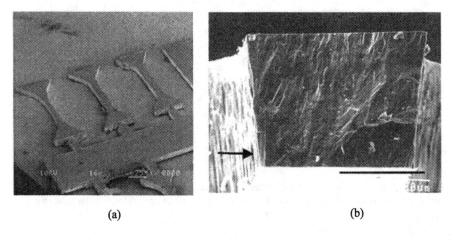

(a)   (b)

Figure 14 (a) shows the miniature 3.1 mm long SiC tensile specimens. From the left: curved, straight and small notched gage sections. The notches are too small to be seen in this view. (b) shows the fracture surface of a straight section specimen. The arrow shows the origin is a 25 μm x 50 μm large grain at the root of an etch groove. (Ref. 23)

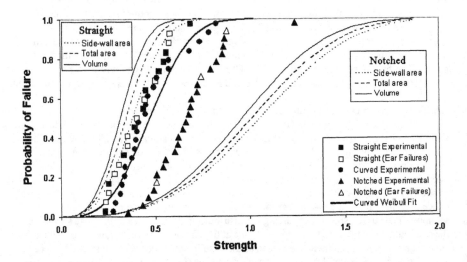

Figure 15 Strength distributions for straight, curved, and notched specimens. Using the curved specimen data set as the baseline, the straight specimen strengths agreed with the Weibull scaling prediction if sidewall area scaling was used. Correlations were poor for notched specimens.[23]

hackle lines that showed the local direction of crack propagation. This enabled the observer to backtrack to the origin site.

Figure 15 shows the strength results for the straight, curved, and notched specimens. The Weibull moduli were quite low (3 to 4) and the standard deviations quite large attesting to significant material or flaw variability. The fracture origins in all three specimen types usually were a combination of a deep etch groove that combined with a large columnar grain with a favorable orientation so that a crack could pop in on a preferred cleavage plane. Thus, the critical flaws were in many cases a combination of the sidewall grooves and a large grain: a hybrid surface-volume type flaw.

Nevertheless, the fractographic results supported the Weibull area scaling for comparing the straight and curved specimens, provided that only the etched *sidewall* areas were used. Weibull scaling to predict the notched specimen strengths was not successful and fractography provided an explanation. The notches were quite small (15 μm to 25 μm radius) and not much larger than some of the grains and the notch itself. Hence, a Weibull analysis based on continuum mechanics is questionable.

CONCLUSIONS

Ceramics are unforgiving materials when loaded in tension. The track record of reliability analysis of structural ceramics is mixed in large part due to the many assumptions implicit in the reliability analysis. Violation of any one of these "weakest links" incurs considerable risk. Nevertheless, there are a number of very successful applications, some of which are cataloged in the review paper by Quinn and Morrell.[11] Judicious fractographic analysis of both laboratory strength coupons and destructive prototype test coupons can dramatically improve models. It is best to let the material reveal how it is apt to fail than rigidly adhere to a preconceived model.

ACKNOWLEDGEMENT

Support for this work was furnished by the U. S. Department of Energy, Office of Transportation Technologies, with Oak Ridge National Laboratory.

REFERENCES

[1] W. Weibull, "A Statistical Distribution Function of Wide Applicability," *J. Appl. Mech.*, **18,** 293–297 (1951).

[2] R. W. Davidge, *Mechanical Behavior of Ceramics*, Cambridge Univ. Press, London, 1979.

[3] D. Munz and T. Fett, *Ceramics, Mechanical Properties, Failure Behavior, Materials Selection*, Springer, Berlin, 1999.

[4] W. E. C. Creyke, I. E. J. Sainsbury, and R. Morrell, *Design with Non-ductile Materials*, Applied Science Publ., London, 1982.

[5] J. B. Wachtman, *Mechanical Properties of Ceramics*, Wiley, New York, 1996.

[6] *Life Prediction Methodologies and Data for Ceramic Materials*, ASTM Special Technical Publication 1201, eds, C. R. Brinkman and S. F. Duffy, ASTM, West Conshohocken, PA, 1994.

[7] J. Menčik, *Strength and Fracture of Glass and Ceramics*, Elsevier, Amsterdam, 1992.

[8] A. Paluszny and W. Wu, "Probabilistic Aspects of Designing with Ceramics," ASME paper 77-GT-41 (1977).

[9] A. De S. Jayatilaka, *Fracture of Engineering Brittle Materials*, Applied Science Publ., London, 1979.

[10] R. Morrell, "Mechanical Properties of Engineering Ceramics: Test Bars Versus Components," *Mat. Sci. and Eng.*, **A109,** 131–137 (1989).

[11] G. D. Quinn and R. Morrell, "Design Data for Engineering Ceramics: A Review of the Flexure Test," *J. Am. Ceram. Soc.,* **74,** [9] 2037–2066 (1991).

[12] R. R. Baker, L. R. Swank, and J. C. Caverly, "Ceramic Life Prediction Methodology - Hot Spin Disk Life Program," AMMRC TR 83-44, Army Materials and Mechanics Research Center, Watertown, MA, August, 1983.

[13] L. R. Swank, R. R. Baker, and E. M. Lenoe, "Ceramic Life Prediction Methodology," pp. 323 –330 in Proceedings of the 21st Automotive Technology Development Contractors Coordination Meeting, Society of Automotive Engineers, Warrendale, PA, 1984.

[14] G. D. Quinn, K. Xu, R. J. Gettings, J. A. Salem, and J. J. Swab, "Does Anyone Know the Real Fracture Toughness? SRM 2100: The World's First Ceramic Fracture Toughness Reference Material," pp. 76-93 in *Fracture Resistance Testing of Monolithic and Composite Brittle Materials, ASTM STP 1409,* J. A. Salem, G. D. Quinn, and M. G. Jenkins, eds. ASTM, West Conshohocken, PA, 2002.

[15] G. D. Quinn, "Fracture Mechanism Maps for Advanced Structural Ceramics, Part I, Methodology and Hot Pressed Silicon Nitride," *J. Mat. Sci*, **25,** 4361-76 (1990).

[16] G. D. Quinn and J. B. Quinn, "Slow Crack Growth in Hot Pressed Silicon Nitride," pp. 603 – 636 in *Fracture Mechanics of Ceramics 6*, Eds. R. Bradt, A. Evans, D. Hasselman, F. Lange, Plenum Press, NY, (1983).

[17] K. D. Mörgenthaler and E. Tiefenbacker, "Bauteile Aus Keramik Für Gasturbinen," (Ceramic Components for Gas Turbines), Final Report of Phase III, Daimler-Benz AG, Technical Report, Stuttgart, 1983.

[18] E. J. Bunning, D. R. Claxton, and R. A. Giles, "Liners for Gun Tubes – A Feasibility Study," *Ceram. Eng. and Sci. Proc.*, 2 [7-8] 509–519 (1981).

[19] G. D. Quinn, "Characterization of Turbine Ceramics After Long Term Environmental Exposure," AMMRC TR 80-15, Army Materials and Mechanics Research Center, Watertown, MA, May l980.

[20] G. D. Quinn, "Review of Static Fatigue in Silicon Nitride and Silicon Carbide," *Ceram. Eng. and Sci. Proc.*, **3,** [1-2] 77– 98 (1982).

[21] G. D. Quinn and R. N. Katz, "Time Dependent High Temperature Strength of Sintered Alpha SiC," *J. Amer. Ceram. Soc.*, **63,** [1-2] 117– 119 (1980).

[22] J. Coppola, M. Srinivasan, K, Faber, and R. Smoak, "High Temperature Properties of Sintered Alpha Silicon Carbide," presented at *the International Symposium on Factors in Densification and Sintering of Oxide and Non-Oxide Ceramics*, Hakone, Japan, October 1978.

[23] W.N. Sharpe, Jr., O. Jadaan, G.M. Beheim, G.D. Quinn, N.N. Nemeth, "Fracture Strength of Silicon Carbide Microspecimens," accept by *J. Micromechanical Systems,* 2004.

THE EFFECTS OF INCORPORATING SYSTEM LEVEL VARIABILITY INTO THE RELIABILITY ANALYSIS FOR CERAMIC COMPONENTS

Robert Carter
Ordnance Materials Branch, Materials Division
Army Research Laboratory
Aberdeen Proving Ground, MD 21005

Osama Jadaan
University of Wisconsin-Platteville
College of Engineering, Mathematics and Science
Platteville, WI 53818

ABSTRACT

Probabilistic failure analyses evaluate the effects of strength variability in brittle materials on the design of ceramic components. However, strength is not the only stochastic parameter in ceramic components and making such an assumption can result in failed designs. Other variable parameters such as geometry, fabrication, materials, and service load histories should also be considered when evaluating the total probability of failure for the component. Slight variations in these variables can alter the stress distribution within the ceramic, and change the probabilities of failure for the system. This effort demonstrates an approach for incorporating the statistical scatter in material properties and system conditions, as well as the Weibull properties of the ceramic in the probability of failure calculations. The evaluated system is an internally pressurized ceramic tube with a steel sheath that imparts a pre-stress through a shrink-fit operation. Two analyses have been performed where the initial analysis is deterministic in nature, in that there is no variability in the inputs to the Weibull analysis. The first analysis calculates the optimal geometry and pre-stress levels for the lowest probability of failure. The second analysis was performed using ANSYS PDS (Probabilistic Design System) and CARES/Life software to evaluate the total probability of failure for a more realistic system with variability in the different material properties, Weibull parameters, pre-stress levels, pressure loading, and geometric inputs. The results of the PDS analysis illustrate that the probability of failure is not a single value, as defined by the first analysis, but a range spanning several orders of magnitude.

INTRODUCTION

Designers of ceramic components use Weibull statistics to describe the variability in the tensile strength to predict the reliability of the part. These analyses calculate a probability of failure for a stressed component using a given set of Weibull parameters. This approach assumes that only one stochastic parameter (strength) exists. In reality, variability in parameters other than the material strength also exists. For example, the loading history, geometry and material properties are not exact and are subject to the accuracy of measurement method. In the case of sheathed ceramic tubes, other parameters such as pre-applied compressive stresses from sheathing, elastic moduli, and Poisson's ratio are subject to scatter. Each of these conditions can cause changes in the stress state, which will in turn change the probability of failure.

This paper presents the results of two analyses for the development of a pressurized silicon nitride tube with a steel sheath. An analytic model has been used to determine the geometry and pre-stress levels for optimal performance. A second model uses ANSYS finite element analysis software and CARES/Life probabilistic analysis software to determine the effects of system level variability on the calculated probability of failure.

ANALYTIC MODEL

An analytic model is used to develop a ceramic lined tube design capable of sustaining large internal pressures. The model combines an analytic solution for a multilayered anisotropic tube coupled with the probability of failure calculations for a ceramic tube subjected to internal and external pressure loads[1,2]. It calculates the stress, strain, and displacement in every layer for a given internal and external pressure, axial force, torsion, and uniform temperature change. With the addition of different modifications to the strain term, the model can represent the effects of shrink-fit, press-fit, shape memory alloy contractions, and high tension filament winding[3].

The Weibull calculations are based solely on the hoop stress in the ceramic, and it does not consider multi-axial stress effects. It is intended as a design tool to perform parametric searches of which design parameters are of significance. Searches of the variation of the wall thickness ratio and shrink fit temperature for optimal survival rates are performed to generate the best gross geometry and pre-stress levels.

To generate a design window plot, the total wall thickness is left constant and only the radial location of the interface between the ceramic and metal sheath is varied. The material properties for the silicon nitride tube (silicon nitride composition labeled as SN47), the steel sheath, and the geometry of the total tube are in Table I. The wall thickness ratio (ceramic thickness/ total thickness) ranges from 0 to 100% ceramic, where 0% is an all steel design and 50% is equal thicknesses of ceramic and steel. The shrink fit generates compressive stresses in the ceramic tube due to the thermal mismatch between the dissimilar materials. By increasing the shrink-fit temperature differential, the sheathing generates higher compressive stress fields in the ceramic tube, which in turn, increases its burst pressure level and reliability. An investigation of the shrink-fit temperature differential (ΔT) and wall thickness ratio is illustrated in Figure 1. Three different regions exist in the plot, and are labeled sheath failure, ceramic failure, and design success. Sheath failure occurs when the calculated Von Mises stress in the steel exceeds the yield strength of the material (827 MPa or 120 ksi). Ceramic failure occurs when the probability of failure for the ceramic is above 1 in 10,000. Design success occurs when the probabilities of failure below 1 in 10,000 and the sheath is not loaded above its yield strength.

By evaluating the results in Figure 1, the safe design region ranges from below 100° to above 450°C for the shrink-fit temperature differential, and from near zero percent ceramic to nearly 85% ceramic wall ratio. A safe design would be to choose a value near the center of the region, so that it would have the greatest tolerance to variability. A value of 33% wall thickness with a shrink-fit temperature differential of 190°C was chosen since it fits this description.

Table I - Geometry, loading, and material parameters used to evaluate failure probability for the sheathed internally pressurized SN47 cylinder.

Geometry	Loading	Material properties
SN47 tube ID = 2.28 mm OD = 4.91 mm Wall thick = 1.315mm L = 500 mm **Steel sheathing** ID = 4.91 mm OD = 10.2 mm Wall thick = 2.645 mm L = 500 mm	P_i = internal pressure = 380 MPa $\Delta T = -190$ °C	**SN47 Silicon Nitride** E = 310 GPa ν = 0.27 α = 3.2e-6 1/°C m = Weibull modulus = 10.9 σ_{0v} = scale parameter = 248.7 MPa. $m^{3/m}$ **Steel** E = 200 GPa ν = 0.32 α = 12.8e-6 1/°C

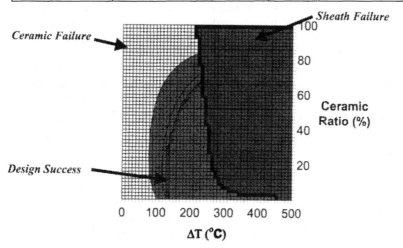

Figure 1 - Design space for varying ceramic wall ratio and shrink-fit temperature

ANSYS/CARES/PDS ANALYSIS:

An ANSYS finite element model was created using the geometric, material, loading parameters listed in Table I. The model consists of 16,600 eight-noded axisymmetric elements (ANSYS Plane 82 elements). A CARES/Life reliability analysis was performed on this tube and the probability of failure was determined to be 5.93×10^{-4} (i.e., approximately 6 out of 10000 ceramic tubes are predicted to fail). The difference between the probability of failure from analytic solution ($<10^{-4}$) and FEA/CARES value (5.93×10^{-4}) is due to the fact that the analytic model does not consider multiaxial stress states while the CARES analysis does.

At this point, the analysis has assumed that only one stochastic parameter (strength) exists, and all other parameters are deterministic in nature. In reality, variability in parameters other than the material strength also exists. For example, the loading history is not exact and is subject to the accuracy of the measurement method. Also, the geometry of the ceramic tube is not exact (wall thickness can vary) and is dependent on the accuracy of the manufacturing process. Other parameters such as the pre-applied compressive stresses due to sheathing, material's elastic modulus, and Poisson's ratio are subject to scatter.

The reliability evaluations to be presented next were all based on volume flaw analysis. Monte Carlo simulation and the Latin Hyper Cube sampling methods were used to compute the following two random output variables:

1) Total probability of failure (computed using CARES and ANSYS/PDS)
2) Maximum stress (computed using ANSYS/PDS only)

The probabilistic design system (PDS) within ANSYS requires that random input variables be defined as statistical distributions with the corresponding parameters. For example, if the ceramic tube ID is considered a random input variable that obeys a Gaussian distribution, then the ANSYS PDS analysis requires the ID to be defined as a Gaussian random variable with the corresponding mean and standard deviation. In this analysis, the statistical distributions for several random variables were not known, so statistical distributions were assigned to them based on the amount of data provided. This information is listed in Table II.

Table II - Statistical distribution functions and parameters for the random input variables.

Random input variable	Statistical Distribution	Parameters
Pressure (MPa)	Triangular	$P_{\text{max likelihood}} = 380$
Shrink-fit temperature (°C)	Triangular $\pm 10\%$	$T_{\text{min}} = 171$ $T_{\text{max likelihood}} = 190$ $T_{\text{max}} = 209$
Weibull modulus	Triangular	$m_{\text{min}} = 5.89$ $m_{\text{max likelihood}} = 10.9$ $m_{\text{max}} = 15.26$
Scale parameter (MPa. m$^{3/m}$)	Triangular	$\sigma_{0v,\text{min}} = 231.3$ $\sigma_{0v,\text{max likelihood}} = 248.7$ $\sigma_{0v,\text{max}} = 266.1$
ID of ceramic tube (mm)	Triangular ± 0.0001 in	$ID_{\text{min}} = 2.27746$ $ID_{\text{max likelihood}} = 2.28$ $ID_{\text{max}} = 2.28254$
OD of sheathing tube (mm)	Triangular ± 0.001 in	$OD_{\text{min}} = 10.1746$ $OD_{\text{max likelihood}} = 10.2$ $OD_{\text{max}} = 10.2254$
SN47 Elastic modulus (GPa)	Gaussian $\pm 5\%$	Mean = 310 Standard dev = 15.5
Steel Elastic modulus (GPa)	Gaussian $\pm 5\%$	Mean = 200 Standard dev = 10

Eight parameters were assumed to be uncertain. These random input variables are: elastic modulus for silicon nitride, elastic modulus for steel, thermal mismatch temperature, internal diameter for the ceramic tube, outer diameter for the steel tube, internal pressure, the scale parameter, and the Weibull modulus. Figure 2 and Figure 3 show the probability density functions and the cumulative distribution functions for the triangular and Gaussian distributions for two of the eight random input variables. The maximum likelihood value, as seen in Table II, corresponds to the peak in the distribution density, while the max and min values define the ranges of the distributions. The variability for the Weibull modulus and the scale parameter were based on the number of samples tested[4]. The sample population size was only ten samples, as that was the number of uniaxial tensile specimens that failed due to volume flaws during the characterization of this material.

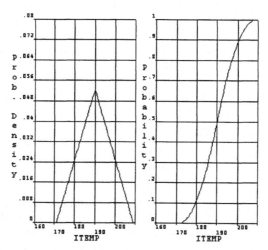

Figure 2 - Triangular statistical distribution for the shrink-fit temperature.

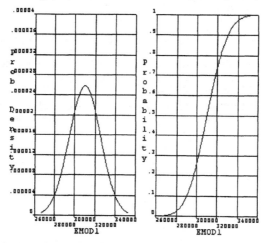

Figure 3 - Gaussian statistical distribution for the SN47 elastic modulus.

257

Table III lists the PDS results for both random output parameters, the total probability of failure and the maximum stress. As can be seen from this table, the mean total probability of failure for the sheathed ceramic tube is computed to be 3.92×10^{-3} while the mean maximum stress equals 696.5 MPa (computed within the stress concentration zone of the steel tube). **However, due to loading, geometric and material uncertainties, the ceramic tube's total probability of failure could be anywhere between 7.7×10^{-6} and 9.4×10^{-2}.** This means that the probability of failure could be more than two orders of magnitude higher than the 5.94×10^{-4} probability of failure computed assuming all parameters other than strength are deterministic. This result highlights the importance of taking into account all input random variables thought to affect the reliability of the system.

Table III - Statistics for the total probability of failure and maximum stress as generated by ANSYS PDS and CARES.

Name	Mean	Standard Deviation	Minimum	Maximum
PFAIL	3.92×10^{-3}	8.98×10^{-3}	7.68×10^{-6}	9.40×10^{-2}
MAXSTRES (MPa)	696.5	35.94	585.8	790.1

In the PDS analysis, 600 samples were used to perform the Monte Carlo simulation. Figure 4 shows the sample history plots for the total probability of failure. This figure seems to indicate convergence, within 95% confidence, after 600 samples. Of course more samples would yield tighter confidence bounds but at the expense of solution time. Such figures can be used to assess the quality of results for the output random variables based on the number of samples used for the simulation. Figure 5 shows similar plots, but for the maximum stress in the steel sheath as an output variable.

Increasing stress in the sheath indicates a larger pre-stress in the ceramic, which would normally lead to a decreasing probability of failure. By inspecting Figures 4 and 5, the peaks and valleys of the probability of failure values do not necessarily correspond to those in the maximum stress value. This is due to the influence of variations on the other six variables in the analysis. Future work will focus on developing a sensitivity analysis to determine which of the variables have the greatest effect on the probability of failure. Thus, more attention can be focused on the most important parameters for improving system-level reliability.

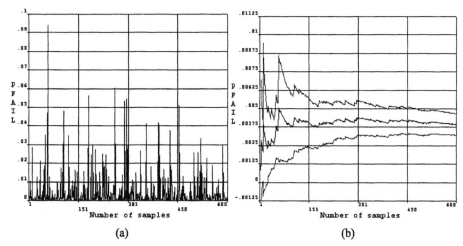

Figure 4 - Sample history plots for the total probability of failure. (a) – Probability of failure history (b) – Average value with 95% confidence interval history

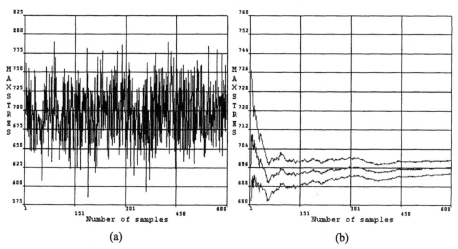

Figure 5 - Sample history plots for maximum stress in the steel sheath. (a) – Maximum sheath stress history (b) – Average value with 95% confidence interval history

CONCLUSIONS

Large changes in the predicted probability of failure are computed when system-level variabilities are incorporated into the analysis of a sheathed ceramic tube. In this example, incorporating uncertainty into eight geometric, material, and service load parameters lead to a

259

wide range of total predicted failure probabilities. At the high end of this range, the probability of failure increased to 9.2% from 0.06% when only strength variability was taken into account. This illustrates the benefit of representing all forms of stochastic behaviors into the probabilistic model.

[1] R. Carter, "Model Development for Parametric Design of Pressurized Ceramic Tubing" *Ceramic Engineering and Science Proceedings*, Vol. 24, Issue 4, pp 477-482 (2003).

[2] C. Rousseau, Master's Thesis: "Stresses and Deformations in Angle-Ply Composite Tubes," Virginia Tech, Blacksburg, VA, 1987.

[3] R.F. Eduljee and J. W. Gillespie, "Elastic Response of Post- and In Situ Consolidated Laminated Cylinders"; *Composites: Part A*, **27**A, 437-446 (1996).

[4] Arthur McLean, and Dale Hartsock, "An Overview of the Ceramic Design Process," Engineered Materials Handbook, Volume 4, p. 676-689 (1991).

FINITE-ELEMENT-BASED ELECTRONIC STRUCTURE CALCULATION IN METAL/CERAMIC INTERFACE PROBLEMS

Yoshinori Shiihara
The University of Tokyo
4-6-1, Komaba
Meguro-Ku, Tokyo, 153-8505 Japan

Osamu Kuwazuru
Institute of Industrial Science, The University of Tokyo
4-6-1, Komaba
Meguro-Ku, Tokyo, 153-8505 Japan

Nobuhiro Yoshikawa
Institute of Industrial Science, The University of Tokyo
4-6-1, Komaba
Meguro-Ku, Tokyo, 153-8505 Japan

ABSTRACT

The Finite Element Method (FEM) is implemented for atomic-scale analyses of metal/ceramic interface problems based on the Density Functional Theory (DFT), which is indispensable to prediction of ultimate bond strength in various combinations of metals and ceramics. The DFT simulation contributes much for the design of the new breed of functional materials without laborious trial and error experiments. For interface problems, an inconveniently large number of atoms should be included in a unit cell. The traditional DFT scheme of the plane-wave basis is not effective for such a large-scale calculation, since its scheme requires quadratic-scaling operation for matrix-vector product. We propose to reduce inherently prolonged computational time by employing finite discretization of real space to perform the DFT scheme. The sparseness of the Hamiltonian matrix improves computational efficiency so effectively as to increase the matrix-vector product linearly in proportion to number of atoms. Additional merit of the DFT by the FEM is in parallel computation caused by real space discretization, which makes decomposition of the matrix systematic. This is not the case with the traditional DFT using Fourier transforms. The finite element formulation is summarized and a trial computation with a Si dimer is demonstrated to verify the applicability of the FEM.

INTRODUCTION

Ceramic coatings are a promising technology for alloy-based machinery to enhance heat tolerance, wear resistance, and chemical protection. The interface problems between metal and ceramic always appear in the lifetime estimation of the coating. A lifetime model by continuum mechanics is useful for engineering,[1] however, its validity should be proven by numerous

261

experiments. Physical phenomena associated with fracture are hardly elucidated by continuum mechanics.

Nowadays there have been calculations based on the Density Functional Theory (DFT) for analyses of ideal mechanical strength of metal/ceramic interfaces,[2] since the interface adhesion strongly depends on the electron distribution on the interface. Köstlmeiler et al.[3] showed that a thin interlayer of titanium can significantly enhance the bonding at a metal/ceramic interface by the DFT. Zhang et al.[4] also discussed the effect of sulfur in the adhesion of Ni/Al_2O_3 by the DFT.

Plane wave method with the norm-conserving pseudopotential has been established and applied to various kinds of problems including metal/ceramic interfaces.[5,6] Its sophistication principally consists of periodicity of the basis function and the Fast Fourier Transform (FFT), which effectively work for periodic systems with a unit cell of a small number of atoms. In metal/ceramic interface problems, the large atomic systems are necessarily concerned so as to include defects, misfits and impurities governing the strength.[2-4,7] However, in the case of a large-scale system of thousands of atoms, computational cost increases greatly in a conventional plane wave method due to the matrix-vector product operation attributed to the dense Hamiltonian matrix. The employment of parallel computer is common to deal with large-scale systems, however, it is not easy to parallelize the Hamiltonian estimation in the reciprocal space. In estimation of a term of Hamiltonian in the reciprocal space, the FFT requires all the values of the electron density defined in the real space. Therefore each node on a parallel computer must keep large memory space for the electron density data. This induces memory overflow in the large-scale calculation, and reduces the advantage of the parallel computation.

We investigate a DFT scheme based on finite element discretization of real space to overcome the obstacles of dense Hamiltonian and complicated FFT. Element-wise integral in real space gives rise to sparse Hamiltonian matrix. The element-wise procedure also allows us systematic decomposition of the governing equation matrix and a concise procedure for parallel computation. These advantages are explained through the finite element formulation. However, there have been less previous studies of the FEM scheme, especially with the norm-conserving pseudopotential, than that of the plane wave method.[8-9] In this paper, we formulate the FEM scheme with the norm-conserving pseudopotential DFT scheme and an analysis of a Si dimer is tentatively carried out to confirm the ability of the FEM.

DENSITY FUNCTIONAL THEORY

The Density functional theory (DFT) was constructed by Kohn et al.[10]. The Kohn-Sham equation is the governing equation of the DFT. In the DFT scheme, the electron density is treated as a basis variable, and the quantum effects are put into the potential-called the exchange-correlation potential. The DFT makes it easier to calculate the nonlinear potential included in the one-particle Schrödinger equation and allows us to solve the electron structure faster than the other methods such as Hartree-Fock method. In addition, the norm-conserving pseudopotential method is an important technique to reduce the computational tasks required in the DFT calculation.[6] This method regards the effect from the core electrons to the valence electrons as the nonlocal pseudopotential. The local pseudopotential is employed as the ionic core potential, which is smoother than the true core potential. The valence electrons are only considered, and the valence wave functions become smoother, so that the number of degrees of freedom in the system becomes less than full-electron calculation.

The Kohn-Sham equation based on the norm-conserving pseudopotential method is described by the nonlinear eigenvalue problem as

$$\left(-\frac{1}{2}\nabla^2 + v_{KS} - \varepsilon_i\right)\psi_i = 0 \tag{1}$$

where ψ_i is the one-particle electron wave function of i-th orbital, ε_i is the Kohn-Sham eigenenergy of i-th orbital. v_{KS} is Kohn-Sham effective potential which causes the nonlinearity of the Kohn-Sham equation,

$$v_{KS} = \hat{v}_{nonlocal} + v_{local} + \mu_{XC} + v_{H} \tag{2}$$

where $\hat{v}_{nonlocal}$ is the nonlocal pseudopotential operator and v_{local} is the local pseudopotential. The exchange-correlation potential μ_{XC} is a function of electron density ρ, which is defined as

$$\rho = \sum_i^{N_{orbital}} \psi_i^* \psi_i \tag{3}$$

where $N_{orbital}$ is the number of electron orbitals and the asterisk means the conjugation of the function. Hartree potential v_H is Coulomb potential between electrons and given by the Poisson equation,

$$\nabla^2 v_H = -4\pi\rho \ . \tag{4}$$

FINITE ELEMENT DISCRETIZATION

For finite element discretization we transform Equation (1) to the weak form by Galerkin weighted residual method. According to the procedure of Galerkin weighted residual method,[11] we integrate the product of Equation (1) and the weighted function $\delta\psi_i^*$ over the analysis region Ω.

$$\int \delta\psi_i^* \left(-\frac{1}{2}\nabla^2 + v_{KS} - \varepsilon_i\right)\psi_i d\Omega = 0 \tag{5}$$

In order to reduce the derivative order due to Laplacian, we apply Green's theorem to Equation (5) and obtain

$$\frac{1}{2}\int \nabla\delta\psi_i^* \cdot \nabla\psi_i d\Omega - \int \delta\psi_i^* \nabla\psi_i \cdot \mathbf{n} d\Gamma + \int \delta\psi_i^* v_{KS}\psi_i d\Omega = 0 \tag{6}$$

where Γ is the surfaces of region Ω and $\mathbf{n}$ is the outward unit normal on Γ. The term of the surface integral vanishes under the periodic boundary condition or the Diriclet boundary condition.

$$\frac{1}{2}\int \nabla \delta \psi_i^* \cdot \nabla \psi_i d\Omega + \int \delta \psi_i^* (v_{KS} - \varepsilon_i) \psi_i d\Omega = 0 \tag{7}$$

The Kohn-Sham effective potential term is considered here. All the potential terms except for the nonlocal pseudopotential are defined on the point of region Ω, since they are handled without any difficulty. By using the Kleinmann-Bylander form as the nonlocal pseudopotential,[12] the term of nonlocal pseudopotential is given by

$$\int \delta \psi_i^* \hat{v}_{\text{nonlocal}} \psi_i d\Omega = \int \delta \psi_i^* \sum_{l,m} G_{l,m} u_{l,m} v_{\text{NL},l} \psi_i d\Omega \tag{8}$$

where

$$G_{l,m}\psi_i = \frac{\int u_{l,m}(|\mathbf{x}|) v_{\text{NL},1}(|\mathbf{x}'|) \psi_i(\mathbf{x}') d\Omega'}{\int u_{l,m}(|\mathbf{x}|) v_{\text{NL},1}(|\mathbf{x}'|) u_{l,m}(\mathbf{x}') d\Omega'} \tag{9}$$

l is the angular momentum quantum number, m is the magnetic quantum number and $u_{l,m}$ is the radial pseudo wave function obtained by solving radial Kohn-Sham equation of an isolated atom. $v_{\text{NL},l}$ is the nonlocal pseudopotential obtained by the same manner as $u_{l,m}$. Equation (8) is written as the following after substitution of Equation (9),

$$\int \delta \psi_i^* \sum_{l,m} G_{l,m} u_{l,m} v_{\text{NL},1} \psi_i d\Omega = \sum_{l,m} \frac{1}{C} \left[\int \delta \psi_i^* v_{\text{NL},1} u_{l,m} d\Omega \right] \left[\int u_{l,m} v_{\text{NL},1} \psi_i d\Omega' \right] \tag{10}$$

with

$$C = \int u_{l,m}(|\mathbf{x}|) v_{\text{NL},1}(|\mathbf{x}'|) u_{l,m}(\mathbf{x}') d\Omega' \tag{11}$$

ψ_i is discretized with the shape function defined in a finite element.

$$\psi_i(\mathbf{x}) = \sum_n^{N_{\text{node}}} N^n(\mathbf{x}) \psi_i^n$$

$$\delta \psi_i^*(\mathbf{x}) = \sum_n^{N_{\text{node}}} N^n(\mathbf{x}) \delta \psi_i^{*n} \qquad \forall \mathbf{x} \in \Omega_e \quad , \tag{12}$$

where N_{node} is the number of nodes in an element, $N^n(\mathbf{x})$ is a shape function and ψ_i^n is a nodal wave function value. Ω_e is the divided region by a finite element. Substituting Equation (8) to Equation (7) and writing the term of the Kohn-Sham effective potential explicitly, the matrix form of Equation (7) is obtained as

$$\sum_e \delta \psi_i^* \left[\int_{\Omega_e} \frac{1}{2} \mathbf{B}^{\mathrm{T}} \mathbf{B} + \left(v_{\mathrm{local}} + \mu_{\mathrm{XC}} + v_{\mathrm{H}} - \varepsilon_i \right) \mathbf{N} \mathbf{N}^{\mathrm{T}} d\Omega_e \right.$$
$$\left. + \sum_{l,m} \frac{1}{C} \int V_{\mathrm{NL},1} u_{l,m} \, \mathbf{N} d\Omega_e \int u_{l,m} v_{\mathrm{NL},1} \mathbf{N}^{\mathrm{T}} d\Omega_e' \right] \psi_i = 0 \quad , \tag{13}$$

where ψ_i is the array of nodal function value ψ_i^n in an element, $\mathbf{N}$ is the element-wise vector of $N^n(\mathbf{x})$ and $\mathbf{B}$ is the matrix of partial derivative of shape function with respect to coordinates $\partial N^i / \partial x_j$. Superscript T means the transpose operation applied to the matrix or vector. $\sum_e$ represents merging all the elements. By denoting $\sum_e \psi_i$ by the Ψ_i, Equation (13) becomes

$$\delta \Psi_i^* \left[\mathbf{H} + \sum_{l,m} \mathbf{p}_{lm} \mathbf{p}_{l,m}^{\mathrm{T}} \right] \Psi_i = 0 \tag{14}$$

where

$$\mathbf{H} = \sum_e \int_{\Omega_e} \frac{1}{2} \mathbf{B}^{\mathrm{T}} \mathbf{B} + \left(v_{\mathrm{local}} + \mu_{\mathrm{XC}} + v_{\mathrm{H}} - \varepsilon_i \right) \mathbf{N} \mathbf{N}^{\mathrm{T}} d\Omega_e \tag{15}$$

$$\mathbf{p}_{l,m} = \sum_e \frac{1}{C^{1/2}} \int V_{\mathrm{NL},1} u_{l,m} \, \mathbf{N} d\Omega_e \quad . \tag{16}$$

Taking account of the arbitrariness of $\delta \Psi_i^*$, the finite-element discretized Kohn-Sham equation is obtained as

$$\left[\mathbf{H} + \sum_{l,m} \mathbf{p}_{l,m} \mathbf{p}_{l,m}^{\mathrm{T}} \right] \Psi_i = 0 \quad . \tag{17}$$

Note that the matrix $\mathbf{H}$ is a sparse matrix because the shape functions are locally defined at an element. The sparseness allows us to perform the matrix-vector product with the number of operations in proportion to the system size linearly. Nevertheless, the matrices $\mathbf{p}_{l,m} \mathbf{p}_{l,m}^{\mathrm{T}}$ are not sparse matrices. Therefore the inner products $\mathbf{p}_{l,m}^{\mathrm{T}} \Psi_i$ are calculated in advance without explicitly constructing the matrices. In this procedure, all the matrix-vector product operations are performed in linear proportion to system size. In parallelization it is great advantage that the terms of Hamiltonian matrix can be estimated in each element independently. It enables us to divide the region Ω to subregions of a small number of elements and to calculate the components of the Hamiltonian matrix independently.

TEST CALCULATION OF SILICON DIMER

We test the validity of our formulation by the DFT calculation of a simple atomic system, Si dimer. We employ an octahedral isoparametric finite element which is generally used in structure analysis.[11] Because the order of Kohn-Sham effective potential is higher than that of wave function, we interpolate the wave function and the potential by the first-order shape function and the second-order shape function, respectively. We perform calculations only on the Γ point, however this scheme is easily expanded to a multi k-point scheme. The BKL method is employed in solving the Kohn-Sham equation.[13] The preconditioning scheme proposed by Gan et al. is employed.[14] The Poisson equation is solved by preconditioned-conjugate-gradient method. The super-cell size is set to 10.6^3 angstrom[3]. For the norm-conserving pseudopotential we use the GTH pseudopotential,[15] which is given by analytical form in real space. By this virtue, the pseudopotential value on an arbitrary integral point can be analytically obtained. The exchange-correlation function given in Ref. 15 is employed.

Figure 1. shows the convergence of the eigenenergies as a function of the FEM grid size. In order to reduce the number of degrees of freedom and calculate efficiently, the grid size should be set so appropriately as to give acceptable accuracy. In this case, we set the grid size to 0.35 angstrom. This adjusted grid size can be applied to more complex Si systems, such as Si crystal including defects, than dimer since the appropriate grid size strongly depend on the variation of wave function localized around the atomic core. In Table 1 we show the result of Si dimer and compare the eigenenergies with the result by all-electron DFT of *ab-initio* software package, Gaussian98.[16] In all-electron calculation, 6-31d(G) is chosen as the gaussian basis set. The exchange-correlation function is set as the same one with the FEM. The difference between them falls in the range of 0.092-2.1 % in spite of using pseudopotential on the FEM schemes. In Figure 2 the pseudocharge densities of Si dimer for the different bond length are presented. These show that our FEM scheme can simulate the collapse of covalent bonding between the dimer when the bond length stretches.

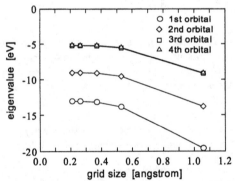

Fig.1. Convergence of the eigenenergies as a function of the FEM grid size. The eigenenergies start to converge at the grid size 0.35 angstrom.

Table 1. The eigenenergies of Si dimer from pseudopotential method of FEM and from all-electron DFT of Gaussian98. The difference between them is also shown. The grid size of FEM is set to 0.2 angstrom.

Eigenenergy [eV]	1st orbital	2nd orbital	3rd orbital	4th orbital
FEM	-13.040	-9.0742	-5.2025	-5.1827
Gaussian98	-12.967	-9.0826	-5.3040	-5.2983
Difference [%]	0.56	0.092	1.9	2.1

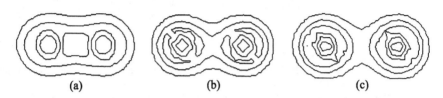

(a)   (b)   (c)

Fig. 2. Pseudocharge densities of Si dimer for the different bond length : (a)1.11 angstrom, (b)1.38 angstrom, and (c).1.64 angstrom.

CONCLUSION

We have studied the FEM method with the norm-conserving potential method for the large-scale electronic-structure calculation such as the metal/ceramic interface problem. Two suitable features for large-scale simulation, the sparseness of Hamiltonian matrix and the decomposition of subregions, are derived from the locality of finite element basis.

ACKNOWLEDGEMENTS

One of the authors (Y. S.) was supported through the 21st Century COE Program, "Mechanical Systems Innovation", by the Ministry of Education, Culture, Sports, Science and Technology.

REFERENCES
[1]F. Traeger, M. Ahrens, R. Vaβen, and D. Stöver, "A life time model for ceramic thermal barrier coatings", *Mater. Sci. Eng. A*, **358**, 255-265, (2003).
[2]M. W. Finnis, "The theory of metal-ceramic interfaces", *J. Phys.: Condens. Matter.*, **8**, 5811-5836, (1996).
[3]S. Köstlmeier and C. Elsässer, "Density functional study of the 'titanium effect' at metal-ceramic interfaces", *J. Phys.: Condens. Matter.*, **12**, 1209-1222, (2000).
[4]W. Zhang, J. R. Smith, X. G. Wang, and A. G. Evans, "Influence of sulfur on the adhesion of the nickel/alumina interface", *Phys. Rev. B.*, **67**, 245414-1-245414-12, (2003).

[5]J. C. Slater, "Wave functions in a periodic potential", *Phys. Rev.*, **51**, 846-851, (1937).

[6]G. B. Bachelet, D. R. Hamann, and M. Schlüter, "Pseudopotentials that work: From H to Pu", *Phys. Rev. B*, **26**, 4192-4228, (1982).

[7]R. Benedick, A. Alavi, D. N. Seidman, L. H. Yang, D. A. Muller, and C. Woodward, "First Principles Simulation of a Ceramic/Metal Interface with Misfit", *Phys. Rev. Lett.*, **84**, 3362-3365, (2000).

[8]E. Tsuchida and M. Tsukada, "Large-scale electronic-structure calculations based on the adaptive finite-element method", *J. Phys. Soc. Jpn.*, **67**, 11, 3844-3858, (1998).

[9]J. E. Pask, B. M. Klein, C. Y. Fong, and P. A. Sterne, "Real-space local polynomial basis for solid-state electronic-structure calculations: A finite-element approach", *Phys. Rev. B*, **59**, 12352-12358, (1999).

[10]W. Kohn and L. J. Sham, "Self-consistent Equations Including Exchange and Correlation Effects", *Phys. Rev.*, **140**, A1133-A1188, (1965).

[11]O. C. Zienkiewicz, R. L. Taylor, The Finite Element Method, 4th edn., McGraw-Hill, New York, (1988).

[12]L. Kleinman and D. M. Bylander, "Efficacious Form for Model Pseudopotentials", *Phys. Rev. Lett*, **48**, 1425-1428, (1982).

[13]D. M. Bylander, L. Kleinman, and S. Lee, "Self-consistent calculations of the energy bands and bonding properties of $B_{12}C_3$", *Phys. Rev. B*, **42**, 1394-1403, (1990).

[14]C. K. Gan, P. D. Haines, and M. C. Payne, "Preconditioned conjugate gradient method for the sparse generalized eigenvalue problem in electronic structure calculations", *Comp. Phys. Commun.*, **134**, 33-40, (2001).

[15]S. Goedecker, M. Teter and J. Hutter, "Separable dual-space Gaussian pseudopotentials", *Phys. Rev. B*, **54**, 1703-1709, (1996).

[16]*Gaussian98*, M. J. Frisch, *et al.*,Gaussian, Inc., Pittsburgh PA, (1998).

3D FEM SIMULATION OF MLCC THERMAL SHOCK

Yang Ho Moon, Hyuk Joon Youn
Samsung Electro-Mechanics Co. Ltd.
314 matan-dong
Suwon-city, Kyung-gi-do, 443-743

ABSTRACT

MLCC(multiplayer ceramic capacitors) for automotive applications need higher thermal cycle reliability than other general purpose MLCC because the automobile environment is more severe. In the paper, we use 3 dimensional FEM simulations try to reduce the stress in MLCC and prevent cracks in MLCC during thermal cycle reliability tests. We calculated the residual stress during production process of MLCC and during reflow soldering. Thermal cycle stress is simulated considering strain hardening and creep effect.

Stress in MLCC is calculated for various MLCC chip design parameter and surface mounting parameter on PCB(like dimension of MLCC, external electrode, and solder height etc) and the parameters were optimized to reduce the stress during thermal cycle including satisfying board flex reliability.

INTRODUCTION

With increasing use of MLCC(multilayer ceramic capacitors) in surface mount applications such as telecommunications, automotive engineering, or aeronautics, the reliability is becoming increasingly important because of the high safety demand of electric system and the large sum of integrated parts. Specially MLCC for automotive applications need higher thermal cycle reliability than other general purpose MLCC because the automobile environment is more severe.

This paper examines thermal load of MLCC for automotive application during thermal cycle reliability tests. Thermal cycle reliability tests are done until 1000cycles on two different MLCC thickness, 0.65mm and 1.25mm (length×width = 2.0mm×1.2mm, X7R) on alumina PCB, and after analyzing the hazard function, it is found that the hazard is very high in the initial 10 thermal cycles in the thinner MLCC. Therefore, the stress field was calculated via 3 dimensional finite element method using ABAQUS commercial software for the initial 10 thermal cycles in the 0.85mm-thickness MLCC, including residual stress of production process and soldering ,and considering mechanical propertied change according to temperature, and creep effect of solder.

To prevent cracking of MLCC during thermal cycles, the factors that effect the crack initiation are optimized after understanding the production process (termination firing, soldering) and parameter study that include MLCC chip design parameters(external electrode thickness, chip thickness, band width, etc.) and surface mounting parameters(solder height, solder thickness, land distance, PCB(FR4, alumina), asymmetric position of MLCC).

RESULTS

Thermal cycle test.

For thermal cycle testing, the specimen of Fig.1 is used. 20 MLCC are soldered in PCB (FR4, thickness 1.6mm) board and thermal cycle tests are performed. A single thermal cycle specimen is shown in the red box in Fig. 1. Finite element analysis was performed in that section and symmetry boundary conditions are applied on symmetry planes. Fig.2 shows the temperature profile of the thermal cycles. To pass this reliability test, MLCC sample must satisfy electrical specification and not contain any cracks.

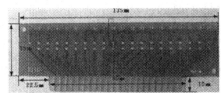

Fig. 1. Thermal cycle specimen that can be mounted 20 MLCC by reflow soldering

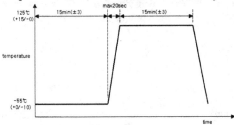

Fig. 2. Thermal cycle temperature profile.

Experimental observations

Table 1 shows thermal cycle reliability test results until 1000cycles in two different MLCC thickness', 0.65mm and 1.25mm (length×width = 2.0mm×1.2mm, X7R) on alumina PCB. After each 10, 49, 80, 1000 thermal cycles, 80 samples (40 sample(0.65 mm) + 40 samples(1.25 mm)) were inspected by SEM to inspect for cracks. Fig.3 shows the hazard function acquired from Table1. From Fig 3, it is found hazard is very high in the initial 10 thermal cycle in thin MLCC. The stress in the 0.85mm thickness MLCC during the initial 10 thermal cycles was calculated via 3 dimensional FEM simulation using ABAQUS software. Typical crack shape during thermal cycle test when FR4 PCB used was a crack at the end of external electrode of MLCC top surface like Fig. 4. When analyzing the thermal cycle, termination firing and soldering is simulated to calculate the residual stress and stress analysis is performed during the thermal cycle like Fig.5.

Results of initial thermal cycle simulation

As a result of the first 1 thermal cycle simulation, the stress distribution during 1 thermal cycle is predicted Fig.6. The maximum value of maximum principle stress occurred at the place

Table 1.
Number of cracked sample after each thermal cycle (MLCC chip thickness: 0.65mm, 0.85mm)

Thermal cycle	no of test samples	no of cracked sample	
		0.65t	0.85t
10	10	1	4
49	10	1	9
80	10	0	7
1000	10	2	8

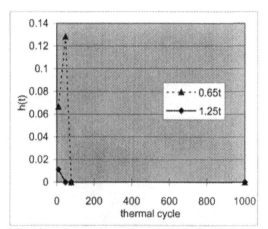

Fig. 3. Hazard function of thermal cycle reliability(chip thickness 0.65mm and 1.25mm)

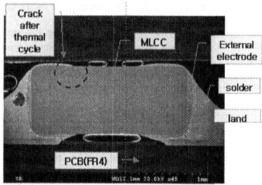

Fig. 4. Typical crack shape during thermal cycle test when FR4 PCB is used(PCB thickness =1.6mm)

271

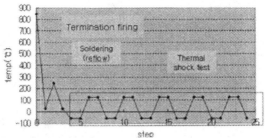

Fig. 5. Analysis step of thermal cycle

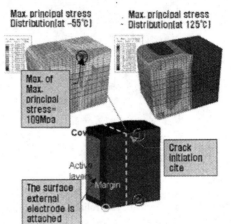

Fig. 6. Stress distribution at –55 ℃ and 125 ℃ of thermal cycle(PCB: FR4)

crack occurred(on the top edge of MLCC at the end of external electrode; Fig.4) . The ceramic material is so brittle that usually a crack occurred where maximum principal stress is large. For the initial thermal cycle, Table 3 shows the maximum value of max. principal stress as a function of MLCC chip thickness and external electrode band width. The results in Table 3 indicate that the stress can be reduced by increasing MLCC chip thickness and band width.

Results of 10 thermal cycles simulation

To get the trend according to the number of thermal cycles, in Case 3 of Table 3 the change of stress at the area marked on Fig.7 is compared for each thermal cycle. Fig.8 shows the comparison of stress distribution on the line of Fig.7 according to the step of Fig.5. The maximum value of the maximum principal stress occurred at the end of external electrode on the top of MLCC at −55℃. As the number of thermal cycles increases, the stress at −55℃ increases.

Table 3.

Maximum principal stress at the end of external electrode

case	case1	case2	case3	case4
Chip thickness	1.25t	0.85t	0.85t	0.85t
Chip edge	rectangle	rectangle	fillet	fillet
Band with	560μm	590μm	590μm	450μm
model	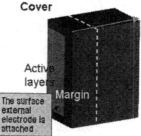			
①Max. of max. principal stress	90.29	118.6	109.23	119.11
②Max. of max. principal stress	83.9	82.7	75.87	50.38

Cover

Active layers

Margin

The surface external electrode is attached

Fig. 7. The line of stress distribution data acquired during thermal cycle test.

After 10 cycles the stress is increased by 3.4%, as shown in Fig.9, due to the increase of the plastic strain of the external electrode and solder. Fig. 10 shows the maximum value of the maximum principal stress during 10 thermal cycles as a function of band width. Increasing the band width reduce the stress at the first thermal cycle. So we can consider if maximum value of maximum principal stress of first cycle is decreased, the probability of forming cracks is also decreased. Therefore when parameter study is performed, the maximum value of max principal stress at the first thermal cycle is compared.

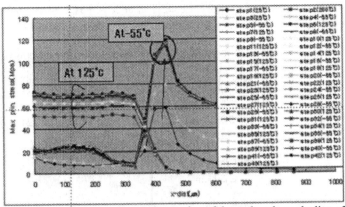

Fig. 8. Stress distribution according to the number of thermal cycle on the line of Fig. 7

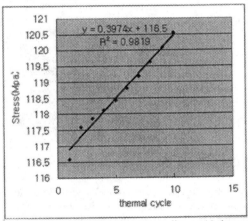

Fig. 9. Maximum value of max. principal stress according to 10 thermal cycles

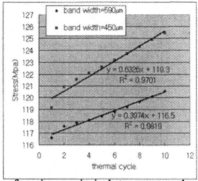

Fig. 10. maximum value of maximum principal stress versus thermal cycle for band width 590μm and 450μm

Parameter study

To reduce the occurrence of cracks during the thermal cycle reliability tests, the maximum value of the maximum principal stress should be reduced through parameter study and optimized the value of factors (Fig. 11) that is estimated to effect the stress during thermal cycle. Table 4 and Fig.12 show the parameter study results(max. value of max. principal stress difference when each parameter level is changed, where max.. value of maximum principal stress shows. Fig. 12 shows that solder height, external electrode thickness, and chip thickness effect stress more than other factors.

Fig.13 shows the max. principal stress distribution when alumina is used as the material of PCB instead of FR4. The maximum value of the maximum principal stress occurred at the end of the external electrode on the bottom edge of MLCC at –55℃. And Fig. 14 shows the crack initiation site when alumina is used as the material of PCB. It is different from the crack initiation site when the material of PCB is FR4(Fig.4). The difference in behavior is likely due to the thermal expansion coefficient difference between alumina PCB and FR4 PCB.

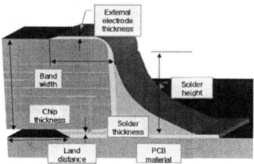

Fig. 11. Factors that is estimated to effect the stress during thermal cycle

Table 4.
Maximum principal stress difference according to factors

fator	level 1	level2	Max. prin. stress diffirence(Mpa) (level2-level1)	crack initiation cite (level1/level2)
band width	450㎛	590㎛	-2,5	①/①
solder thickness	20㎛	50㎛	-3,7	①/①
external electrode	13㎛	22㎛	12,6	①/①
asymmetry	0㎛	144㎛	0,2	①/①
land distance	1200㎛	1460㎛	-1,1	①/①
chip thickness	0,75mm	0,85mm	-9,4	①/①
PCB material	Fr4	alumina	1,9	①/②
solder height	334㎛	610㎛	17,7	①/①

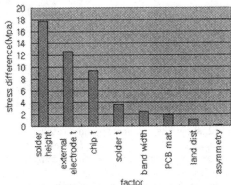

Fig. 12. Maximum principal stress difference according to factors

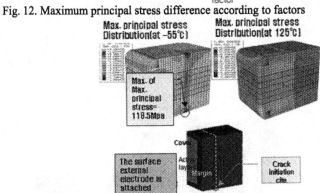

Fig. 13. Stress distribution at −55 ℃ and 125 ℃ of thermal cycle on the alumina PCB, PCB thickness =0.8mm)

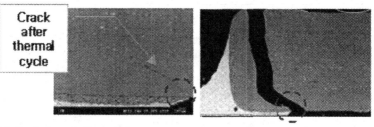

Fig. 14. Typical crack shape during thermal cycle test when alumina PCB is used(PCB thickness =0.8mm)

Comparison between experiment and simulation

Table 5 shows comparison between experimental results and simulation that we mentioned above. Simulation results well matched with experimental results. So we can optimize parameters via simulation.

Table 5
Comparison between experiment and simulation at thermal cycle test

	experiment	Simulation
Effect of MLCC thickness	Increasing Chip T improve thermal cycle reliability	Increasing Chip T reduce the max. value of max. principal stress
Crack initiation cite	1) FR4 PCB : At the end of external electrode on the top edge of MLCC 2)alumina PCB: At the end of external electrode on the bottom edge of MLCC	The place max. value of maximum principal stress is appeared is same as crack initiation cite of test when two PCB is used
Effect of Band width	small effect	0.85T: effect is smaller than other factors

Comparison with board flex reliability test

Board flex reliability test is to see how MLCC endure when PCB bended. Many factors that effect board flex reliability also effect thermal cycle reliability. So we compare them in Table 6 and optimize the factors to satisfy board flex test and thermal cycle reliability tests simultaneously at the next paragraph.

Table 6.
Comparison between board flex and thermal cycle reliability test

Reliability test	Board flex	Thermal cycle
The place max. value of max. principal stress is	At the end of external electrode on the bottom edge of MLCC	1) FR4 PCB : At the end of external electrode on the top edge of MLCC 2)alumina PCB: the same as board

		flex
appeared(simulation result)		flex
Crack initiation cite(SEM)	At the end of external electrode on the bottom edge of MLCC	1) FR4 PCB : At the end of external electrode on the top edge of MLCC 2)alumina PCB: the same as board flex
Crack mode	Instant crack	cyclic fatigue crack (few shows instant crack)
Vital few X's (from simulation and experiment)	1)MLCC:MLCC thickness, external electrode thickness, band width 2)Mounting :asymmetry	1)MLCC: external electrode thickness > MLCC thickness □ band width 2)Mounting : solder height □ solder thickness
Effect of asymmetry mounting	Crack occur on the side of the distance between external electrode band end and land of PCB is closer than the other	No effect
External fore that makes crack	Residual stress(termination firing, soldering) + board flex	Residual stress(termination firing, soldering) + -55□ cooling + thermal fatigue

Optimization

Using DOE by finite element analysis we optimized the value of design factors of MLCC to reduce the maximum value of maximum principal stress. Table 7 shows the optimized value of factors. Because external electrode thickness and chip thickness among the MLCC design parameter effect maximum principal stress more than other factors, they are optimized. The band width was optimized from the board flex reliability test. Even though solder height effects max. principal stress, it is specified by user requirements so we fixed the value of solder height to represent a severe condition. Reducing external electrode thickness is not so easy because of limits of the process capability. The sigma level is estimated using MonteCarlo simulation from the regression equation from the result of DOE. Table 8 shows that the 6-sigma level in thermal cycle reliability quality after optimization and bend reliability quality become higher too.

Table 7.
Optimized value of factors

factor	Before optimized (measured)		After optimized	
	mean	Sigma level	mean	Sigma level
band width(μm)	590	50	590	32.5
external ele. t(μm)	24.4	2.49	13	2.49
Chip t (mm)	0.834	0.00647	0.88	0.00647

Table 8.
Sigma level of board flex and thermal reliability after optimized

Reliability test	Before optimized (measured)		After optimized (simulation)	
	mean (spec.)	Sigma level	mean (spec.)	Sigma level
bending	2mm	2.17	4.3mm	5.4
Thermal cycle	(1000cycle)	2.0	(1000)	6.0

SUMMARY AND CONCLUSION

Using 3-dimensional FEM simulation we tried to understand how MLCC cracking occurs during thermal cycle, accounting for temperature dependent Young's moduli, plastic properties, creep of solder, and the residual stress during external electrode termination firing and reflow soldering. From the result we can reach the following conclusions:

1) From the thermal cycle experimental result, it is found hazard of crack is very high in initial 10 thermal cycle in thin MLCC. So we calculated stress via 3 dimensional finite element analysis using ABAQUS software during initial 10 thermal cycle in 0.85mm thickness MLCC.

2) Cracks during thermal cycle(FR4 PCB) occurred on the top edge of MLCC at the end of external electrode. The place that crack occurred is the same as the place of maximum of maximum principal stress at $-55\,°C$ from the result of thermal cycle simulation.

3) From the simulation of the first thermal cycle, increasing MLCC chip thickness and band width reduces the maximum value of maximum principal stress.

4) From the simulation of the first 10 thermal cycle, as thermal cycle proceed, the maximum value of maximum principal stress is increased by 3.4%. Increasing of plastic strain of external electrode and solder according to thermal cycle makes the stress of ceramic part of MLCC increased. During the steady state of temperature for 15 minutes of thermal cycle the stress decreases due to the creep of solder.

5) From the simulation of the first 10 thermal cycles, we can consider if maximum value of maximum principal stress of first cycle is reduced, we can reduce the probability of cracking. Therefore when a parameter study is performed, the maximum value of maximum principal stress of the first thermal cycle is compared.

6) As a result of the parameter study via simulation, solder height, external electrode thickness, and chip thickness effect stress more than other factors. Low solder height, thin external electrode and thick chip can reduce the stress and crack during thermal cycle.

7) From the comparison between experimental result and simulation, simulation results are well matched with experimental results. So we could optimize parameters via simulation.

8) Board flex reliability and thermal cycle reliability is compared. Many factors that effect thermal cycle reliability also effect board flex reliability. So we optimize the factors that effect board flex and thermal cycle reliability to satisfy bend and thermal cycle reliability simultaneously via simulation.

Acknowledgment
The author acknowledges, with thanks, the useful comments made by Jin Woo Park.

REFERENCES

[1]Klaus Franken, "Finite-Element Analysis of Ceramic Multilayer Capacitors : Failure Probability Caused by Wave Soldering and Bending Loads", *J. Am. Ceram. Soc.*,. 83[6] p.1433-40 (2000)

[2]John H. Lau, "Creep Analysis of Solder bumped direct chip attach(DCA) on Microvia Build-up Printed Circuit board with Underfill", *Int'l Symp on Electronic Materials & Packaging*,. p.127-135 (2000)

[3]Abaqus V6.4 Analysis user's manual Vol III : materials p11.2.4-41.

[4]Prume, K., K. Franken, and e. al., "Modeling and numerical simulation of the electrical, mechanical, and thermal coupled behavior of multilayer capacitors", *Journal of the European Ceramic Society*, 22: p. 1285-1296. (202)

[5]Franken, K., et al., "Finite-Element Analysis of Ceramic Multilayer Capacitors: Failure Probability Caused by Wave Soldering and Bending Loads,. *Journal of the American Ceramic Capacitors*,. 83(6): p. 1433-1440. (2000)

[6]John H.L.Pand and Tze_Ing Tan, "Thermo-Mechanical Analysis of Solder Joint Fatigue and Creep in A Flip Chip On Board Package Subjected to Temperature Cycling Loading", *Electronic Components and Technology Conference*,. : p.878~883. (1998)

[7]Bharat S. Rawal, "Reliability of Multilayer ceramic capacitors after thermal shock", *IEEE*,. p.371-375 (1988)

[8]L.A. Mann, "The Influence of Termination Properties on the Thermal Shock and Thermal Cycle Performance of Surface Mount Multilayer Ceramic Capacitors", *IEEE*, , p.437-441 (1991)

[9]GeoFFrey C. Scott, "Thermal Stresses in Multilayer Ceramic Capacitors: Numerical Simulations", *IEEE Transactions on Components, Hybrids, and Manufacturing Technology*, Vol.13, No.4, December 1990. p. 1135-1145 (1999)

ANALYSIS OF FIRING AND FABRICATION STRESSES AND FAILURE IN CERAMIC-LINED CANNON TUBES

J.H. Underwood, M.E. Todaro, M.D. Witherell
US Army Armament Engineering and Technology Center
Watervliet, NY 12189, USA

A.P. Parker
Royal Military College of Science, Cranfield University
Swindon, SN6 8LA, UK

ABSTRACT

Four aspects of service loading of ceramic lined cannons are considered, based upon recent work with ceramics under cannon thermomechanical loads and upon experience with cannons made from more conventional materials. First, candidate ceramics are evaluated by comparing models of thermal compressive firing stress with their measured compressive strength at elevated temperatures. The strength of each of the ceramics exceeds typical near-bore firing stress, with SiAlON having more margin between high temperature strength and near-bore firing stress than SiC and Si_3N_4. Second, an upper bound analysis is performed of the shear failure of a typical ceramic "island" created by thermal cracks and subjected to further compressive firing stress. The analysis shows that failure of cracked ceramic islands is possible. Third, a generic design concept of a thin ceramic liner is considered, supported by a thicker, intermediate, shrink-fitted, steel liner, and further supported by a shrink-fitted steel or tension-wrapped carbon fiber/epoxy jacket. Radial and hoop fabrication and firing stresses are calculated and axial fabrication stresses are estimated for typical cannon steel and composite properties, showing radial crushing of composites to be a concern. Fourth, a fracture mechanics analysis is used to calculate the cannon firing pressure at which a thermal crack would grow to full depth of the ceramic liner. The low fracture toughness of SiC is a limitation.

INTRODUCTION

Ceramic liners for gun tubes are being considered[1] because of recent increased performance requirements for guns, particularly those that increase temperature and duration of the propellant gasses at the tube bore. Metals, such as the electroplated chromium often used at a gun bore, have limited hot strength and melting temperature. The hot strength of chromium has become a critical limiting factor in the role of chromium as a thermal barrier[2], and once the barrier cracks and pieces break away, melting of the underlying steel cannot be avoided.

Recent work to evaluate ceramic liners in guns has involved experimental simulation of the transient heating at a gun bore using a laser[3,4]. In addition, analysis has been described of the actual and simulated gun firing using finite difference calculation of near-bore temperatures[2,4] and mechanics models of coating and bulk transient thermal stresses[5,6]. A brief outline of the sequence of critical near-bore damage processes considered is: one-dimensional transient convective-conductive heating near the bore; near-bore thermal expansion and compressive yielding of thermally softened ceramic (or metal); development of tensile residual stress and cracks upon cooling; further softening, compressive yielding, opening of cracks and failure of cracked islands during subsequent heating cycles.

This work continues the evaluation of ceramic-lined gun tubes, concentrating on four aspects of the critical loading, both thermal and mechanical, expected for a ceramic lined cannon. First, refined descriptions of laser heating and ceramic properties are used for finite difference calculation of near-bore temperatures in four candidate ceramics. This improves the critical comparison of thermal compressive stress with compressive strength for ceramics at the elevated near-bore temperatures in a cannon. Second, the improved stresses are used for an upper bound analysis of the shear loading on a typical ceramic "island" created by thermal cracks and subjected to further compressive firing stress. Third, a generic concept for fabrication of a thin ceramic liner is considered, in which the ceramic is supported by a thicker, intermediate, shrink-fitted, steel liner, and further supported by a shrink-fitted steel or tension-wrapped carbon fiber/epoxy jacket. Radial and hoop stresses from fabrication and firing are calculated for typical high-strength steel and composite properties, with emphasis on the hoop stresses near the ceramic bore surface. Fourth, a fracture mechanics analysis is used to calculate the cannon firing pressure at which a given depth of thermal crack in the various ceramics would grow to full depth of the ceramic liner. Cannon pressures that would cause ceramic crack growth are calculated for representative near-bore hoop stresses arising from typical liner-jacket fabrication stresses and for expected values of ceramic fracture toughness.

MATERIALS AND ANALYSIS
Four ceramics are considered here, as listed in Table I. The first two listed ceramics, SiAlON and Si_3N_4-1 are the same as in other recent work[7], and Si_3N_4-2 and SiC-4 are new. The temperature dependent conductivity, diffusivity and compressive strength expressions were based on measurements over the range 300-1270 K. In Ref. 7 the upper limit of strength data was 1070 K, so the 1270 K limit of the data here is an improvement, since it is closer to the bore temperatures indicated in upcoming results.

Table I. Ceramics investigated and selected properties

ceramic	Elastic Modulus E; GPa	Poisson's Ratio ν --	Thermal Expansion α; K^{-1}	Thermal Conductivity k; W/m K	Thermal Diffusivity δ; cm^2/s	Compressive Strength S_C; GPa
SiAlON	320	0.25	3.3E-6	$28.3\ T^{-0.170}$	$0.714\ T^{-0.503}$	$37.8\ T^{-0.303}$
Si_3N_4-1	310	0.27	3.2E-6	$339\ T^{-0.442}$	$14.1\ T^{-0.844}$	$30.7\ T^{-0.315}$
Si_3N_4-2	294	0.28	3.3E-6	$522\ T^{-0.484}$	$20.9\ T^{-0.879}$	$39.8\ T^{-0.371}$
SiC-4	410	0.14	4.5E-6	$15800\ T^{-0.823}$	$66700\ T^{-1.229}$	$518\ T^{-0.698}$

The analysis of laser heating was significantly improved in this work by incorporating the rise and fall of the flux density of the laser with time (which peaks at about 1 ms and falls to one half the peak value at about 4 ms) in a manner similar albeit slower than that of cannon heating (which peaks in about 0.5 ms and falls to one half of the peak at about 1 ms). In prior work the simple expedient of a constant laser heat flux was used. The varying flux was used in the finite difference calculations of laser heating described here, while keeping the total heat input (over 6 ms) for laser heating the same as for cannon gas heating. An example of the resulting surface temperature for Si_3N_4-1 for these two types of heating is shown in Figure 1. The cannon gas heating has a peak convection coefficient of about 400,000 W/m^2 and peak gas temperature of 3000 K; details have been described [7]. Note that the peak laser surface temperature is slightly lower and occurs much later in time that that for gas heating. These features are each consistent with the slower rise and fall of the laser heating compared with gas heating. Thus, although the temporal nature of the laser heating is not yet a close match to gas heating, laser heating is well described by the analysis, and the laser heating results for ceramics can thereby be directly related to cannon gas heating of ceramics – the topic of main concern. A laser with shorter and more adjustable pulse duration will soon be available to the authors.

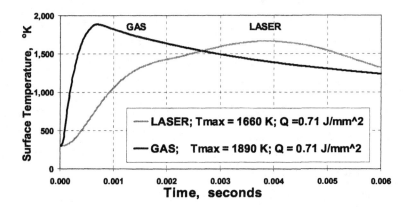

Figure 1. Surface temperature for cannon gas and laser heating of Si$_3$N$_4$-1 for the same heat input

NEAR-SURFACE TEMPERATURES AND STRESSES

The calculation methods were, apart from the improved analysis of the laser pulse described above, the same as described in recent work[7]. Near-bore temperatures were determined from a spread-sheet finite difference method using the temperature dependent thermal properties of the ceramics from Table I. Cannon gas temperature and convection coefficient versus time data from an interior ballistics model were used to model cannon heating; the laser heating has been discussed. In-plane transient compressive stresses, S$_T$ were determined from the finite difference temperatures using conventional[6] solid mechanics relations and properties, including elastic modulus, E, Poisson's ratio, ν and thermal expansion, α, as follows

$$S_T = E\,\alpha\,(T - 300\ ^\circ K) / (1 - \nu) \tag{1}$$

When the transient compressive stress from Eq. 1 exceeds the elevated temperature compressive strength, S$_C$ determined from hot hardness tests, or S$_T$ > S$_C$, a permanent compressive displacement takes place, resulting in tensile residual stress and crack formation upon subsequent cooling.

The peak finite difference temperatures versus depth at any time in the heating cycle are shown in Figure 2 for the four ceramics with either cannon or laser heating. The total heat input, Q in J/mm^2 calculated for cannon heating (shown in the legend) was used as input for laser heating analysis, so that the temperatures and stresses could be compared on a common basis.

283

Note that surface temperatures are higher for cannon heating, whilst below the surface (at 0.05 mm for example) the temperatures are higher for laser heating. This is consistent with the faster rise of cannon heating compared with laser heating, discussed earlier. Also, calculations using the conductivity expressions in Table I show that surface temperatures for either type of heating are inversely related to conductivity, as expected. Next, the finite difference temperatures are used (with Eq. 1) to calculate transient compressive stress, for comparison with elevated temperature compressive strength.

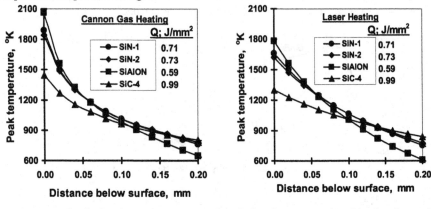

Figure 2. Peak calculated temperature at any time during the cannon gas or laser heating cycle

The critical comparison of applied stress to material strength for a ceramic cannon liner is controlled by the transient heating at and near the bore surface. Figures 3 and 4 show this comparison for the four ceramics with cannon or laser heating. The lower curves in all the plots are the transient compressive stress, S_T resulting from thermal expansion; it increases near the bore surface in response to an increased ΔT. The upper curves are the compressive strength, S_C of the ceramics, determined from hot hardness tests[7]; strength decreases near the bore as the temperature increases, as is true for all materials. It is clear for all of the results in Figures 3 and 4 that the compressive stress does not exceed the strength, even at the bore surface. But the margin between stress and strength varies among the ceramics. For both cannon and laser heating the ratio of strength-to-stress, S_C/S_T, for SiAlON shows the largest margin between strength and stress. The noticeable difference between the S_C/S_T behavior of Si_3N_4-1 and Si_3N_4-2 is due to their different compressive strengths at elevated temperature.

The S_C/S_T ratios for the cannon and laser heating analyses of Figs. 3 and 4 are listed in Table II, along with analysis results for the laser heating tests that had heat inputs close to those from analysis. Comparing first the surface temperatures and S_C/S_T ratios from analysis of cannon and laser heating, the laser temperatures are lower whilst the laser S_C/S_T ratios are higher. As has been discussed, this is expected due to the slower laser heating. The last group of data in Table II shows an unexpected, and as yet unexplained, result for the laser heating test of SiAlON. This test had the lowest heat input (0.56 J/mm^2) of the four ceramics and yet cracking was observed. The other laser test resulting in cracking, ceramic Si_3N_4-1, had a much higher heat input (0.90 vs. 0.56 J/mm^2) and the S_C/S_T ratio (1.18) more closely approached the critical value of 1, so the cracking observed in this test can be understood. And the remaining two laser tests had high

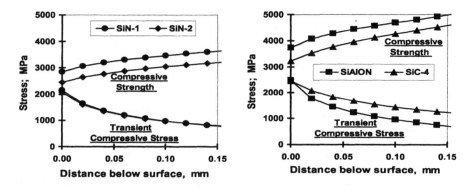

Figure 3. Transient compressive stress and strength for *cannon gas heating* of four ceramics

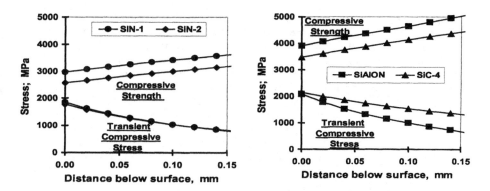

Figure 4. Transient compressive stress and strength for *laser heating* of four ceramics

Table II. Heat, temperatures and strength/stress for cannon and laser heating of four ceramics

ceramic	Cannon Heating Analysis			Laser Heating Analysis			Laser Heating Tests		
	heat input J/mm²	surface temperature deg K	S_C/S_T ---	heat input J/mm²	surface temperature deg K	S_C/S_T ---	heat input J/mm²	surface temperature deg K	S_C/S_T ---
SiAlON	0.59	2070	1.51	0.59	1780	1.87	0.56*	1720	1.28
Si₃N₄-1	0.71	1890	1.32	0.71	1660	1.60	0.90*	2040	1.18
Si₃N₄-2	0.73	1840	1.18	0.73	1620	1.44	0.78	1710	1.33
SiC-4	0.99	1450	1.31	0.99	1300	1.62	0.98	1280	1.67

* cracking observed

S_C/S_T ratio and no cracking, as would be expected. The most likely rationale for the unexpected cracking in the SiAlON laser tests is the higher surface temperature expected for either gas or laser heating (Fig. 2) due to lower conductivity. Thus the SiAlON surface temperature is well above the 1270 K maximum temperature of input data for the analysis, and the results from analysis become less certain.

FAILURE OF CRACKED CERAMICS

It is useful to analyze how a ceramic "island" surrounded by surface cracks might fail completely and break away from the surface, since this is the final, critical event that leads to rapid erosion failure of a chromium plated gun tube[2]. Such a final failure of a ceramic island is considered in Figure 5, showing an idealized sketch of island failure and a plot of the shear stress along the base of a ceramic island. An expression for the shear stress, τ on the base of an island of length, L, width, b, and depth, h, can be written using the force-balance concept of Evans and Hutchinson[6], as follows.

$$\tau\, b\, L = [E\, \alpha\, (T_{h/2} - 300)\, /\, (1-\nu)]\ b\, h \qquad (2)$$

The open crack (on the left) relieves the thermal compression, S_T applied to the island, and the compression force once carried by the island must be balanced by the shear force along the base of the island. S_T is calculated as in Eq. 1, with the finite difference temperature at the half depth of the island, $T_{h/2}$ used as an upper bound estimate of the mean temperature from the surface to depth h. Using the ceramic properties of Table I and the typical L/h ratio of 3 from prior results[7], plots of shear stress versus time are calculated using Eq. 2 and shown in the plot of Figure 5. As earlier, the effects of the faster cannon gas heating can be seen. However, the slower laser heating results in higher shear stress at later times, because slower heating more effectively raises an interior temperature such as $T_{h/2}$. The important result is that the overall levels of shear stress are significant for each type of heating and each ceramic. It is suggested that these levels of shear stress are high enough to cause complete failure and separation of a ceramic island

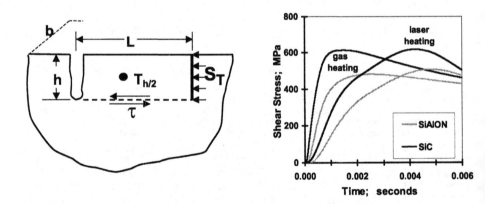

Figure 5. Analysis of shear failure of a cracked ceramic island upon subsequent heating

adjacent to an open crack. The mode of failure may well be tensile for ceramics, given their limited tensile strength. A significant tensile principal stress can accompany the shear stress, and bending and its accompanying tensile stresses can also be envisioned from the sketch in Fig. 5. Finally, note that open cracks, associated here with significant shear and tensile stresses, are often observed in thermal damage of ceramics[4,7].

CERAMIC-LINED CANNON CONCEPT

Given the preceding discussion of high near-bore temperatures and cracking in a ceramic-lined cannon tube, it is prudent to consider the firing and fabrication stresses that will be present. Other current work addresses this[8] and will be used here to evaluate a ceramic-lined tube concept. Exact spread-sheet calculations give hoop and radial applied and residual stresses at all locations of a three-layer thick-wall pressurized tube of any absolute size made from any elastic materials (E and v) with any bore pressure (r = a) and any interference or ΔT at the layer interfaces (r = c and d). Figure 6 shows the concept to be discussed here, with b/a = 2.0, d/a = 1.5, c/a = 1.1. The concept is to first heat-shrink a thin (10% of total wall thickness) ceramic liner into a thicker, intermediate steel liner (50% of wall thickness) and then either heat-shrink or tension-wrap the intermediate liner and ceramic assembly into a steel or carbon/epoxy composite jacket, respectively.

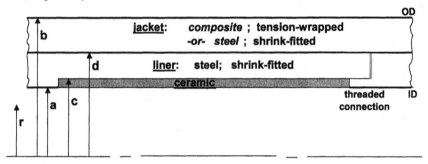

Figure 6. Concept for ceramic-lined, steel or carbon/epoxy -jacketed gun tube

Stress calculations were performed for the four ceramics of Table I, the two types of jacket, an internal pressure, p_a of 500 MPa (typical mid-range of modern cannons), a ΔT_c of 400 K between ceramic and steel intermediate liner, and an interface pressure, p_d, of 120 MPa. The mean elastic properties of a $[\pm 30/0_4]_s$ P75 graphite fiber/1962 Epoxy, 62 % fiber volume, aerospace composite laminate[9] were used for the carbon/epoxy jacket, that is, E=225 GPa, v=0.28. Figure 7 shows results for a Si_3N_4-1 liner and carbon/epoxy jacket, which are typical of all results. A significant hoop residual compressive stress of –1660 MPa is produced at the ceramic inner surface (r = a) that counteracts the firing stress, so the total stress remains compressive, at – 500 MPa for a Si_3N_4-1 liner (–440 to –500 MPa for all results).

The most critical stress at other tube locations is, somewhat counter-intuitively, the relatively low total radial compressive stress at the steel liner-carbon/epoxy jacket interface, – 250 MPa for a Si_3N_4-1 liner (–230 to –250 MPa for all results). This liner-jacket interface stress is critical because of the fundamentally limited strength of a fibrous composite material in a

287

direction unsupported by fibers. It is not clear if radial crushing of a carbon-epoxy composite can be avoided in the concept tube discussed here. For example, the 90° (axial) crushing strength of the [±30/0₄] laminate above, whose fibers provide a known albeit small amount of 90° support, is –190 MPa[9]. Its radial crushing strength, with no supporting fibers, may be lower still. This indicates that radial crushing of a carbon/epoxy jacket at the liner-jacket interface (r = d here) is a potential limitation that becomes more severe the closer the interface is to the tube bore. Crushing at the interface will cause loss of the critical hoop compressive residual stress. The use of a steel jacket would avoid this critical limitation, but add weight to the tube.

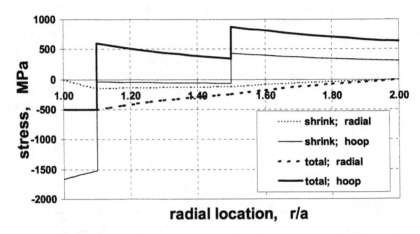

Figure 7. Stresses for Si_3N_4-1 lined, steel-intermediate lined, carbon/epoxy jacketed vessel; b/a=2.0, d/a=1.5, c/a=1.1, p_a=500 MPa, p_d=120 MPa, ΔT_c=400 K

Finally, the tube concept can be evaluated as to how well it provides the critically important axial compressive residual stress support[1] to the ceramic liner. An approximate, upper-bound estimate of the axial compression, S_{MAX} imparted to a thin ceramic liner when it is shrink-fitted to the steel intermediate liner can be written as follows,

$$S_{MAX} = E_C (\alpha_S - \alpha_C) (T_C - T_S) \qquad (3)$$

where the elastic modulus and expansion coefficients for the ceramics, E_C and α_C, are from Table I; the expansion coefficient for steel is α_S =12 E-6 /K; and $\Delta T = T_C - T_S$ = 400 K, as mentioned earlier. This gives compressive stresses of: SiAlON, 1110 MPa; Si_3N_4-1, 1090 MPa; Si_3N_4-2, 1020 MPa; and SiC-4, 1230 MPa. These values are for the ideal limiting case of a thin ceramic liner with negligible cross-section compared to that of the steel intermediate liner, and also for the case of no slippage between ceramic and steel. For these conditions, it appears that protective axial compression can be maintained in a ceramic liner shrink-fitted into a steel intermediate liner.

FRACTURE MECHANICS EVALUATION

An evaluation of the critical crack depth that can be tolerated near the inner surface of the ceramic liner can be made using the shallow crack stress intensity factor expression from prior work[4,8]. At fracture the applied stress intensity factor $K = K_{Ic}$, and the near-bore applied K is determined from: critical crack depth, a_c; applied pressure, p; and the total bore hoop stress (sum of applied and residual hoop stresses) , S_{TOT} described in Fig. 7. The expression is

$$K_{Ic} = 1.12 \, (\pi \, a_c)^{1/2} \, (S_{TOT} + p) \qquad (4)$$

It was used to calculate the pressures at which liner cracks will grow for given values of K_{Ic}, a_c, S_{TOT} and p, as seen in Fig. 8. The addition of p to S_{TOT} in Eq. 4 accounts (exactly, for shallow cracks) for the increase in K due to the pressure loading on the crack faces. The highest pressure considered here is 700 MPa, corresponding to the upper end of modern cannon pressures. The important overall result of Fig. 8 is that the low K_{Ic} (3.0 MPa m$^{1/2}$) and high E of SiC results in an alarmingly low $a_c = 0.004$ mm, whilst the higher K_{Ic} of SiAlON and Si$_3$N$_4$ (7.6 and 8.7 MPa m$^{1/2}$, respectively) and their lower E results in significantly higher a_c. Higher amounts of prestress, at both the jacket-intermediate liner interface and the intermediate liner-ceramic interface, would allow larger critical cracks, a_c without failure. But, as has been discussed, there are practical limits to the amount of prestress that can be achieved.

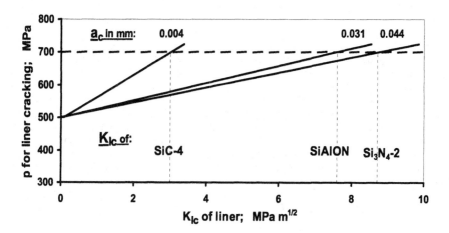

Figure 8. Analysis for ceramic lined, steel-intermediate-lined, carbon/epoxy jacketed vessel; b/a=2.0, d/a=1.5, c/a=1.1, p_d=120 MPa, ΔT_c=400 K

A final fracture mechanics consideration of a ceramic-lined cannon tube is the assessment of the overall reliability and associated safe lifetime of the tube under service loading conditions. A reliability and safe life determination for chromium plated cannon tubes has been described[10] that uses full-scale cannon firing tests followed by continued laboratory firing simulation. These methods have a proven history of success with chromium-plated tubes and are suggested here for ceramic-lined tubes. Admittedly, a chromium-plated tube is different from a ceramic-lined tube, but the failure mechanisms have much in common. Thermal loads cause full-thickness chromium cracks after one or two severe firings, but this is only the start of overall failure of the tube. A similar scenario could play out for a ceramic-lined tube. The suggested reliability and safe life methods, as typically used for chromium plated tubes, are outlined as: a few hundred full-scale cannon firings (at a proving ground) until near-bore thermal damage is well developed; continued hydraulic-pressure-simulated firing (in a laboratory) until a fatigue crack grows completely through the tube wall; reliability and safe life assessment of replicate test results (typically six). The assessment of a safe life, N*, is in terms of the mean and standard deviation of the natural logs of the test lives, N, as:

$$N^* = exp[\ (ln\ N)_{MEAN} - k\ (ln\ N)_{SD}] \tag{5}$$

where k is the normal distribution tolerance factor (from statistical tables) for a given confidence that the safe life will be reached in service loading. Example results for chromium-plated tubes are[10]: six test lives varying from 8,501 to 13,800 load cycles to failure; k = 4.24 for 90% confidence, 0.99 reliability; a safe life N* = 5,337 cycles. Actual safe life results for specific cannon components vary, depending on the confidence and reliability requirements of intended service. Nevertheless these methods of reliability and safe life determination, including as they do the complete assessment of the full-size fabricated condition of the entire cannon tube system, may have great utility for ceramic-lined cannons.

CONCLUSION

A final evaluation and rating of the structural integrity of a ceramic-lined cannon tube can be made by using quantitative comparisons of four critical factors discussed here: gas heating surface temperature; resistance to crack formation; resistance to crack growth; level of axial compression. Table III compares these factors. A lower surface temperature for cannon gas heating improves structural integrity, assuming all other other things being constant. If this is true, SiC-4 is the best, the two Si_3N_4 are intermediate, and SiAlON is the worst in this regard. The ratio of elevated temperature strength to transient stress, S_C/S_T, from Fig. 3, shows superior resistance to thermal crack formation for SiAlON, compared with perhaps adequate resistance for the other ceramics. Resistance to crack growth, measured by a_{Ic} and thereby including effects of hoop firing and fabrication stresses and fracture toughness, is alarmingly low for SiC-4 and at best adequate for the other ceramics. Regarding axial compressive stress, for a thin liner with no

Table III. Comparison of structural integrity of ceramic-lined cannon tubes

ceramic	Surface Temperature deg K	Crack Formation S_C/S_T; --	Crack Growth a_{Ic}; mm	Axial Stress S_{MAX}; MPa
SiAlON	2070	1.5	0.03	−1110
Si_3N_4-1	1890	1.3	0.04	−1090
Si_3N_4-2	1840	1.2	0.04	−1020
SiC-4	1450	1.3	0.004	−1230

slippage it seems possible to obtain a suitable amount of compression in all ceramics by shrink-fitting. Combining these measures of structural integrity, Si_3N_4-1 is the preferred material for the concept of a ceramic-lined, steel intermediate lined, steel or carbon/epoxy jacketed cannon tube considered here.

ACKNOWLEDGEMENT

The authors are pleased to acknowledge the help of Dr. Robert Carter and Mr. Jeffery J. Swab of the Army Research Laboratory for supplying the elastic, thermal and mechanical properties and the test samples of the ceramic materials described here.

REFERENCES

[1] R.N.Katz, "Ceramic Gun Barrel Liners: Retrospect and Prospect," *Proceedings of Sagamore Workshop on Gun Barrel Wear and Erosion*, US Army Research Laboratory report, 66-84 (1996).

[2] J.H. Underwood, M.D. Witherell, S. Sopok, J.C. McNeil, C.P. Mulligan and G.N. Vigilante, "Thermo-Mechanical Modeling of Transient Thermal Damage in Cannon Bore Materials", *Wear*, 257 992-998 (2004).

[3] P.J.Cote, G.Kendall and M.E.Todaro, "Laser Pulse Heating of Gun Bore Coatings", *Surface and Coatings Technology*, 146-147 65-69 (2001).

[4] J.H.Underwood, P.J.Cote and G.N.Vigilante, "Thermo-Mechanical and Fracture Analysis of SiC in Cannon Bore Applications", *Ceramic Engineering and Science Proceedings*, 24, 3, 503-508 (2003).

[5] J.H. Underwood, A.P. Parker, G.N. Vigilante and P.J. Cote, "Thermal Damage, Cracking and Rapid Erosion of Cannon Bore Coatings", *J. Pressure Vessel Technology*, 125, 299-304 (2003).

[6] A.G. Evans and J.W. Hutchinson, "The Thermomechanical Integrity of Thin Films and Multilayers," *Acta. Metal. mater.*, 43, 7, 2507-2530 (1995).

[7] J.H. Underwood, M.E. Todaro and G.N. Vigilante, "Modeling of Transient Thermal Damage in Ceramics for Cannon Bore Applications", *Ceramic Engineering and Science Proceedings*, 25 189-194 (2004).

[8] J.H. Underwood and A.P. Parker, "Stress and Fracture Analysis of Ceramic Lined, Composite or Steel Jacketed Pressure Vessels," *J. Pressure Vessel Technology*, 126, 485-488 (2004).

[9] S.P. Rawal and J.W. Goodman, "Space Applications," in *ASM Handbook, Volume 21, Composites*, ASM International, Material Park, OH, 1033-1042 (2001).

[10] J.H. Underwood and E. Troiano, "Critical Fracture Processes In Army Cannons: A Review," *J. Pressure Vessel Technology*, 125, 287-292 (2003).

Characterization Tools for Materials Under Extreme Environments

ON THE COMPARISON OF ADDITIVE-FREE HfB$_2$-SiC CERAMICS SINTERED BY REACTIVE HOT-PRESSING AND SPARK PLASMA SINTERING

Frédéric Monteverde
National Research Council – Institute of Science and Technology for Ceramics
Via Granarolo 64
48018 Faenza - Italy

Alida Bellosi
National Research Council – Institute of Science and Technology for Ceramics
Via Granarolo 64
48018 Faenza - Italy

ABSTRACT

Ultra-high-temperature HfB$_2$-SiC ceramics were produced by reactive hot-pressing (RHP) and spark plasma sintering (SPS) successfully. In the former case, a mixture of Hf/Si/B$_4$C, mechanically mixed in molar ratio 2.2/0.8/1, was "in-situ" converted into HfB$_2$ and SiC, and then directly hot-pressed until full density was achieved. In the SPS case, a powder mixture of HfB$_2$ + 30vol% SiC was fully densified at 2,100°C, 100°C/min heating rate and 2 min dwell time. The microstructure in both the materials consisted of faceted diboride grains, finer in the RHP case, with SiC particles evenly distributed intergranularly. The combination of some thermo-mechanical properties was of considerable significance. Flexural strength of the RHP fabricated material measured at 25 °C and 1,500 °C in ambient air was 770±35 and 310±15 MPa, respectively. A relevant merit characterized fracture toughness and flexural strength of the material produced by SPS: the values measured at room temperature (3.9±0.3 MPa√m and 590±50 MPa, respectively) did not decrease appreciably at 1500 °C (4.0±0.1 MPa√m and 600±15 MPa, respectively).

INTRODUCTION

The design and processing of materials with enhanced high temperature capabilities represent one of the most challenging tasks of modern engineering. Transition metal diborides and carbides are naturally selected for ultra-high-temperature structural applications because of their melting points exceeding 3,000 °C, coupled to an overall thermo-structural stability at very high temperatures [1]. Great attention is currently addressed towards the engineering of ultra-high-temperature ceramics (UHTCs) capable of increasing the toleration of heat on sharp leading profiles of space vehicles re-entering the Earth's atmosphere [2]. During recent decades, ceramic matrix composites (CMCs) have been largely produced by densification of mechanically-mixed powders. Since the melting points of the UHTCs are among the highest known, processing them into CMCs with final complex shapes and full density required prolonged exposure in atmosphere-controlled conventional furnaces at sintering temperatures in excess of 2,000 °C. Different approaches were tried for improving the fabrication procedures and performances of these UHTCs. Metals or other additives were incorporated into the basic mixtures to control the microstructure development. Such an intentional introduction of impurities, in connection to conventional hot-consolidation techniques like hot-pressing and gas-pressure sintering, often deteriorates the high-temperature strength [3,4].

Processing routes such as reactive hot-pressing (RHP) from powdered precursors or spark plasma sintering (SPS) offer alternative opportunities to fabricating strongly covalent UHTCs. Unlike self-propagating high-temperature synthesis (SHS), which basically exploits the exothermicity of uncontrolled chemical reactions, RHP can be accomplished gradually via solid-state diffusion at temperatures below those typically generated during SHS processes. It ensures an high chemical compatibility of the "in-situ" formed individual phases, a better control over the final microstructures, and the isotropy of properties. SPS is a processing technique recently applied for densifying poorly sinterable ceramics [5]. The SPS process is similar to conventional hot-pressing in which the precursors are loaded in a mould, and a mechanical uniaxial pressure is applied during hot-consolidation. However, instead of an external heating, a (pulsed) direct current is allowed to pass through the (uniaxially) pressing dies and, depending on the electrical conductivity of the processed material, also through the cold-compacted powdered sample inside the mould. This implies that the mould itself acts as an heat source, and that the sample is heated from outside and inside. In contrast to widely used furnaces, the heat transfer is surprisingly efficient, and heating rate up to 600 °C/min or more can be applied. The improved densification rate of the powder particles was frequently argued being obtained by the application of direct current pulses of high energy, which in turn, activated spark and plasma discharges among the powder particles. The presence of activated surfaces, along with the cited spark/plasma discharges, are expected to enhance surface and boundary diffusion, which will promote mass transfer, and therefore also densification and grain growth.

The present study compares the microstructure development and some thermo-mechanical properties of HfB_2-SiC materials sintered by RHP and SPS.

EXPERIMENTAL

The Reactive Hot-Pressing

Three commercial precursors were selected: Hf (99.8% pure, Neomat Co. – Latvia,), Si (99.9% pure, grade AX05, FSSS 3.5μm, H.C. Starck – Germany,) and B_4C (grade HS, FSSS 0.8 μm, H.C. Starck – Germany). In accordance with the following reaction

$$(2+x) Hf + (1-x) Si + B_4C \Rightarrow 2 HfB_2 + (1-x) SiC + x HfC \qquad (1)$$

the corresponding stoichiometric precursors for x=0.2 were wet-milled, dried and sieved. The formulation above reported was adjusted by adding limited amounts of HfB_2 (99.5% pure, 325 mesh, Cerac Inc. – USA) and α-SiC (grade UF25, FSSS 0.45μm, H.C. Starck – Germany). The expected final composition (vol%) reads HfB_2 + 22.1 SiC + 5.9 HfC. Reaction 1 was investigated through a campaign of heat treatments hereafter labelled PLSHT-n, (10°C/min heating rate, dwell time 60 min, graphite heating elements and crucibles) on the as-ground powder mix from 1,000 °C up to 1,650 °C in flowing argon.

The RHP experiment was carried out at low vacuum (0.5 mbar) using a BN-lined induction-heated graphite die, into which the as-ground powder mix was directly loaded and heat-treated. The reactive hot-pressed material will be labelled RHP_HFSC for the rest of the text.

The Spark Plasma Sintering

The following starting mixture, hereafter indicated as SPS_HFSC, was selected: HfB_2 + 30vol% SiC. This powder mixture of HfB_2 (99.5% pure, 325 mesh, Cerac Inc. – USA,) and β-

SiC (grade BF12, 11-13 m^2/g, H.C. Starck – Germany,) was wet-milled, dried and sieved. The as-treated mixture was processed using a spark plasma sintering furnace (FCT Systeme GmbH, Rauenstein - Germany) loading it into a graphite mould lined with a 0.75 mm thick graphitised paper. The following parameters were used: inner pressure 0.02 mbar, uniaxial mechanical pressure 30MPa, heating rate 100°C/min, 2,100°C set-point and 2 min dwell time. The temperature was measured by means of an optical pyrometer focused on the bottom of the upper graphite punch, about 4 mm far from the sample.

Characterization of the as-sintered materials

The crystalline phases were identified by means of an X-ray diffractometer (XRD). Size, shape and distribution of the phases present in the sintered materials were investigated on fracture and polished surfaces using a scanning electron microscope (SEM) equipped with an energy dispersive X-ray microanalyzer (EDX). Polished cross-sections were prepared applying the widely-accepted metallographic method. Young's modulus (E) was measured on a 28.0 x 8.0 x 0.8 mm^3 plate using the resonance frequency method. Flexural strength (σ) in a 4-pt. configuration was tested in ambient air at room temperature (5 specimens tested) and 1,500 °C (3 specimens tested) on 25.0 x 2.5 x 2.0 mm^3 chamfered bars using 20 mm and 10 mm as outer and inner span, respectively, and a cross-head speed of 0.5 mm/min. Fracture toughness (K$_{iC}$) was measured at room temperature and 1,500 °C through the chevron-notched beam method (3 specimens tested each) using 25.0 x 2.5 x 2.0 mm^3 bars, cross-head speed 0.05 mm/min.

RESULTS

The RHP_HFSC material

Table 1 and Fig. 1 show processing parameters and some results from the campaign of the heat treatments PLSHT-n. In the (untreated) sample PLSHT-0, α-Hf, Si and B$_4$C were separate phases. X-ray diffraction identified also a cubic modified HfH$_{1.5}$ (ICDD 05-639), and HfB$_2$. The former phase was initially present in the as-received hafnium powder because of the wet storage. Such a hafnium hydride proved to be stable up to 1,100 °C. Reaction 1 is highly exothermic (-695kJ/mol enthalpy of formation at 25 °C) and satisfies the thermodynamic conditions for a self-sustaining reaction process (T$_{ADIAB}$ $\cong$2,900 °C).

Table 1. PLSHT-n tests: temperature T, dwell time t, relative weight loss WL, and crystalline end-products.

PLSHT-n Test	T (°C)	t (min)	WL (%)	Crystalline end-products[*]	
				Main	Minor
-1	1,000	60	1.3	Hf[♦], Si, HfB$_2$, HfC, HfH$_{1.5}$, B$_4$C	(Hf,Si)
-2	1,100	60	1.5	Hf[♦], (Hf,Si), HfB$_2$, HfC	-
-3	1,200	60	2.3	HfB$_2$, SiC, HfC	m-HfO$_2$
-4	1,300	60	1.0	HfB$_2$, SiC, HfC	m-HfO$_2$
-5	1,450	60	0.5	HfB$_2$, SiC, HfC	m-HfO$_2$
-6	1,500	60	1.0	HfB$_2$, SiC, HfC	m-HfO$_2$
-7	1,500	120	1.1	HfB$_2$, SiC, HfC	m-HfO$_2$
-8	1,550	60	1.4	HfB$_2$, SiC, HfC	m-HfO$_2$[°]
-9	1,600	60	1.2	HfB$_2$, SiC, HfC	-
-10	1,650	60	1.5	HfB$_2$, SiC, HfC	-

[*]hexagonal type (ICDD 38-1478) but longer lattice parameters; [♦]pulverized sample; [°]uncertain

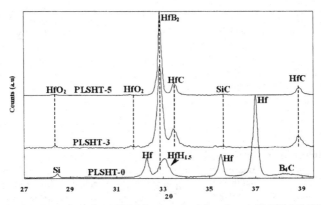

Fig. 1 XRD (raw) patterns of sample PLSHT-0 (as-ground powder mixture), PLSHT-3 (1,200 °C for 60 min) and PLSHT-5(1,450 °C for 60 min).

The formation of HfB_2 during mixing/milling clearly highlights how reaction 1 is prone to have succeeded even at room temperature (Gibbs free energy of formation -665.5 kJ/mol at 25 °C). In accordance with the XRD outcomes, reaction 1 proceeded to a certain extent at 1,000 °C, until the consumption of the initial precursors was virtually completed at 1,200 °C and 1 hour of dwell time. Within the temperature range of 1,000 1,200 °C, several chemical reactions took place, giving rise to stable intermediate solid by-products, specifically a mixture of hafnium silicides (Hf,Si). Raising the processing temperature of the PLSHT-n test up to 1,650 °C, the end-products of the synthesis vary slightly in composition and relative percentages.

Considered the processing conditions of the PLSHT-5 test, the as-ground powder mixture was loaded straight into a graphite mould of the hot-pressing equipment and heat-treated, in accordance with the thermal programme shown in Fig. 2. Also, to prevent the molten silicon (or silicon containing compounds) from squeezing out of the dies at high temperatures, the external pressure was stepwise applied only once the thermal treatment at 1,450°C for 60 min was completed. The heating rate from about 900 up to 1,450 °C was kept as low as 10 °C/min in order to quench any emergence of spontaneous self-combustion. The terminal stage was 1,900°C for 10 min, 50 MPa of applied pressure. In accordance with the full completion of reaction 1

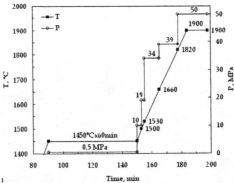

Fig. 2 Temperature (T) and mechanical pressure (P) variation vs. time during the RHP experiment

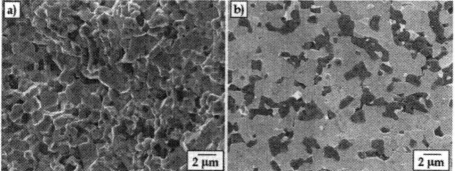

Fig. 3 SEM micrographs of material RHP_HFSC: a) fracture surface; b) polished cross-section:
the dark features corresponds to SiC particles

the predicted bulk density of the composite is 9.51 g/cm^3. The bulk density, measured with the Archimedes' method, was 9.43 g/cm^3, 0.992 of the theoretical relative density. The XRD analysis of RHP_HFSC confirmed the formation of HfB$_2$, SiC, HfC, and m-HfO$_2$ as secondary phase. The analysis by SEM (Fig. 3) agrees with a microstructure free of porosity. The RHP process developed a ceramic composite with faceted diboride grains (from 0.5 to 4μm as grain size), and a SiC particulate (about 2μm max as grain size) located intergranularly. The HfB$_2$/HfB$_2$ grain interfaces appear depleted of secondary phases. Micro-cracking of HfB$_2$ grains and debonding of SiC/HfB$_2$ interfaces were observed in little amount (Fig. 4). The emergence of tensile or compressive residual strained fields in the diboride matrix and SiC particulate, respectively, was considered the main reason of such a phenomenon. Table 2 summarises some mechanical properties. The limited dispersion of strength data points out the suitability of the processing procedures which enable some control over manufacture flaws. At 1,500 °C, a reduction in strength, 60% of the room temperature value occurred.

The SPS_HFSC material

Fig. 5 plots the densification curves during spark plasma sintering. The onset of some measurable shrinkage of the green body was recorded at around 1,300 °C. A marked increase in

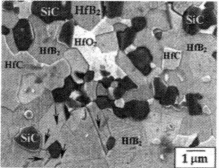

Fig. 4 SEM micrograph of material RHP_HFSC (polished area):
the arrows indicate micro-cracking of HfB$_2$ grains.

Table 2   Relative density rd, Young's modulus E, fracture toughness K_{iC}, and
4-pt. flexural strength σ.

Sample	Density		E (GPa)	K_{iC} (MPa√m)		σ (MPa)	
	g/cm³	rd*		25°C	1500°C	25°C	1500°C
RHP_HFSC	9.51	1	520	na	na	770±35	310±15
SPS_HFSC	8.72	1	512	3.9±0.3	4.0±0.1	590±50	600±15

*: estimated by SEM; na: not measured

the densification rate occurred only for temperatures higher than 1,700 °C, and suddenly fell down once the temperature had reached the set point of 2,100 °C. The whole process, cooling down included, lasted about half an hour. The XRD analysis, beside HfB_2 and SiC, detected limited amounts of monoclinic HfO_2. The examination by SEM (Fig. 6) typically shows a grained structure, with uniformly faceted HfB_2 grains and the SiC particles dispersed within the diboride skeleton intergranularly. The size distribution of the HfB_2 grains is narrow, with an average value of about 2μm. Also in this case, some residual micro-cracking occurred. Table 2 contains the experimental data for the measured thermo-mechanical properties. The fracture toughness (K_{iC}) and the flexural strength (σ) showed an interesting combination of data. In fact, neither K_{iC} nor σ, tested at 1500 °C, demonstrated to decrease appreciably, if compared to the values at room temperature.

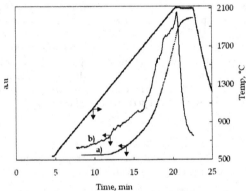

Fig. 5 Linear shrinkage (a) and shrinkage rate (b) vs. time of sample SPS_HFSC during spark plasma sintering

DISCUSSION

The "in-situ" synthesis of the HfB_2-SiC mixture in material RHP_HFSC proceeded via decomposition and further mass exchange processes at high temperatures. Of course the solid precursors (i.e. Hf, Si and B_4C) were selected such that the resulting reaction was highly exothermic. The real driving force of such solid-state exchange reactions definitely arises from the formation of thermodynamically very stable end-products. For temperatures above 1,600 °C, the conversion yield of HfC starts dominating over that of SiC. Therefore, 1,450 °C was deemed an acceptable temperature for synthesizing as fine as possible solid constituents for the final composite and, at the same time, for promoting the formation of HfB_2 and SiC against the competing raise of HfC. The RHP_HFSC material had a viable thermal history, in particular terminal processing conditions like 1,900 °C for 10 minutes.

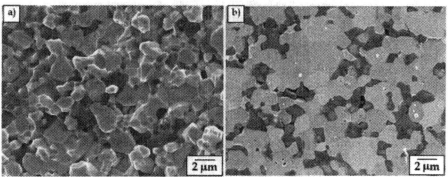

Fig. 6 SEM micrographs of material SPS_HFSC: a) fracture surface; b) polished cross-section: the dark features corresponds to SiC particles

This represents a very important issue because the intrinsic poor sinterability of a mechanically mixed HfB$_2$/SiC powder system was overturned by a correct selection either of the precursors and of the heat treatment. The initial fineness of the powder mixture was decisive in obtaining well dispersed and fine-grained solid phases constituting the RHP_HFSC composite. In addition, the intimate contact between oxygen-depleted reacting interfaces of HfB$_2$ and the simultaneous application of high temperature and pressure meet readily the conditions for surface/boundary diffusion assisting densification of the diboride particles framework, the boron activity being no further depressed. Once B$_4$C decomposes, boron and carbon readily diffuse and interact with the metallic precursors. This is the main reason why the synthesized solid HfB$_2$ and SiC phases kept the dimensional features of the starting precursors.

Regarding the flexural strength behaviour, the lower value in material SPS_HFSC was primarily due to its coarser grain size, if compared with RHP_HFSC. Even though the SPS process was manifestly quicker than RHP, its final processing temperature at 2,100 °C activated grain growth and thus coarsening of the HfB$_2$ grains.

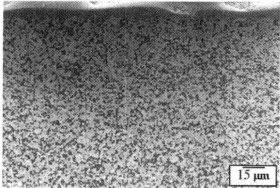

Fig. 7 SEM micrograph from material SPS_HFSC tested at 1500 °C: the polished cross-section in proximity of the external faces of the specimen

The observation by SEM of the specimens fractured at 1,500 °C of both the materials revealed that material RHP_HFSC had a more serious microstructure degradation than material SPS_HFSC. The key-feature which adversely affected the strength behaviour of material RHP_HFSC was its (limited) fraction of HfC. Most of the HfC grains directly facing the oxidizing atmosphere transformed into an oxide form during the heating stage of the test, impairing partially the initial compactness of the base material. On the contrary, the SPS_HFSC material retained its original strength and fracture toughness owing to the effective protection for the faces exposed to air against oxidation (Fig.7). The initial compositional design of material RHP_HFSC aimed at introducing a controlled level of HfC in a HfB_2+SiC framework. This incorporation was reported to have succeeded in improving oxidation/ablation resistance when subjected to heating regimes simulating hypersonic re-entry space missions [6]. However, with reference to the strength behaviour in air at 1,500 °C (conditions that differ markedly from those just mentioned), the presence of HfC negatively affects the virtual ability of HfB_2–SiC materials to retain the original strength effectively.

SUMMARY

This work highlighted interesting advances in the microstructure and mechanical properties of an "in-situ" synthesized ultra-high-temperature HfB_2-SiC composite. Solid powder precursors like Hf, Si and B_4C were properly processed via reactive hot-pressing, and a full dense HfB_2-SiC composite thus obtained. In addition, a powder mixture of $HfB_2+30vol\%$ SiC was fully densified by spark plasma sintering at 2,100 °C for 2 min (100 °C/min heating rate) successfully. The entire SPS process lasted about 30 minutes. The microstructure developed in both the materials was uniform and rather fine, with HfO_2 as the main secondary phase. The combination of the mechanical properties was of remarkable significance. The material processed by SPS for instance retained its original strength (590±50 MPa) and fracture toughness (3.9±0.3 MPa√m) up to 1500 °C. The processing routes herein proposed are an alternative answer to fabricating poorly sinterable HfB_2-SiC systems.

Acknowledgements
    The authors wish to acknowledge the helpful contribution of their colleagues, including A. Balbo (powder processing), D. Dalle Fabbriche (thermal treatments), and C. Melandri (mechanical tests).

References
[1] K. Upadhya, J-M Yang, W.P. Hoffman, "Materials for ultrahigh temperature structural applications", *Am. Ceram. Soc. Bull.* **58**, 51-56 (1997).
[2] M.M. Opeka, I. Talmy, J.A. Zaykoski, "Oxidation-based materials selection for 2000°C + hypersonic aerosurfaces: theoretical considerations and historical experience", *J. Mat. Sci* **39**, 5887-5904 (2004).
[3] F. Monteverde, A. Bellosi, S. Guicciardi, "Processing and properties of zirconium diboride-based composites", *J. Eur. Ceram. Soc.* **22**, 279-288 (2002).
[4] F. Monteverde, A. Bellosi, "Microstructure and properties of an HfB_2-SiC composite for ultra high temperature applications", *Advanced Engineering Materials* **6(5)**, 331-336 (2004).
[5] M. Nygren, Z. Shen, "On the preparation of bio-, nano- and structural ceramics and composites by spark plasma sintering, "*Solid State Sciences* **5**, 125-131 (2003).
[6] J. Bull, J. White, L. Kaufman, "Ablation resistant zirconium and hafnium ceramics", US Patent 5,750,450 (1998).

DYNAMIC ANALYSES OF THE THERMAL STABILITY OF ALUMINIUM TITANATE BY TIME-OF-FLIGHT NEUTRON DIFFRACTION

I.M. Low, D. Lawrence, A. Jones
Materials Research Group, Department of Applied Physics, Curtin University of Technology, Perth. WA 6845, Australia

R.I. Smith
ISIS Facility, CLRC Rutherford Appleton Laboratory, Chilton, Didcot, OX11 OQX, UK.

ABSTRACT

The nature and processes of decomposition of aluminium titanate (Al_2TiO_5) in vacuum from 20-1100°C have been characterized by time-of-flight neutron diffraction to study the temperature and time-dependent structural changes in real time during thermal decomposition. Results show that the thermal stability of Al_2TiO_5 in vacuum is strongly influenced by temperature, heating rates, phase purity, grain size, and additives. Possible mechanisms of structure stabilization or instability of Al_2TiO_5 under different conditions are discussed.

INTRODUCTION

In the Al_2O_3-TiO_2 system, aluminum titanate (Al_2TiO_5) is the only thermodynamically stable compound above 1280°C, up to its melting point. It is one of several materials that is isomorphous with the mineral pseudobrookite (Fe_2TiO_5).[1,2] Each Al^{3+} or Ti^{4+} cation is surrounded by six oxygen ions forming distorted oxygen octahedral. These AlO_6 or TiO_6 octahedra form (001) oriented chains weakly bonded by shared edges. Such structure is responsible for the thermal expansion anisotropy which leads to generation of localized internal stresses and microcracking during cooling from the sintering temperature.[3] This display of microcracking is believed to be grain-size dependent,[4,5] and is responsible for imparting poor mechanical strength and low apparent thermal expansion coefficient (~1 × 10^{-6} K^{-1}) but outstanding thermal shock resistance.

In addition, Al_2TiO_5 is thermodynamically unstable and undergoes a eutectoid-like decomposition to α-Al_2O_3 and rutile-TiO_2 over the temperature range of 900-1280°C in air but peaks at 1100°C. Hitherto, the origins or causes of this instability have been a subject of on-going debate and many explanations have been proposed.[2,6-8] Experimental evidences suggest that decomposition of Al_2TiO_5 is a nucleation and growth controlled process and residual alumina particles might act as preferred nucleation sites.[2] The thermal instability of Al_2TiO_5 has also been related to high internal stresses caused by the anisotropic thermal expansion of the individual grains,[6] presence of vacancies and anti-site defects,[7] as well as cation disorder in the crystal structure.[8]

However, the thermal stability of Al_2TiO_5 in other ageing environment such as vacuum or argon has not been investigated. Our initial study showed that the rate of decomposition of Al_2TiO_5 in vacuum at 1100°C is much faster than in air.[9] In contrast to a bulk-initiated decomposition in air, the decomposition of Al_2TiO_5 in vacuum is surface-initiated forming a graded surface layer rich and the process is controlled by nucleation and growth.[10]

In this paper, we present results on the decomposition behaviour and isothermal stability of Al_2TiO_5 in the temperature range 20-1100°C. The temperature-dependent thermal stability and isothermal decomposition of Al_2TiO_5 have been dynamically monitored and characterized using time-of-flight neutron diffraction to study the structural changes of phase decomposition in real time. The effects of phase purity, stabilizers, temperature, grain size, ageing environment and heating rates on thermal stability have been investigated.

EXPERIMENTAL

Sample Preparation

Alumina and rutile powders of both commercial (<95%) and high purity (>99%) were used for the processing of pure AT samples. Several cylindrical bars of length 20 mm and diameter 12 mm were uniaxially-pressed from the powder mixture, followed by sintering at 1600°C for 4 h. In order to study the effect of stabilizers on the thermal stability, AT samples containing 1-10 wt% MgO and 20 wt% mullite were also fabricated. Some samples were sintered at 1500°C for 2 h to study the effect of finer grain size on the thermal stability. The microstructures of a sintered high purity Al_2TiO_5 sample before and after thermal decomposition are shown in Figure 1. Fine microcracks within certain grains can be seen to form as a result of thermal expansion anisotropy which induces high localized internal stresses. Following thermal decomposition, both needle-like and angular particles can been seen to form on the surface of grains. Based on the EDS results, it is believed that these nano-sized particles are the surface by-products (ie. Al_2O_3 and TiO_2) of thermally decomposed Al_2TiO_5. This may indicate that the initial process of thermal decomposition is surface-initiated and the kinetics are both temperature and time dependent.

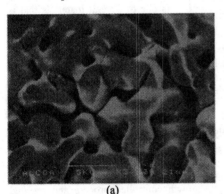
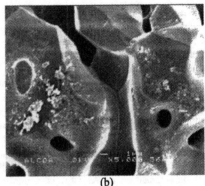

(a)   (b)

Figure 1: Scanning electron micrograph showing the microstructure of a sintered Al_2TiO_5: (a) before (b) after annealing at 1000°C for 14 h.

Neutron Diffraction (ND)

High temperature time-of-flight neutron diffraction was used to monitor the structural evolution of phase decomposition from 20 to 1100°C in real time. Neutron diffraction data were collected using the Polaris medium resolution, high intensity powder diffractometer at the UK pulsed spallation neutron source ISIS, Rutherford Appleton Laboratory.[11] Two diffraction patterns were collected at 700°C and at every 50°C between 800 and 1050°C with a rate of 20°C/min. A total of 12 patterns were collected at 1100°C. The data acquisition times were 1 h

(~200μAh) for the room temperature diffraction pattern, and 15 min (~50μAh) for each of the diffraction patterns collected at elevated temperatures. Normalised data collected in the highest resolution, backscattering detector bank over the d-spacing range of ~0.4-3.2Å were analysed by the Rietveld method using the General Structure Analysis System, GSAS.[12] The variation of phase abundances has also been qualitatively analysed using the relative peak intensity method (ie. I/I_{max}).

RESULTS AND DISCUSSION
Effect of Temperature and Annealing Time

The structural changes of commercial-purity Al_2TiO_5 (AT) during thermal dissociation into corundum and rutile at the temperature range 20-1100°C are shown in Fig. 2. Below a critical temperature of 950°C, AT is quite stable and shows absence of phase degradation. As soon as that temperature is attained, the process of phase decomposition commences within the first 15 min. Further rises in temperature cause the rate of decomposition to increase rapidly so that by the time the temperature reaches 1100°C the sample contains only 10wt% AT, and after less than 30 minutes at 1100°C this has reduced to ~3wt%. The decomposition process is completed in less than 2 h.

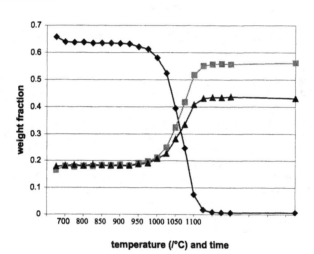

Fig. 2: Effect of temperature and time on the phase decomposition of AT. Legend: Al_2TiO_5 (♦), Al_2O_3 (■) and TiO_2 (▲)

Effect of Phase Purity

Phase purity has a strong influence on the thermal stability of AT (Fig. 3). When the sample was made from high purity Al_2O_3 and TiO_2 powders (ie. >99%), the decomposition of AT commences only at ~950°C and finishes only after very prolonged isothermal ageing. In contrast, the process of decomposition commences as early as 900°C and finishes soon after 1100°C when the sample was made from powders of lower purity. The presence of impurities can act as a catalyst to accelerate the process of phase decomposition. These impurities may offer an explanation for the rather large range of decomposition temperature (ie. 750-1300°C) reported in the literature.[2,6]

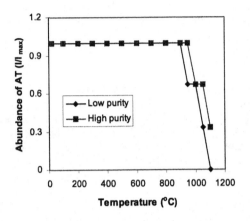

Fig. 3: Effect of phase purity on the decomposition of AT.

Effect of Additives or Stabilisers

The improved thermal stability of AT due to MgO and mullite stabilization is clearly depicted in Figs. 4 & 5. However, the addition of 3-wt% MgO is insufficient to achieve full stabilization but when compared to un-doped AT, it has delayed the start of decomposition temperature from 950° to 1000°C. In addition, the decomposition rate of Al_2TiO_5 has also been significantly reduced where ~80% AT was retained after 30 min at 1100°C. As the content of MgO increases to 10-wt%, full stabilization of AT is evident and this concurs with the results reported in literature.[13-15] The presence of 20-wt%mullite is also effective in stabilizing AT against thermal decomposition. Again, this observation is consistent with the work of Tsetsekou,[13] Huang et al.,[16,17] and Melendez-Martinez et al.[18] The plate-like mullite grains also impart strength improvement to AT. In passing, it is worth-noting that besides MgO and mullite, various other oxide additives that form solid solutions with AT have also been found to have a stabilization effect and these include Fe_2O_3, SiO_2, ZrO_2, $ZrSiO_4$, La_2O_3 and feldspar.[2,13,19-21]

Hitherto, the precise mechanism of structure stabilization in Al_2TiO_5 has been poorly understood. It is probable that the presence of an additive may help to limit the level of internal stresses caused by the anisotropic thermal expansion of the individual grains.[6] This structure stability has also been explained in terms of preferred charge compensation effects, which in the case of MgO, is by interstitial Mg ions while for TiO_2, compensation by oxygen interstitials or aluminium vacancies is preferred.[7] Atomistic simulation calculations have shown that although aluminium titanate is reluctant to form Schottky or Frenkel defects, anti-site defects or cation disorder are easily formed. The thermal stability of aluminium titanate may also be related to the degree of cation disorder in the crystal structure.[8] It is also possible that the incorporation of doping cations such as Mg^{2+}, Fe^{3+} or Zr^{4+} which have a relatively stable oxidation state can help to suppresses oxygen nonstoichiometry changes in Al_2TiO_5 and, as a result, increases the resistance to thermal instability by virtue of the tendency of Ti ions to undergo changes in the oxidation state from 4+ to 3+ especially in reducing and inert environments.[22]

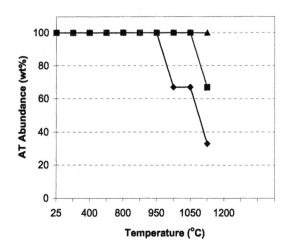

Fig. 4: Effect of additives on the thermal stability of AT. Legend: 0-wt% MgO (♦), 3-wt% MgO (■), 10-wt% MgO (▲) and 20-wt% mullite (▲)

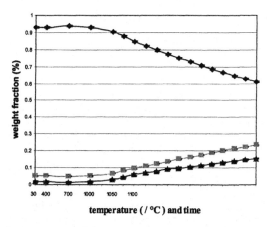

Fig. 5: Effect of temperature and time on the thermal stability of 3wt% MgO-Al_2TiO_5. Legend: Al_2TiO_5 (♦), Al_2O_3 (■) and TiO_2 (▲)

Effect of Heating Rate

The heating rate has been found to have an effect on the temperature at which the decomposition process commences. A slow heating rate (5°C/min) appears to enhance the decomposition process by triggering it on at 900°C as compared to 1050°C with a fast heating rate of 20°C/min. It appears that a slowing heating rate may have accumulated sufficient energy over a long period of time to activate the phase decomposition at a lower temperature. This heating-rate dependence behaviour has also been observed to control the formation temperature of AT where a slow heating rate caused the AT to form at a lower temperature when compared to a fast heating rate.[6]

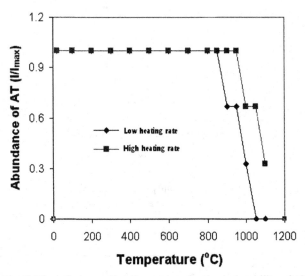

Fig. 6: The influence of heating rate on the thermal stability of AT.

Effect of Grain Size

The influence of grain size on the thermal stability of AT is shown in Fig. 7. Coarse-grained AT exhibits a greater rate of thermal decomposition when compared to its fine-grained counterpart. To the best of our knowledge, this is the first time that grain size has been shown to affect the propensity of thermal degradation in AT. However, it is unclear whether there is a critical grain size associated with this phenomenon. The reason for this grain-size effect is unclear at this stage although it may be closely related to its greater tendency for microcracking as the grain size increases. The microcracking phenomenon is closely related to the material microstructure and thermal expansion anisotropy.[23-25] Below a critical grain size, the elastic energy of the system is insufficient to nucleate microcracks during cooling and thus causing no degradation to the mechanical strength. The density of microcracks increases drastically with grain size once the critical value is exceeded. A closely related mechanism is believed to be also responsible for the observed grain size dependence on the thermal stability in AT.

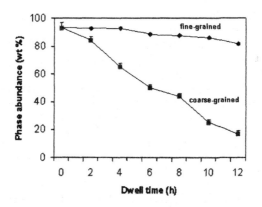

Fig. 7: The influence of grain size on the thermal stability of AT.

CONCLUSION

The thermal stability of Al_2TiO_5 in vacuum is strongly influenced by various factors which include temperature and dwell time, heating rates, phase purity, grain size, and additives. Possible causes of thermal instability and sources of stabilization have been discussed.

ACKNOWLEDGMENTS

The neutron diffraction data for this work were acquired at ISIS (Proposal RB12443), Rutherford-Appleton Laboratory, U.K. with financial support funded by the Commonwealth of Australia under the major National Research Facilities Program. We thank our colleagues (P. Manurung, Z. Oo and B. O'Connor) for providing the MRPD results in Figure 7.

REFERENCES

1. Austin, A.E. & Schwartz, C.M., "The crystal structure of aluminium titanate," *Acta Cryst.* **6**, 812-13 (1953).
2. H.A.J. Thomas, & R. Stevens, "Aluminium titanate – a literature review," *Br. Ceram Trans. J.* **88**, 144-90 (1989).
3. Y. Ohya, Z. Nakagawa & K. Hamano, "Grain boundary microcracking due to thermal expansion anisotropy in aluminium titanate ceramics," *J. Am. Ceram. Soc.* **70**, C184-86 (1987).
4. J.J. Cleveland & R.C. Bradt, "Grain size – microcracking relationships for pseudobrookite oxides," *J. Am. Ceram. Soc.* **61**, 478-81 (1978).
5. F.J. Parker & R.W. Rice, "Correlation between grain size and thermal expansion for aluminium titanate materials," *J. Am. Ceram. Soc.* **72**, 2364-66 (1989).
6. E. Kato, K. Daimon, & Y. Kobayashi, "Factors affecting decomposition temperature of β-Al_2TiO_5," *J. Am. Ceram. Soc.* **63**, 355-56 (1980).
7. R.W. Grimes, & J. Pilling, "Defect formation in β-Al_2TiO_5 and its influence on structure stability," *J. Mater. Sci.* **29**, 2245-49 (1994).
8. Dean, G.A., "An investigation of reaction-sintered aluminium titanate," *Ph.D Thesis*, University of WA. 2000.

9. P. Manurung, "Microstructural design and characterisation of alumina/ aluminium titanate composites." *Ph.D Thesis,* Curtin University, 2001.

10. I.M. Low, P. Manurung, R.I. Smith & D. Lawrence, "A novel processing method for the microstructural design of functionally-graded ceramic composites." *Key Engineering Materials* **224-226**, 465-70 (2002).

11. S. Hull, R.I. Smith, W. David, A. Hannon, J. Mayers, & R. Cywinski, "The POLARIS powder diffractometer at ISIS," *Physica B*, Vol.180-181, 1992, pp. 1000-1002.

12. A.C. Larson & R.B. Von Dreele, "General Structure Analysis System GSAS," *Report LAUR 86-748*, Los Alamos National Laboratory (1984).

13. A. Tsetsekou, "A comparison study of tialite ceramics doped with various oxide materials and tialite-mullite composites: microstructural, thermal and mechanical properties," *J. Europ. Ceram. Soc.* **25**, 335-48 (2005).

14. V. Buscaglia, F. Caracciolo, M. Leoni, P. Nanni, M. Viviani & J. Lemaitre, "Synthesis, sintering and expansion of $Al_{0.8}Mg_{0.6}Ti_{1.6}O_5$: a low thermal expansion material resistant to thermal decomposition," *J. Mater. Sci.* **32**, 6525-31 (1997).

15. V. Buscaglia, F. Caracciolo, M. Leoni, P. Nanni, M. Viviani & J. Lemaitre, "Micriostructure and thermal expansion of Al_2TiO_5-$Al_{0.8}Mg_{0.6}Ti_{1.6}O_5$ solid solutions obtained by reaction sintering," *J. Europ. Ceram. Soc.* **22**, 1811-22 (2002).

16. Y.X. Huang, A.M.R. Senos & J.L. Baptista, " Thermal and mechanical properties of aluminium titanate – mullite composites," *J. Mater. Res.* **15**, 357-63 (2000).

17. Y.X. Huang, A.M.R. Senos & J.L. Baptista, "Thermal and mechanical properties of aluminium titanate – mullite composites," *J. Europ. Ceram. Soc.* **17**, 1239-46 (1997).

18. J.J. Melendez-Martinez, M. Jimenez-Melendo, A. Dominguez-Rodriguez & G. Wotting, "High temperature mechanical behaviour of aluminium titanate – mullite composites," *J. Europ. Ceram. Soc.* **21**, 63-70 (2001).

19. U.S. Bjorkert, R. Mayappan, D. Holland & M.H. Lewis, "Phase development in La_2O_3 doped Al_2TiO_5 ceramic membranes," *J. Europ. Ceram. Soc.* **19**, 1847-57 (1999).

20. M. Takahashi, M. Fukuda, H. Fukuda & T. Yoko, "Preparation, structure and properties of thermally and mechanically-improved aluminium titanate ceramics doped with alkali feldspar," *J. Am. Ceram. Soc.* **85**, 3025-30 (2002).

21. H.A.J. Thomas, R. Stevens & E. Gilbert, "Effect of zirconia additions on the reaction sintering of aluminium titanate," *J. Mater. Sci.* **26**, 3613-16 (1991).

22. A. Duran, H. Wohlfromm, & P. Pena, "Study of the behaviour of Al_2TiO_5 materials in reducing atmosphere by spectroscopic techniques," *J. Europ. Ceram. Soc.*, **13**, 73-80 (1994).

23. K. Hamano,Y. Ohya & Z. Nakagawa, "Crack propagation resistance of aluminium titanate ceramics," In Int. Journal of of High Tech. Ceram. Elsevier Science Publishers Ltd., UK., pp. 129-37 (1985).

24. Y. Ohya, Z. Nakagawa & K. Hamano, "Crack healing and bending strength of aluminium titanate ceramics at high temperature," *J. Am. Ceram. Soc.* **71**, C23-33 (1988).

25. Y. Ohya, & Z. Nakagawa, "Grain boundary microcracking due to thermal expansion anisotropy in aluminium titanate ceramics at high temperature," *J. Am. Ceram. Soc.* **70**, C184-86 (1987).

CHARACTERIZING THE CHEMICAL STABILITY OF HIGH TEMPERATURE
MATERIALS FOR APPLICATION IN EXTREME ENVIRONMENTS

Elizabeth Opila
NASA Glenn Research Center
21000 Brookpark Rd.
Cleveland, OH 44135

ABSTRACT

The chemical stability of high temperature materials must be known for use in the extreme environments of combustion applications. The characterization techniques available at NASA Glenn Research Center vary from fundamental thermodynamic property determination to material durability testing in actual engine environments. In this paper some of the unique techniques and facilities available at NASA Glenn will be reviewed. Multiple cell Knudsen effusion mass spectrometry is used to determine thermodynamic data by sampling gas species formed by reaction or equilibration in a Knudsen cell held in a vacuum. The transpiration technique can also be used to determine thermodynamic data of volatile species but at atmospheric pressures. Thermodynamic data in the Si-O-H(g) system were determined with this technique. Free Jet Sampling Mass Spectrometry can be used to study gas-solid interactions at a pressure of one atmosphere. Volatile $Si(OH)_4(g)$ was identified by this mass spectrometry technique. A High Pressure Burner Rig is used to expose high temperature materials in hydrocarbon-fueled combustion environments. Silicon carbide (SiC) volatility rates were measured in the burner rig as a function of total pressure, gas velocity and temperature. Finally, the Research Combustion Lab Rocket Test Cell is used to expose high temperature materials in hydrogen/oxygen rocket engine environments to assess material durability. SiC recession due to rocket engine exposures was measured as a function of oxidant/fuel ratio, temperature, and total pressure. The emphasis of the discussion for all techniques will be placed on experimental factors that must be controlled for accurate acquisition of results and reliable prediction of high temperature material chemical stability.

INTRODUCTION

The aim of this paper is to describe capabilities for determining materials chemical stability at high temperatures. In general, experimental equipment which is unique to NASA Glenn Research Center is described[1]. These facilities include equipment for fundamental thermodynamic property determination as well as those which allow testing in complex combustion environments. Midway between fundamental property determination and actual combustion environments, a varied array of test rigs are also used to determine degradation mechanisms for materials at high temperatures. These include vacuum furnaces, box furnaces, controlled atmosphere furnaces, thermogravimetric analysis furnaces, furnaces for cyclic oxidation in air and controlled atmospheres, ultra high temperature furnaces for use in air environments (2300K capability) and Mach 0.3 burner rigs which operate at one atmosphere. While these capabilities are not discussed within this report, they are an important bridge between the fundamental and application-oriented equipment discussed here.

Experimental techniques rather than the results are featured in this paper. The information that can be obtained by each method, and the range of achievable temperatures, pressures, gas velocities, and reactive environments are described. Limitations and lessons learned for each technique are also discussed. Examples of results are presented. Many examples describe the Si-O-H system, since a large body of work has been conducted in this area at the NASA Glenn Research Center in the effort to develop silicon-based ceramics for propulsion applications and water vapor-containing combustion environments.

KNUDSEN EFFUSION MASS SPECTROMETRY

Knudsen Effusion Mass Spectromety (KEMS) is a powerful technique used to identify vapor species and their relative pressures[2,3]. A condensed sample is placed in a Knudsen cell (inner dimensions ~10 mm dia., 7 mm height) and uniformly heated until equilibrium between the vapor and the condensed phase is achieved. The vapor is continuously sampled by effusion through a small orifice (~1.5 mm dia., 4 mm height) in the cell which is held in a vacuum. A molecular beam is formed from the effusion vapor, ionized and directed into a magnetic sector mass spectrometer for identification and pressure measurement. A schematic diagram of this process is shown in Figure 1. The vapor species is identified by the mass-to- charge ratio (m/e). The species pressure is directly proportional to the intensity of the signal obtained. Measurements of the vapor species pressure as a function of temperature allow a number of thermodynamic quantities to be determined. The enthalpy of sublimation of the condensed phase or the enthalpy of vapor species reaction can be determined from the slope of a plot of log measured pressure versus inverse temperature. Component activities can be determined by comparing the relative vapor pressure of a characteristic species in equilibrium with a solution phase to that of a pure component. The temperature dependence of measured activities enables the partial enthalpies and entropies of mixing to be determined. In addition, plateaus in vapor pressure as a function of temperature allow the accurate determination of phase transformation temperatures.

The temperature range of measurements is about 1000 to >2000K. The lower limit is determined by the ability to make pyrometer measurements as well as thermal gradients in the furnace. The upper limit is determined by the temperature limit of the furnace components. The range of pressures that can be measured by this technique is between 10^{-12} and 10^{-4} atm. The lower limit is established by the sensitivity of the mass spectrometer while the upper limit is determined by the transition from molecular to hydrodynamic flow. Molecular flow is desired for accurate measurements and occurs when molecule – molecule collisions are negligible compared to molecule – cell wall collisions.

The primary concern for accurate thermodynamic measurements is the accurate measurement of temperature. This problem is overcome in a multiple cell KEMS in which the melting point of gold, or another primary temperature standard, is measured in one cell as part of the same experiment in which the vapor species of interest are studied in another cell. Container compatibility is a second important issue. In practice, the cell material becomes part of the system under study. A non-reactive cell material that can reach equilibrium with the sample quickly is desired. Typically metallic alloys are studied in refractory ceramic cells (Al_2O_3, ZrO_2, Y_2O_3) and ceramics are studied in refractory metal cells (W, Mo, Ir). Another issue is the repeatable formation of a well-defined molecular beam. Molecular beam formation is controlled by the effusion orifice, apertures between the effusion cell and the ion-source, as well as system alignment. Variation in instrument calibration, an issue for single cell instruments, is mostly overcome by the use of multiple cells.

As an example, the multiple cell KEMS was used to determine the effect of dissolved oxygen on the activities of aluminum and titanium in Ti₃Al(O). These results show that dissolved oxygen acts to increase the activity of aluminum (see Figure 2) and decrease the titanium activity. These effects are important to the understanding of how dissolved oxygen affects protective aluminum oxide formation on Ti₃Al at high temperatures.

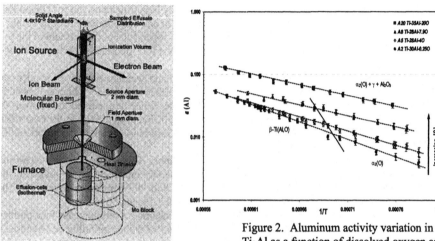

Figure 1. Schematic drawing of ion beam formation for analysis in the KEMS.

Figure 2. Aluminum activity variation in Ti₃Al as a function of dissolved oxygen and temperature.

THE TRANSPIRATION TECHNIQUE

The transpiration technique has been reviewed by Merten and Bell[4]. This is a relatively simple technique for determining thermodynamic data for gas phase species as well as indirectly identifying the gas species. In this method, shown schematically in Figure 3, a carrier and/or reactive gas mixture flows over a heated condensed phase sample at rates low enough for equilibrium to be maintained. Any product gases formed from the reaction between condensed phase and reactant gas are carried downstream to a cooler portion of the gas train where they condense. The condensate is collected and the amount of condensate is determined quantitatively by chemical analysis techniques, such as atomic emission spectroscopy. This technique is a direct measurement of condensable gas species pressure. Thermodynamic data for the reaction can be determined from the temperature dependence of the reaction. The species identity can be determined indirectly from the dependence of the product on the reactant gas pressure.

This technique is a useful analogue to KEMS since it allows the determination of thermodynamic data of gas species, but at relatively high pressures: 0.1 to 1 atm. The temperature range is limited at low temperatures by the formation of measurable product for the reaction of interest (about 20 μg) and at high temperatures by either the temperature capability of the furnace or by reactions between the condensed phase and the container. In our laboratory, transpiration experiments have been conducted at temperatures as high as 1725K for the $SiO_2(c)$-$H_2O(g)$ system. Gas flow rates must be low enough so that equilibrium in the transpiration cell is maintained. About 50 to 500 ccm are typical flow rates.

Experimental concerns include the following. First, equilibrium between the solid and gas must be maintained. This is attained by placing a constriction in both the gas inlet and outlet of the reaction cell. At very low flow rates, diffusion of the products out of the reaction cell can contribute significantly to transport through the cell. At very high flow rates, the gas stream may not reach the equilibrium product saturation. In the course of any transpiration study, it should be demonstrated that experiments were conducted in a flow regime where the pressure of the product species was independent of flow rate, and the product flux increases linearly with flow rate.

Another area of concern is the complete collection and accurate analysis of condensate. Complete collection of condensates can be difficult for oxides such as Al_2O_3 which are insoluble in acids. In such cases, techniques such as pyrosulfate fusion are needed to recover the condensate[5,6]. Collection of liquid condensates can also be difficult, especially in the case where water vapor, a reactant gas, is condensing also. Finally, care must be taken that standards for spectroscopic analysis are accurate.

Accurate temperature measurement is of utmost importance when measuring thermodynamic data. In this case, a thermocouple is typically immersed in the middle of the condensed phase sample. Reactions of the thermocouple with the sample are a concern. This concern is typically addressed by replacing the thermocouple at regular intervals.

Undesired reactions between the sample and the transpiration cell may also be a problem for obtaining accurate thermodynamic data. Careful choice of the transpiration cell material is required, as previously discussed for Knudsen cells.

In our laboratory, we have studied the Si-O-H system extensively[7]. It is now accepted that the following reaction occurs in water vapor at pressures near one atmosphere[7,8]:

$$SiO_2(c) + 2H_2O(g) = Si(OH)_4(g) \tag{1}$$

Examination of this reaction shows that the formation of $Si(OH)_4$ should have a square dependence on the water vapor partial pressure. The pressure dependence of volatile species in the Si-O-H system is shown in Figure 4. At temperatures of 1473K and below, Reaction (1) is found to be occur. At higher temperatures, the power law exponent for the water vapor pressure dependence is lower than two, suggesting that other Si-O-H species such as $SiO(OH)(g)$ may be important. Temperature dependent results for Si-O-H volatility have also been determined. The slope of a plot of $log(P_{Si(OH)4})$ versus inverse temperature can be used to calculate the second law heat of reaction. Third law heats of reaction can be determined from each pressure-temperature data point when the Gibbs free energy function for the vapor molecule is known or can be estimated based on its structure.

The transpiration method is a conceptually simple technique for determining thermodynamic data and vapor species identity at higher gas pressures and with oxidizing gases where KEMS is not possible.

314

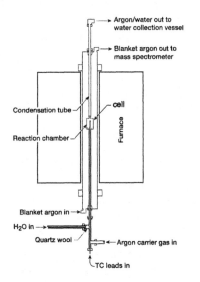

Figure 3. Schematic drawing of the transpiration furnace.

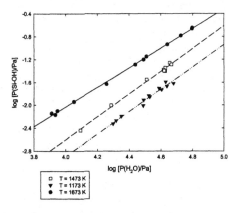

Figure 4. Water vapor partial pressure dependence of volatile Si-O-H species formation: Slopes equal (1.95 ± 0.01), (1.99 ± 0.06), and (1.71 ± 0.02) at 1173, 1473, and 1673K respectively. The theoretical slope for $Si(OH)_4$ formation is 2.

FREE JET SAMPLING MASS SPECTROMETRY

Volatile species formed at one atmosphere can be directly identified using the specialized Free Jet Sampling Mass Spectrometer (FJSMS) at NASA Glenn Research Center. The mass spectrometer has been described in detail[9] and is shown schematically in Figure 5. This instrument is a nice complement to the transpiration method, since it can sample the gas phase at the same conditions under which the transpiration experiments are conducted. Both simple vaporization reactions and reactions which form gaseous products can be studied. This system is composed of a high temperature furnace containing a porous solid or powder sample and a flowing reactant/carrier gas mixture at one atmosphere total pressure that is positioned directly beneath a platinum-rhodium sampling cone. The volatile species of interest, entrained in the gas mixture at one atmosphere pressure, enters an 0.2 mm diameter hole in the sampling cone. A free jet expansion of the gas stream occurs as it flows through a series of differentially pumped chambers. The expansion leads to a well defined molecular beam with essentially collisionless flow. This sampling system preserves the chemical and dynamic integrity of the volatile species even if they are condensable or reactive. The molecular beam is then ionized and directed into a quadrupole mass spectrometer for identification by mass-to-charge (m/e) ratio. Because of the nature of the quadrupole mass spectrometer, quantitative data can not be obtained, precluding thermodynamic data determination. However, important qualitative trends in vapor pressures can be observed.

The temperature range of measurement is limited at low temperatures by the formation of measurable amounts of product for the reaction of interest and the quadrupole sensitivity. By vaporizing NaCl and comparing to known thermodynamic data[10], the lower detection limit for vapor pressure of a condensable species in this instrument has been estimated to be about 1×10^{-6} atm. The measurements are limited at high temperatures by the temperature capability of the furnace which is about 1400°C. Carrier gas flow rates must be high enough so that the condensable species is carried to the sampling cone, but not so high as to cause the cone to plug with condensate. A typical flow rate is about 200 ccm.

Experimental concerns include the following. Typically, sample temperature is measured by immersing a thermocouple in the sample, or by placing a thermocouple in direct contact with the sample container. Sample temperature is not easily controlled in experiments with the FJSMS because variations in gas flow rate and carrier/reactive gas composition cause significant changes in both sample temperature and the baseline in the mass spectra. Because condensable vapor species are often of interest, alignment of the cone, collimators, and mass spectrometer are critical. Permanent gases are relatively easily detected with small amounts of misalignment, but condensable species are not. The optimization of the quadrupole sensitivity varies with m/e. For a given tuning of the instrument, mass discrimination of the larger masses occurs relative to the smaller masses. The instrument should therefore be tuned at the approximate m/e of the unknown volatile species using a known vapor species. The inert gases helium, neon, argon, krypton, and xenon are useful standards which cover a range of m/e.

For example[11], Figure 6 shows results for mass spectra of SiO_2 in dry oxygen compared to results obtained in water vapor plus oxygen as per Reaction 1. In the presence of water vapor, the peak at m/e=79 is attributed to $Si(OH)_3^+(g)$, a fragment of $Si(OH)_4(g)$. The peak at m/e=96 is attributed to $Si(OH)_4^+(g)$. The loss of an OH group during the ionization process is typical. Note that krypton found naturally in small amounts in the oxygen carrier gas serves as a good standard for m/e in this mass region. The distinctive isotopic distribution of krypton provides a fingerprint for accurate identification of m/e.

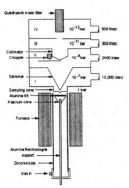

Figure 5. Schematic drawing of the FJSMS.

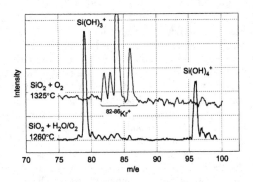

Figure 6. Mass spectra for vapor species observed from SiO_2 plus dry oxygen compared to SiO_2 plus wet oxygen.

316

HIGH PRESSURE BURNER RIG

The High Pressure Burner Rig (HPBR) at the NASA Glenn Research Center was designed specifically to test materials in combustion environments that simulate a gas turbine engine[12]. Water vapor makes up about 10% of the combustion products when a hydrocarbon fuel is burned[13]. Formation of volatile hydroxides, such as $Si(OH)_4$, is a great concern in combustion environments, especially at high pressures. The HPBR allows testing of materials at high pressures and moderately high gas velocities where volatility is more easily observed. A schematic diagram of the rig is shown in Figure 7. Preheated air is mixed with a hydrocarbon fuel. Combustion is initiated by an igniter. The combustion products flow downstream to a test section, over the material samples to be tested, and then are exhausted. Material durability is monitored by periodic weight change and material recession measurements.

Fuel to air ratios (F/A) between 0.02 and 0.14 are attainable. This corresponds to equivalence ratios (ϕ) of 0.3 to 0.9 for fuel-lean conditions and 1.5 to 2.0 for fuel-rich conditions. These in turn correspond to theoretical gas temperatures between 800 and 2100°C. The stoichiometric condition, $\phi=1$, is avoided since the rig components can not withstand the high gas temperatures. Test section pressures between 4 and 15 atm are typical, although 25 atm conditions were achieved with some damage to the combustor liner. Gas velocities vary inversely with the test section pressure and range from about 10 to 30 m/s. Typical exposure times are for 50 to 100h. Sample configurations vary from uncooled coupons of 1.3 x 8 x 0.3 cm dimensions to cooled subcomponent airfoil designs.

Empirical measurements of recession have been correlated to fundamental oxidation/volatilization mechanisms[14,15]. The biggest issue for quantifying these mechanisms is determination of accurate sample temperatures. The sample temperature is measured using a two-color optical pyrometer in fuel-lean conditions. The gas temperature is also monitored with a thermocouple placed behind the sample position. Concerns with the optical pyrometer measurements include changes in sample emissivity as oxidation of the sample surface occurs[16] and flame luminosity under fuel-rich conditions. A second issue is sample contamination that may affect the oxidation or volatilization characteristics of the sample. Reactions with sample grips as well as contamination from the burner environment, such as hardware cooling water leaks, must be avoided.

Silicon carbide (SiC) and silicon nitride (Si_3N_4) recession rates due to exposure in the HPBR have been measured as a function of sample temperature, gas pressure and gas velocity. The empirical results[14] were found to give excellent agreement with predictions based on a mechanism in which recession is controlled by the rate of volatile $Si(OH)_4$ transport outward from the sample surface through a laminar gas boundary layer[15]. These results are shown in Figure 8. These results demonstrate that material exposures in a high temperature combustion environment as complex as the HPBR can be used to gain both quantitative degradation rates as well as information about chemical mechanisms of degradation.

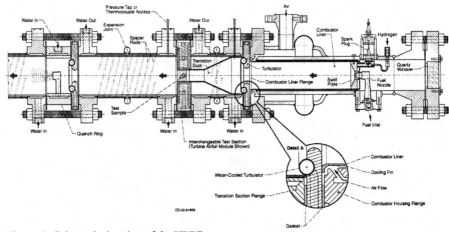

Figure 7. Schematic drawing of the HPBR.

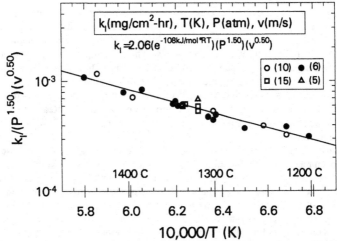

Figure 8. Temperature (T), pressure (P) and velocity (v) dependence of normalized linear recession rates (k_l) for SiC exposed in fuel-lean conditions in the HPBR. The legend specifies total pressure values in atm at which the data were obtained.

ROCKET COMBUSTION LAB, HYDROGEN/OXYGEN ROCKET ENGINE (CELL 22)

A hydrogen/oxygen rocket engine test cell (Cell 22) was developed at NASA Glenn Research Center in the 1980's to test the durability of thermal barrier coatings in a high heat flux environment[17,18]. Since that time, Cell 22 has been used in various programs to screen materials for a number of properties and applications including thermal shock resistance, ablative material performance, nonablative nozzle throat performance, and design viability for cooled ceramic matrix composite components[19,20]. Durability is generally assessed based on visual observations of sample integrity after test. High temperature materials can be tested in two different configurations, shown schematically in Figure 9. Many features are common to both configurations. Gaseous hydrogen and oxygen are supplied to a water cooled injector. The oxygen to fuel ratio (O/F) is controlled by the relative amounts of each gas supplied to the injector. Combustion is initiated with an igniter. The combustion products flow downstream through water cooled hardware. In the first test configuration, the sample to be tested, typically in coupon form, is placed in the pressure chamber upstream of the nozzle. Here pressures can vary from about 6 atm to 60 atm. The gas velocities are relatively low, on the order of 200 m/sec. Gas temperatures depend on the O/F. The stoichiometric condition for the hydrogen/oxygen system is O/F=8 corresponding to a gas temperatures in excess of 3000°C. Fuel-rich conditions with an O/F of about 2.0, corresponding to a gas temperature of 1800°C, are more typically used. In general, fuel rich conditions are used, although some testing has been conducted in the oxygen-rich regime. In the second configuration for material testing, the sample to be exposed is placed downstream of the nozzle. The sample configuration can be of subcomponent complexity since it is outside of the engine. Here the pressure is 1 atm, but gas velocities up to mach 2 (660 m/s) have been achieved. Gas temperature is again dependent on O/F. O/Fs between 1.3 and 7.0 have been used, corresponding to gas temperatures between about 1300 and 3000°C.

The increased complexity of the test engine makes accurate determination of the material exposure parameters difficult. Pressure in the test chamber is measured by strain gauge transducers at a pressure port in the combustion chamber. Gas flow rates are controlled by choosing the appropriately sized sonic venturi that should be operated with the appropriate pressure drop across the venturi. Temperature, again, is very difficult to measure. Thin film thermocouples have been applied to coupons tested within the pressure chamber. Unfortunately, these thin films are not robust enough to withstand the high gas velocities and temperatures typically found in test. Only limited success at the most benign conditions has been attained measuring material temperature by this method. Sheathed thermocouples can also be inserted into the combustion flow allowing temperature profiling of the flowing gas across the pressure chamber. Typically the temperature across the chamber is found to be very nonuniform and depends strongly on the choice of injector. Single element injectors have been found to result in the smallest temperature gradients, about 25°C in 1 cm as measured at the midline of the chamber. Average gas temperatures can also be estimated by monitoring the heat loss to the water cooled chamber walls. There is no view port in the water-cooled pressure chamber precluding any optical pyrometry. Temperature measurements of components downstream of the nozzle have been attempted with optical equipment, such as pyrometers or IR cameras. Difficulties with flame luminosity are severe, so the accuracy of these measurements is uncertain. Generally, the most reliable measurements have been obtained using thermocouples in contact with the cool side of the test article or in a hole just below the hot side surface.

Surface temperatures can then be calculated for materials of well-defined and well-known thermal conductivity.

Based on the success of quantitatively determining SiC and Si_3N_4 recession rates and defining the chemical degradation mechanism of these materials in the HPBR, the same type of test program was conducted in Cell 22 for carbon fiber reinforced SiC composite coupons (1.3 cm x 10 cm x 0.3 cm) as part of a larger program on ceramic matrix composite life prediction[21]. In this case, the recession/volatilization mechanism could differ from that observed in the HPBR because of the fuel-rich conditions, higher pressures, and higher gas temperatures. The injector, combustion chamber, sample holder, and "nozzle" were all configured to allow high pressure, long duration (10 min.) tests with minimal temperature gradient across the sample length. Tests were conducted at O/Fs from 1.4 to 2.3 (theoretical gas temperatures between 1350 and 2000°C), chamber pressures of about 7, 35, and 60 atm. Exposure times varied from 2 minutes to 1 hour per condition. One coupon was tested at each condition. Interrupted weight change and recession measurements were acquired. The recession results are shown in Figure 10. Correlation of recession to chemical mechanistic models is in process.

Problems that occurred during testing and the associated remedies are as follows. Due to the low mass flow testing used to enable long test time, some venturis were operated outside their calibration limits and required additional calibration. The sample surface temperature was unknown for the most part due to failure of thin film thermocouples. Surface temperatures were estimated based on sample surface temperature under more benign conditions, thermocouple data obtained by probing the gas stream, and estimations based on loss of heat to the hardware cooling water. Samples were initially contaminated from city water used to build up back pressure in the engine. The water source was switched to de-ionized water ending this problem. Finally, at the highest pressure (60 atm) samples were contaminated by volatile species originating from degradation of other metal engine components.

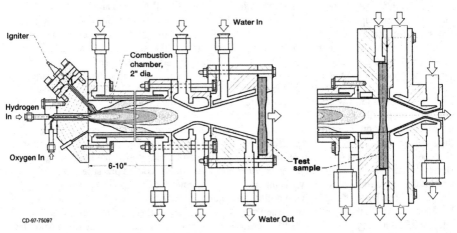

Figure 9. Schematic drawing of material test configurations in a hydrogen/oxygen rocket engine test (Cell 22).

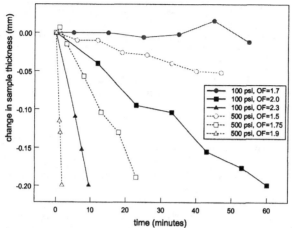

Figure 10. SiC recession measurements as a function of chamber pressure and OF exposures in a hydrogen/oxygen rocket engine test (Cell 22).

CONCLUDING REMARKS

In summary, several statements can be made about the determination of material chemical stability in severe environments. First, accurate temperature measurements are challenging, but critical, for accurate prediction of material stability. Second, experimental determination of thermodynamic data is critical to prediction of material stability in extreme environments. More work needs to be done in this area. Finally, identification of chemical degradation mechanisms and quantitative rate determination can be made, even in complex test environments, if experimental parameters are well controlled and accurately measured.

ACKNOWLEDGMENTS

The author would like to acknowledge Evan Copland (Case Western Reserve University, Cleveland, OH), Nathan Jacobson (NASA-GRC), Dennis Fox (NASA-GRC), Dwight Myers (East Central University, Ada, OK), Craig Robinson (QSS, Inc., Brookpark, OH), Martha Jaskowiak (NASA-GRC) and Andy Eckel (NASA-GRC) for their contributions to this work.

REFERENCES

[1] http://www.grc.nasa.gov/WWW/EDB/index.htm

[2] R.T. Grimley, "Mass Spectrometry" pp. 195-243 in *The Characterization of High Temperature Vapors*, ed. J.L. Margrave, John Wiley & Sons, NY, 1967.

[3] E.H. Copland and N.S. Jacboson, "Thermodynamic Activity Measurements with Knudsen Cell Mass Spectrometry," *Interface* **10** [2] 28-31 (2001).

[4] U. Merten and W.E. Bell, "The Transpiration Method," pp. 91-114 in *The Characterization of High Temperature Vapors*, ed. J.L. Margrave, John Wiley & Sons, NY, 1967.

[5] Z. Sulcek and P. Povondra, Decomposition by Fusion, pp. 167-242 in *Methods of Decomposition in Inorganic Analysis*, CRC Press, Boca Raton, FL , 1989.

[6]L. Brewer and A.W. Searcy, "The Gaseous Species of the Al-Al$_2$O$_3$ System," *J. Am. Chem. Soc.* **73**, 5308 (1951).

[7]N.S. Jacobson, E.J. Opila, D. Myers, and E. Copland, Thermodynamics of Gas Phase Species in the Si-O-H System," accepted for publication in *Journal of Chemical Thermodynamics*.

[8]A. Hashimoto, "The Effect of H$_2$O Gas on Volatilities of Planet-Forming Major Elements: 1. Experimental Determination of Thermodynamic Properties of Ca-, Al-, and Si-Hydroxide Gas Molecules and its Application to the Solar Nebula," *Geochim. Cosmochim. Acta* **56**, 511-32 (1992).

[9]C.A. Stearns, F.J. Kohl, G.C. Fryburg, and R.A. Miller, "A High Pressure Modulated Molecular Beam Mass Spectrometric Sampling System," NASA TM 73720, (1977).

[10] M.W. Chase, Jr., C.A. Davies, J.R. Downey, Jr., D.J. Frurip, R.A. McDonald, and A.N. Syverud, Editors, *JANAF Thermochemical Tables*, 3rd ed., American Chemical Society and American Physical Society, New York, 1985.

[11] E.J. Opila, D.S. Fox, N.S. Jacobson, "Mass Spectrometric Identification of Si-O-H(g) Species from the Reaction of Silica with Water Vapor at Atmospheric Pressure," *J. Am. Ceram. Soc.*, **80** [4] 1009-12 (1997).

[12]R.C. Robinson, "SiC Recession due to SiO$_2$ Scale Volatility Under Combustor Conditions," NASA CR 202331, 1997.

[13]N.S. Jacobson, "Corrosion of Silicon-Based Ceramics in Combustion Environments," *J. Am. Ceram. Soc.*, **76** [1] 3-28 (1993).

[14]R.C. Robinson and J.L. Smialek, "SiC Recession due to SiO$_2$ Scale Volatility under Combustion Conditions. Part I: Experimental Results and Empirical Model," *J. Am. Ceram. Soc.* **82** [7] 1817-25 (1999).

[15] E.J. Opila, J.L. Smialek, R.C. Robinson, D.S. Fox, and N.S. Jacobson, "SiC Recession due to SiO$_2$ Scale Volatility under Combustion Conditions. Part II: Thermodynamics and Gaseous Diffusion Model," *J. Am. Ceram. Soc.* **82** [7] 1826-34 (1999).

[16] E.J. Opila, R.C. Robinson, "The Oxidation Rate of SiC in High Pressure Water Vapor Environments," pp. 398-406 in *High Temperature Corrosion and Materials Chemistry*, eds. M. McNallan, E. Opila, T. Maruyama, T. Narita, The Electrochemical Society, Pennington, NJ, 2000.

[17]J.A. Nesbitt and W.J. Brindley, "Heat Transfer to Throat Tubes in a Square-Chambered Rocket Engine at the NASA Lewis Research Center," NASA TM 102336 – Oct. 1989.

[18]J.A. Nesbitt, "Thermal Modelling of Various Thermal Barrier Coatings in a High Heat Flux Rocket Engine," NASA TM 102418 – Dec. 1989.

[19]M.J. Bur, "A Combustion Research Facility for Testing Advanced Materials for Space Applications," 41st Aerospace Sciences Meeting and Exhibit, Reno, Nevada, Jan. 6-9, 2003, AIAA-2003-282.

[20]K.W. Dickens, D.L. Linne, and N.J. Georgiadis, "Experiment and Modeling of a Rocket Engine Heat Flux Environment for Materials Testing," 41st Aerospace Sciences Meeting and Exhibit, Reno, Nevada, Jan. 6-9, 2003, AIAA-2003-0283.

[21]S.R. Levine, A.M. Calomino, J.R. Ellis, M.C. Halbig, S.K. Mital, P.L. Murthy, E.J. Opila, D.J. Thomas, L.U.J.T. Ogbuji, and M.J. Verrilli, "Ceramic Matrix Composites (CMC) Life Prediction Method Development," NASA TM−2000-210052, 2000.

EFFECT OF OXYGEN PARTIAL PRESSURE ON THE PHASE STABILITY OF Ti$_3$SiC$_2$

I.M Low, Z. Oo, B.H. O'Connor
[1]Materials Research Group, Department of Applied Physics, Curtin University of Technology, GPO Box U1987, WA 6845 Australia

K.E. Prince
ANSTO, PMB 1, Menai, NSW 2234 Australia

ABSTRACT

The thermal stability and topotactic transition of Ti$_3$SiC$_2$ in vacuum, argon and air have been investigated by synchrotron radiation diffraction (SRD), in-situ neutron diffraction (ND) and secondary ion mass spectroscopy (SIMS). In the presence of a low oxygen partial pressure such as in vacuum or argon, Ti$_3$SiC$_2$ undergoes a surface dissociation to form non-stoichiometric TiC and/or Ti$_5$Si$_3$C at 1200°C. In contrast, it forms non-stoichiometric surface oxides of rutile at 1000°C and cristobalite at 1300°C when exposed to an oxygen-rich environment. Near-surface depth profiling of vacuum-treated Ti$_3$SiC$_2$ by SIMS has revealed a distinct gradation in phase composition and confirmed the existence of Ti$_5$Si$_3$C in solid solution with oxygen.

INTRODUCTION

Titanium silicon carbide (Ti$_3$SiC$_2$) is a remarkable ternary compound that defies many of the expected properties of a ceramic.[1-5] It has better thermal and electrical conductivity than titanium metal, is resistant to thermal shock, and is relatively light. Its hardness is exceptionally low for a carbide, and like graphite, it is readily machinable. For this reason Ti$_3$SiC$_2$ is considered the silicon equivalent of graphite. Extensive hardness testing has shown the ability of the material to undergo an extensive plastic deformation without macroscopic fracture. However, unlike the traditional binary carbides (eg. WC, SiC and TiC) which are some of the hardest (>25 GPa), stiffest, and most refractory (>2000°C) materials known, the ternary carbide Ti$_3$SiC$_2$ is relatively soft, not wear resistant, and has lower thermal stability (<1700°C). In view of this, Barsoum and co-workers have successfully improved the surface hardness and oxidation resistance of Ti$_3$SiC$_2$ by using processes such as carburization and silicidation to form surface layers of TiC and SiC.[6] Using a new approach, we have recently shown that it is possible to form a thin surface layer of TiC through dissociation of Ti$_3$SiC$_2$ in the presence of vacuum at elevated temperatures.[7-9] However, the exact surface chemistry of dissociation to form TiC in vacuum remains unclear and it is also unknown whether a similar surface dissociation process will occur in an inert atmosphere such as argon.

Hitherto, mixed and confusing results have been reported for the oxidation resistance and behaviour of Ti$_3$SiC$_2$ in air.[10-14] For instance, the oxidation resistance of Ti$_3$SiC$_2$ was reported by Zhou et al.[10] and Tong et al.[11] to be excellent at temperatures below 1100°C due to the formation of a protective SiO$_2$ surface layer. However, oxidation of Ti$_3$SiC$_2$ was detected by Racault et al.[12] to commence as low as 400°C through the formation of an anatase-like TiO$_2$ film that eventually transformed to rutile at 1050°C. In addition, although the existence of the protective TiO$_2$ (rutile) has been confirmed by all the researchers, the presence of the protective SiO$_2$ film is much more elusive.[13] In a recent study, Li et al.[14] reported the oxidized layers to exhibit a duplex microstructure in the temperature range 1000-1500C with an outer layer of TiO$_2$ (rutile) and an inner layer consisting of SiO_2 and TiO_2. In a similar study, Barsoum *et al.*[15] also found the protective oxide scales that formed to be layered with the inner layer comprised of silica (formed at ~1200°C) and titania and the outer layer comprised of pure rutile (formed at ~900°C). The growth of these oxide layers is both temperature and time-dependent and was thought to occur by the outward diffusion of titanium and carbon and the inward diffusion of oxygen through surface pores or cracks. However, the nature and precise composition of the oxide layers formed during oxidation remain controversial, especially in relation to the presence of SiO_2 and the graded nature of the oxides formed.

In this paper, the in-situ thermal stability of Ti_3SiC_2 in vacuum, argon and air has been investigated by neutron diffraction and synchrotron radiation diffraction in the temperature range 20-1400°C. Rietveld analyses of the data show that the thermal stability of Ti_3SiC_2 is highly sensitive to the conditions of the ageing environment where it decomposes to form a surface layer of TiC at ~1200°C in argon and vacuum, but oxidizes in air to form rutile and cristobalite. The roles of the oxygen partial pressures on the surface chemistry of thermal dissociation and oxidation of Ti_3SiC_2 are discussed.

EXPERIMENTAL METHOD

Sample Preparation

Ti_3SiC_2 samples were fabricated by reaction-sintering and hot-isostatic-pressing of Ti, SiC and C powders. The powder compacts were initially prepared by mixing in the proper molar ratio, cold pressed, followed by reaction-sintered in a vacuum furnace (~10^{-5} torr) at 1500°C for 1 h, and finally hot-isostactically-pressed (HIPed) in argon at 1650°C for 2 h with a pressure of 150 MPa. Vacuum heat-treatment of HIPed samples was conducted in an Elatec™ vacuum furnace (~2×10^{-5} torr) at 1000 – 1500°C for 1 h. Samples for the ex-situ oxidation study were oxidized in an air-ventilated furnace at 1000 – 1500°C for 15 min.

Neutron Diffraction (ND)

Measurements of neutron diffraction of HIPed Ti_3SiC_2 in air and argon in the temperature range 25-1400°C were performed using the medium-resolution powder diffractometer (MRPD) at the Australian Nuclear Science and Technology Organization (ANSTO) with a wavelength of 1.665 Å. The relative abundances of the phases formed were computed using the Rietveld refinement method. The models of Kisi et al. (ICSD #86213) for Ti_3SiC_2, Christensen et al. (ICSD #1546) for TiC, Riley Nu model for Ti_5Si_3C, Maslen et al. (ICSD #73725) for α-Al_2O_3, and Shintani et al. (ICSD #64987) for rutile (TiO_2) were used for the Rietveld analysis.

Synchrotron Radiation Diffraction (SRD)

The phase composition of vacuum-treated and air-oxidized Ti_3SiC_2 in the temperature range 1000-1500°C was analysed at the Photon Factory using the BIGDIFF diffractometer. Imaging plates were used to record the patterns over 2θ of 10-90°. The diffractometer was operated in Debye-Scherrer mode under vacuum with wavelength of 0.8 Å and at an incidence angle of 3°.

Secondary Ion Mass Spectroscopy (SIMS)

SIMS is an analysis technique for gathering compositional information about the surface and near-surface regions of solid materials, and particularly suited to the measurement of concentration profiles. It employs a beam of primary ions to eject secondary ions from near-surface regions of the target. The secondary ions are collected and focussed through a mass spectrometer, and then directed into a detector. Since a sample will be gradually eroded with time, a depth profile can be obtained by recording the detector signal as a function of time. The depth-profiling of vacuum-treated Ti_3SiC_2 samples was conducted through the elemental imaging of titanium, carbon, silicon and oxygen on the near-surface. The enhanced elemental sensitivity of SIMS allows the detection of oxygen (if present) within the structure of the near-surface layer.

The facility used in this work is the Cameca ims-5f SIMS of the Australian Nuclear Science and Technology Organisation (ANSTO). Positive secondary ion depth profiles were taken using 150 nA primary Cs^+ ion current at 3 keV impact energy, rastered over a 250×250 μm^2 area. The ion count rates in all mass channels were normalised to Cs^+ secondary ion counts rate to minimise the effect of variations in the primary ion beam current.

RESULTS AND DISCUSSION
Thermal Stability of Ti_3SiC_2 at Elevated Temperature in Vacuum and Air

The phase relations of Ti_3SiC_2 samples vacuum-treated or oxidized at $1000°$, $1200°$, $1400°$ and $1500°C$ as revealed by SRD have been qualitatively analysed. The existence of various phases at ambient and elevated temperatures is shown in Table 1.

Table I: Relative peak intensity of phases present in Ti_3SiC_2 vacuum-treated or air-oxidized at elevated temperature as revealed by SRD.

Temp (°C)	Heat-Treatment	Ti_3SiC_2	TiC	Ti_5Si_3C	Cristobalite (SiO_2)	Rutile (TiO_2)
25	Nil	vs	w	-	-	-
1000	Vacuum	vs	w	-	-	-
1200	Vacuum	vs	m	-	-	-
1400	Vacuum	s	m	w	-	-
1500	Vacuum	w	s	m	-	-
1000	Oxidization	m	-	-	m	s
1200	Oxidization	t	-	-	-	vs
1400	Oxidization	-	-	-	-	vs
1500	Oxidization	-	-	-	-	vs

Legend: vs = very strong; s = strong; m = medium; w = weak; t = trace

There is clearly no apparent dissociation of Ti_3SiC_2 to form TiC at temperatures below 1200°C. It dissociates initially slowly at 1200-1300°C but the process becomes quite rapid from 1400 to 1500°C. Formation of Ti_5Si_3C is also observed at 1400°C. This transient phase is believed to convert eventually to the stable TiC at elevated temperature. Similarly, oxidation of Ti_3SiC_2 commences at ~1000°C to form oxide layers of rutile and cristobalite. The thickness of these oxide layers increased rapidly (>100μm) as the temperature approached 1400°C. Since the presence of cristobalite was not detected on the surface at 1200-1500°C, this suggests that the thick rutile layer formed on the outer surface with the cristobalite layer underneath. However, it is unclear whether the cristobalite crystallised in-situ or formed when the sample was cooled down. It is quite possible that amorphous SiO_2 initially formed at <1000°C but recrystallised to cristobalite at room temperature when the SRD pattern of the sample was measured. This agrees well with the observed microstructure (Fig. 1) and the literature, which states that rutile forms as an outer layer with the inner oxide layer consisting of both TiO_2 and SiO_2.[10-15]

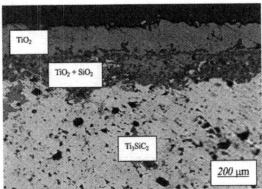

Fig. 1: Optical micrographs of Ti_3SiC_2 showing formation of oxide layers when oxidized at 1300°C for 15 min.

Depth-Profiling of Elemental Composition in Vacuum-Treated Ti_3SiC_2

As would be expected, the control sample showed no variations in the elemental compositions of Ti, Si and O with depth. In contrast, the sample vacuum-treated for 4 h showed an initial drop in the concentration of Si before leveling off after 1000 s. Its concentration then increased again after 5000 s (Fig. 2). This implies that the near-surface of the sample is covered mainly (~ 97 wt%) with a thin layer of TiC (~ 5 μm) to protect the Ti_3SiC_2 bulk. The graded nature of the Ti_3SiC_2-TiC interface has been verified.

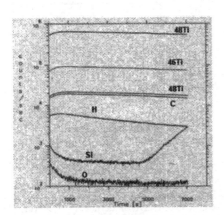

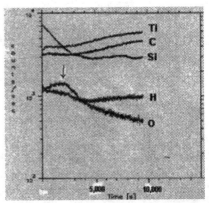

Fig-2: Elemental composition versus time/depth in Ti_3SiC_2 vacuum-treated at 1500°C for 4 hr. Note the rapid rise in Si content after 5000 s.

Fig-3: Elemental composition versus time/depth in Ti_3SiC_2 vacuum-treated at 1500°C for 8 hr. Note the large O peak (see arrow) at 2000 s.

More complex results are oberved for the sample vacuum-treated at 8 h (Fig. 3). High concentrations of Ti, Si, C and O are present at the near-surface (0-3000 s) which indicate that in addition to TiC, a second phase co-exists, ie. Ti_5Si_3C with dissolved O_2. This phase is believed to cause the discolouration of Ti_3SiC_2 following a prolonged vacuum treatment at 1500°C. The Ti_5Si_3C (with dissolved O_2) phase appears brownish and acts as an intermediate or precursor phase for the eventual formation of TiO_2 and SiO_2. This further suggests that Ti_5Si_3C can act as a 'sponge' to soak up oxygen and when a supersaturated state is reached as in during air-oxidation, the Ti_5Si_3C (with supersaturated O_2) will segregate out to form TiO_2-rutile and SiO_2-cristobalite.

Dynamic Phase Dissociation of Ti_3SiC_2 at Elevated Temperature in Argon

Figure 4 shows the temperature dependence of the relative phase abundances formed. From room temperature to 1000°C, the phase concentrations of Ti_3SiC_2 and TiC remain quite stable and constant. At 1000°C, Ti_3SiC_2 dissociates to form TiC. Below 1200°C, the thermal dissociation process is slow but the process becomes quite rapid from 1250°C to 1400°C. In addition, a small amount of Ti_5Si_3C is observed as a transient phase from 20°C to 1400°C. This phase is believed to form during the initial decomposition stage of Ti_3SiC_2 and eventually converts to the stable TiC at elevated temperature. The abundance of TiC increases with temperature whereas the reverse is true for Ti_5Si_3C and Ti_3SiC_2. These findings are consistent with the observations of Wu and Kisi.[16]

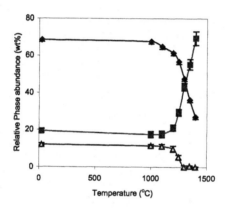

Fig. 4: Relative phase abundances of TiC (■), Ti_5Si_3C (Δ) and Ti_3SiC_2 (♦) present during the heat-treatment of Ti_3SiC_2 in argon from room temperature to 1400°C. Errors bars indicate two estimated standard deviations ±2σ.

From this study, the sintered Ti_3SiC_2 appeared to be thermally unstable in argon from 1000°C and above. The intermediate phase Ti_5Si_3C slowly disappeared at around 1400°C but significant amounts of Ti_3SiC_2 and TiC remained. In the presence of a very low oxygen partial pressure, it was previously postulated that the surface of Ti_3SiC_2 may undergo a high temperature topotactic thermal dissociation process to form TiC as follows:[8]

$$Ti_3SiC_2 \xrightarrow{O_2} 2TiC_{(s)} + TiO_{(g)} + SiO_{(g)} \qquad (1)$$

327

However, this chemical reaction is inconsistent with the above observations and it does not explain the existence of Ti_5Si_3C. According to Wu & Kisi,[16] the initial dissociation of Ti_3SiC_2 to Ti_5Si_3C is more favourable to TiC because silicon has a low diffusion rate in TiC-related systems. Thus the Ti_5Si_3C phase is more likely to nucleate on Ti_3SiC_2 because this mechanism would not contain any long range diffusion of Si atoms. In view of this, it is proposed that the pathways for the topotactic thermal dissociation of Ti_3SiC_2 in the presence of low oxygen partial pressure to form TiC and Ti_5Si_3C are as follows:

$$3Ti_3SiC_2 + O \rightarrow 4\ TiC_{(s)} + Ti_5Si_3C_{(s)} + CO_{(g)} \hspace{2cm} (2)$$

$$Ti_5Si_3C + 7O \rightarrow TiC_{(s)} + 3\ SiO_{(g)} + 4\ TiO_{(g)} \hspace{2cm} (3)$$

The proposed volatility of gaseous of CO, SiO and TiO released is consistent with the observed results of reduced Ti_3SiC_2 mass following the heat treatment in vacuum or argon.[8]

It should be pointed out that carbon non-stoichiometry exits in both TiC_x and $Ti_5Si_3C_x$ and its defect concentration is sensitive to the variation in the oxygen partial pressure and temperature.[16-18] In the case of TiC_x. the value of x has the astonishing range from 0.5 to 0.97, without change in crystal structure. The presence of Ti ions in Ti_3SiC_2 having an unstable oxidation state is believed to enhance carbon content variations with a concomitant susceptibility to surface dissociation. Changes in the oxidation state of titanium from 4+ to 3+ in low oxygen partial pressure give rise to carbon vacancy formation. The change in defect concentrations in TiC_x and $Ti_5Si_3C_x$ with atmosphere may be attributed to the variation in charge compensation mechanisms to form carbon vacancies and other defects. Indeed, non-stoichiometry in TiC has been ascribed to the existence of defects such as stacking faults and Si solid solution.[18] A thorough study of the defect chemistry is essential to gain a complete understanding of the non-stoichiometry and defects associated with the topotactic thermal dissociation of Ti_3SiC_2 in various oxygen partial pressures.

Dynamic Phase Transformation of Ti_3SiC_2 During Oxidation in Air

The initial results of phase evolution of Ti_3SiC_2 during oxidation and relative abundance of oxidised phases formed at various temperatures as revealed by in-situ neutron diffraction is shown in Figure 5. Before oxidation, the phases present in sample were mainly Ti_3SiC_2 with TiC and Ti_5Si_3C as minor phases. At ~750°C, a portion of Ti_3SiC_2 commenced to oxidise to form rutile although the presence of cristobalite was not evident. However, a glassy phase is believed to have formed at <1000°C, as evidenced by the appearance of characteristic "broaden" peaks at $2\theta \approx 35\text{-}40°$, which may indicate the existence of an amorphous phase which subsequently crystallizes to cristobalite at an elevated temperature. The intensity of rutile peaks increased rapidly as the temperature increased beyond 1100°C with a commensurate decrease in peak intensity of Ti_3SiC_2. Interestingly, the abundance of TiC increased as the temperature increases and reached a maximum of ~13wt% at 1000C before it decreased again as the temperature approached 1350°C. In contrast, the abundance of Ti_5Si_3C remained fairly constant below 1200°C before it increased to ~14wt% at 1350°C.

Two possible reactions involving the transitions of Ti_3SiC_2 and TiC during oxidation in ambient oxygen partial pressure may account for the observed phenomenon:

328

$$Ti_3SiC_2 + 2O_2 \rightarrow 2TiC_{(s)} + TiO_{2(s)} + SiO_{2(l)} \tag{4}$$

$$TiC + 3Ti_3SiC_2 + 11O_2 \rightarrow Ti_5Si_3C_{(s)} + 5TiO_{2(s)} + 6CO_{2(g)} \tag{5}$$

Equation (4) would explain the apparent increase in TiC content due to oxidation of Ti_3SiC_2 and the concomitant formation of a glassy phase that will eventually devitrify to cristobalite at the elevated temperature. Similarly, Equation (5) would explain the formation of Ti_5Si_3C but the reduction in TiC. However, it should be pointed out that Ti_5Si_3C has a defect structure that dissolves oxygen readily to form a solid solution as elucidated by the SIMS results mentioned above. Non-stoichiometry is also expected for the oxides formed (ie. TiO_2 & SiO_2) in the atmosphere of high oxygen partial pressure as evidenced by the colour changes at various temperatures.

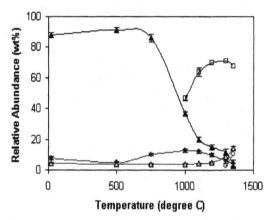

Fig. 5: Relative abundance of phases formed during the in-situ oxidation of Ti_3SiC_2 in air. Errors bars indicate two estimated standard deviations $\pm 2\sigma$. (Legend: ▲ = Ti_3SiC_2; ● = TiC; △ = Ti_5Si_3C; □ = rutile; ○ = cristobalite)

In contrast to the SRD results above, crystallisation of cristobalite was only observed at 1300°C instead of 1000°C. This suggests that amorphous SiO_2 formed initially at 1000°C but remained glassy before it crystallised to cristobalite at ~1300°C. However, this glassy phase can readily devitrify to cristobalite when it is cooled from 1000°C to room temperature as revealed by SRD. The existence of such glassy phase has also been alluded to by Barsoum et al.[15] during SEM observation and confirmed by TEM observation.[19]

CONCLUSION

The phase stability and topotactic transition of Ti_3SiC_2 at elevated temperature are strongly dependent on the oxygen partial pressures. When the oxygen partial pressure is low, it dissociates readily to non-stoichiometric TiC and Ti_5Si_3C. In contrast, it oxidizes readily to non-stoichiometric rutile and cristobalite in environment of high oxygen partial pressure. A solid solution is formed between Ti_5Si_3C and oxygen.

ACKNOWLEDGMENTS

IML is grateful to AINSE (99/030, 00/90P, 01/091, 02/075 & 04/188) to provide fundings for the HIPing of Ti_3SiC_2 samples and the collection of MRPD and SIMS, and to and the ASRP (00/01-AB-31 & 01/02-AB-36) for the collection of SRD data at the Photon Factory. We thank Drs. D. Perera, A. Studer & D. Goossens of ANSTO and Dr. J. Hester of ANBF for experimental assistance in HIPing and the collection of MRPD and SRD data respectively. We are also grateful to Prof. M. Barsoum of Drexel University for providing some samples for initial and comparative studies.

REFERENCES

1. M.W. Barsoum & T. El-Raghy: Room temperature ductile carbides. *Metall. Mater. Trans.*, **30A**, 363 (1999).
2. T. El-Raghy, A. Zavaliangos, M.W. Barsoum & S. kalidinidi, "Damage mechanisms around hardness indentations in Ti_3SiC_2," *J. Am. Ceram. Soc.*, **80**, 13 (1997).
3. I.M. Low, "Vickers contact damage of micro-layered Ti_3SiC_2," *J. Europ. Ceram. Soc.*, **18**, 709 (1998).
4. I.M. Low, S.K. Lee, M. Barsoum & B. Lawn, "Contact Hertzian response of Ti_3SiC_2 ceramics," *J. Am. Ceram. Soc.*, **81**, 225 (1998).
5. Y. Kuroda, I.M. Low, M.W. Barsoum & T. El-Raghy, "Indentation responses and damage in HIPed Ti_3SiC_2," *J. Aust. Ceram. Soc.*, **37**, 95 (2001).
6. T. El-Raghy, & M. Barsoum, "Diffusion kinetics of the carburization and silicidation of Ti_3SiC_2," *J. Appl. Phys.*, **83**, 112 (1998).
7. I.M. Low, P. Manurung, R.I. Smith & D. Lawrence, "A novel processing method for the microstructural design of functionally-graded ceramic composites," *Key Engineering Materials*, **224-226**, 465 (2002).
8. I.M. Low & Z. Oo, "Diffraction studies of a novel Ti_3SiC_2–TiC system with graded interfaces," *J. Aust. Ceram. Soc.*, **38**, 112 (2002).
9. I.M. Low, "Depth-profiling of phase composition in a novel Ti_3SiC_2–TiC system with graded interfaces," *Mater. Lett.*, **58**, 927 (2004).
10. Y.C. Zhou, Z.M. Sun and F.H. Sun, *Z. Metsllkd.*, **91**, 329 (2000).
11. X.H. Tong, T. Okano, T. Iseki and T. Yano, *J. Mater. Sci.*, **30**, 3087 (1995).
12. C. Racault, F. Langlais and R. Naslain, "Solid state synthesis and characterisation of the ternary phase Ti_3SiC_2," *J. Mater. Sci.*, **29**, 3384 (1994).
13. N.F. Gao, Y. Miyamoto and D. Zhang, "Dense Ti_3SiC_2 prepared by reactive HIP," *J. Mater. Sci.* **34**, 4385 (1999).
14. S. Li, L. Cheng and L. Zhang, "The morphology of oxides and oxidation behaviour of Ti_3SiC_2-based composite at high temperature," *Mater. Sci. & Eng.*, **A341**, 112 (2003).
15. M.W. Barsoum, T. El-Raghy and L. Ogbuji, "Oxidation of Ti_3SiC_2 in air," *J. Electrochem. Soc.*, **44**, 2508 (1997).
16. E. Wu, E., E.H. Kisi, "Powder Diffraction Study of Ti_3SiC_2 Synthesis," *J. Am. Ceram. Soc.*, **88**, 81 (2001).
17. W.S. Williams, "Physics of transition metal carbides," *Mater Sci Eng* **A105/106** 1-10 (1988).
18. M.W. Barsoum, "Comment on 'reaction layers around SiC particles in Ti: an electron microscopy study'," *Scripta Materialia*, **43**, 285-286 (2000).
19. I.M. Low and E. Wren, "Characterisation of phase relations in air-oxidised Ti_3SiC_2," *J. Aust. Ceram. Soc.*, **39**, 103 (2003).

MECHANICAL BEHAVIOR CHARACTERIZATION OF A THIN CERAMIC SUBSTRATE AT ELEVATED TEMPERATURE USING A STEREO-IMAGING TECHNIQUE

Sujanto Widjaja
Corning Incorporated
Science & Technology Division, SP-FR-04
Corning, NY 14831

Karen L. Geisinger
Corning Incorporated
Science & Technology Division, SP-FR-04
Corning, NY 14831

Scott C. Pollard
Corning Incorporated
Science & Technology Division, SP-AR-02
Corning, NY 14831

ABSTRACT

High temperature mechanical behaviors of a thin ceramic substrate of 3%yttria-stabilized zirconia (3YSZ) were characterized using ARAMIS™ optical measurement system (GOM Optical Measuring Techniques, Braunschweig, Germany). The technique relies on 3D image correlation photogrammetric principles providing non-contact determination of shape, deformation and full-field strain. Experimentally, a random pattern with good contrast was applied to the surface of a thin substrate, and the deformation of substrate under load was recorded and evaluated using digital image processing. The elastic modulus of the ceramic substrate at room temperature and 725°C was indirectly calculated from the deflection data, and generally it was found to be in agreement with values reported elsewhere. This characterization technique also allows the determination of the maximum value of deflection of the substrate at failure during pressure rupture testing and *in-situ* observation of ceramic membrane response under temperatures.

INTRODUCTION

Ceramic membranes have found various applications in fields such as energy, environmental and chemical industries. In these applications the ceramic membrane is typically dense and must allow the transport of certain species through the oxide lattices. Enormous efforts in research have concentrated on the identification and optimization of materials to achieve functional properties such as electro-catalytic activities. As other materials are incorporated into the membrane-based system targeting a specific application, it is necessary to evaluate the mechanical properties of the ceramic membrane and its design to achieve the desired reliability.

The mechanical reliability of ceramic membrane utilized at elevated temperatures is primarily dictated by stresses, of which the magnitudes are dependent upon material properties, design geometry and operating conditions. The objectives of this work are: (1) to demonstrate the applicability of a stereo-imaging technique for the evaluation of elastic modulus of a ceramic membrane at elevated temperature; and (2) to characterize the rupture strength and maximum out-of-plane deflection at failure of a thin ceramic membrane, which are critical parameters in design geometry for a specific application.

EXPERIMENTAL PROCEDURE

A yttria-stabilized tetragonal zirconia polycrystal (Y-TZP) ceramic substrate of about 20 μm in thickness, prepared by tape casting, was mounted to a stainless steel frame with an adhesive. The specimens considered in this study were of round geometry with the thickness determined on individual samples based on the average of multiple measurements.

The surface of the specimen was sprayed with a random pattern of ink, as shown in Figure 1, which is needed to allocate pixels in the captured images. Prior to characterization at elevated temperatures, measurement was performed on a similar specimen at room temperature secured by epoxy bonding to a stainless steel annular frame. The room temperature results were analyzed and compared to those obtained from other known techniques.

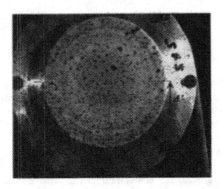

Figure 1   **Illustration of a thin ceramic specimen, mounted on the annular ring, with a random pattern of ink on the surface that is needed to allocate pixels in the captured images.**

Figure 2 illustrates a schematic of the specimen test set-up comprising three ports: an input compressed air port allowing the application of pressurization on specimen, a measurement port for connecting a pressure transducer and an output port for specimen mounting. A given specimen was typically tested at least at two different pressures, where the deformation profile at each pressure was recorded, prior to loading the substrate until failure to obtain its rupture strength. For elevated temperature measurements, the specimen was placed inside an insulated furnace (of embedded SiC heating elements) with a fused silica glass window as a top cover, allowing the observation of deformation and recording of substrate's response due to temperature and pressure loading. Room temperature measurements were performed without the silica glass window. Figure 2 includes the illustration of test set-up where two cameras for stereo-imaging measurements were used. The ARAMIS™ system[1] recognizes the surface of the specimen to be tested in digital images by allocating coordinates to the image pixels. The cameras were calibrated prior to any loading to establish reference condition of the coordinates of pixels. During loading, which deforms the specimen, further images were captured. Based on the captured images, the software compares the digital images obtained at different stages of loading and calculates the displacement and deformation of the specimen.

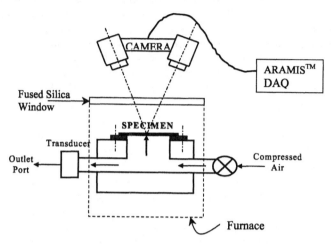

Figure 2   A schematic of test set-up showing a specimen test rig placed within a furnace, and the positioning of stereo-imaging cameras to monitor the deformation of specimen

ANALYSIS OF LARGE DEFLECTION OF MEMBRANE

The solution to a bending of a clamped circular plate under uniform pressure p leading to a large deflection (i.e. deflection >> thickness of the plate) is well-documented elsewhere[2-4], which in some cases require considerable amount of numerical calculation and series solution to solve the integration of non-linear equations. For instance, the equations to describe deflection w and radial stress σ_r profiles obtained by Fichter[4], following Hencky's approach[3], are given as follows:

$$\sigma_r(r) = \frac{E h}{4}\left(\frac{p a}{E h}\right)^{2/3} \sum_{n=0}^{\infty} b_{2n}\left(\frac{r}{a}\right)^{2n} \tag{1}$$

$$w(r) = a \left(\frac{p a}{E h}\right)^{1/3} \sum_{n=0}^{\infty} a_{2n}\left[1-\left(\frac{r}{a}\right)^{2n+2}\right] \tag{2}$$

where E is the elastic modulus of plate, h and a are thickness and radius of plate, respectively, and r is the radial distance from the center of the plate. The coefficients a_n and b_n were calculated numerically. For a known deformation profile obtained from pressurization of a thin ceramic specimen with the known geometry, one can obtain the elastic modulus of substrate by using Eq. (2).

Alternatively, Finite Element (FE) was utilized as a tool to model the problem of large deflection of a membrane under pressure and temperature loading. Large deflection behavior of a thin substrate was taken into account by using the appropriate element, namely SHELL63 provided by ANSYS[5], to mesh the geometry of interest. SHELL63 has both bending and membrane capabilities where both in-plane and normal loads are permitted. The comparison of deflections as a function of radial distance obtained from FE calculation and that of given by Eq. (2) for a chosen geometry of "plate" ($h = 23$ μm and $a = 31.75$ mm) and material properties ($E = 215$ GPa and Poisson's ratio $\nu = 0.25$) summarized in Figure 3. It can be seen that the results of two methods of calculation are in agreement, indicating the validity of the FE model. Thus, FE analyses were performed to simulate the pressurization of thin substrates at room and elevated (725°C) temperatures, respectively. The FE results allowed the evaluation of the effect of elastic modulus on deformation of thin substrate by curve fitting the calculated deflection profile (for a given E) to the deflection data obtained from the stereo-imaging measurement.

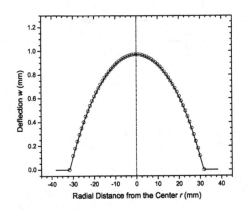

Figure 3   A comparison of out-of-plane deflection profile which is calculated from the series solution (denoted as symbol) to that of given by FEA (solid line).

RESULTS and DISCUSSION

Validation of the Characterization Tool
    Figure 4 shows the experimentally obtained deflection profiles (denoted as symbols) of a YSZ substrate, with a diameter of 63.5 mm and a thickness of 23 μm, along the diagonal section A-A' (as shown in Figure 1) under two pressure loadings (15.2 kPa and 23.4 kPa) at room temperature. A series of FE calculations was performed with the assumed elastic modulus of the substrate to be in the range of 200 – 250 GPa. It can be seen from Figure 4 that the deflection profiles (denoted as solid lines) calculated with an E of 215 GPa fit the measured deflection values well, suggesting that the Young's modulus of the material is 215 GPa. Multiple deflection profile measurements on an additional sample were conducted, resulting in an average elastic modulus of 3YSZ to be (218±8) GPa at room temperature.
    Independent measurement of the same thin ceramic substrate was performed via nano-indentation, where the sample was mounted on a glass slide using a thin layer of epoxy. Figure 5 illustrates the calculated Young's modulus of the YSZ substrate as a function of indentation depth. The modulus was found to be (249±4) GPa for the indentation depth of 100-400 nm,

while at greater indentation depth the measured modulus seemed to drift lower due to the effect of the properties of material beneath the substrate of interest[6]. A higher measured modulus obtained by nano-indentation as compared to that extracted from the stereo-imaging measurement is believed to be attributed to a slight unevenness of thin ceramic membrane under testing. The small waviness and "undulating" surface of substrate requires additional work (in the form of pressure) before transferring all the applied work to deform the substrate to its ultimate shape as captured by the stereo-cameras. However, in general it can be deduced that the Young's modulus calculated from FE based on the experimentally measured deflection profile for an even and flat surface is in agreement with our value measured independently via nano-indentation. Our finding is also comparable to the reported value of 190 GPa by Atkinson et.al.[7] from the test result of 8YSZ substrates of 200 μm in thickness via ring-on-ring method.

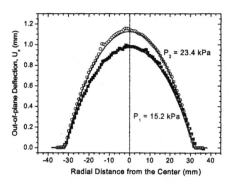

Figure 4   Comparison of the FE calculated deflection profiles with E value of 215 GPa (denoted as solid lines) to the experimentally measured deflection (denoted as symbols) by stereo-imaging technique.

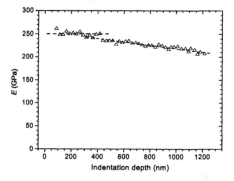

Figure 5   Nano-indentation results showing the calculated elastic modulus of YSZ ceramic substrate of 23 μm in thickness as a function of indentation depth. At smaller indentation depth up to 400 nm the elastic modulus appears to remain constant (i.e. 249 ± 4 GPa).

Evaluation of Mechanical Behaviors of 3YSZ at 725°C

Figure 6 illustrates typical out-of-plane deflection profiles of a ceramic membrane at 725°C obtained from the stereo-imaging measurement (denoted as symbols) as a function of applied pressure. It should be noted that experimentally, the deflection at 725°C with no pressure loading was taken as the reference point of subsequent measurements where pressure was applied. Figure 6 also includes the deflection profiles calculated by FE (solid lines) by taking the elastic modulus of YSZ to be 160 GPa. The FEA result of an E of 160 GPa for 3YSZ shows a good agreement with a recently reported value (157 GPa) of the elastic modulus of 8YSZ at 800°C.[7] Once the deflection profiles of the ceramic substrate were obtained as a function of applied pressure, the specimens were pressurized until rupture. The rupture strength of 3YSZ specimens at 725°C calculated from Eq.(1) was (225 ± 73) MPa.

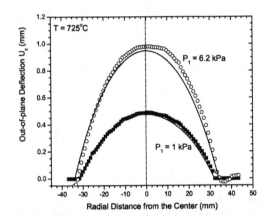

Figure 6   **Comparison of the FE calculated deflection profiles with E value of 160 GPa (denoted as solid lines) to the experimentally measured deflection (denoted as symbols) by stereo-imaging technique.**

An additional benefit of utilizing a stereo-imaging method of measurement in our work is that this technique allows an *in-situ* observation of the ceramic membrane at elevated temperature under pressure loading, which provides useful insights into the design of a particular system where a ceramic membrane is incorporated. For instance, it was of interest to determine the maximum out-of-plane deflection $(U_z)_{max}$ at failure at elevated temperature. Continuous recording or intermittent collections of digital images at various temperatures of the tested samples allows the determination of the evolution of deflection under loading. Figure 7 shows the deflection map U_z of a round membrane at 725°C under increasing pressure loading at a time prior to rupture as calculated by the ARAMIS™ measurement system. Additional feature of the measuring system that is yet to be demonstrated in our work is strain calculations of the tested substrate at elevated temperatures under various pressure loadings.

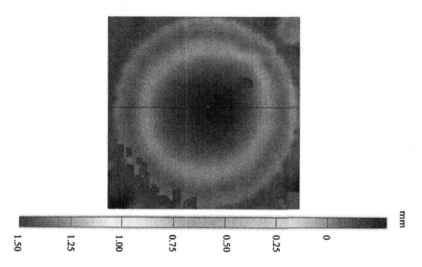

mm

1.50   1.25   1.00   0.75   0.50   0.25   0

Figure 7   Out-of-plane deformation contour map of a round ceramic substrate pressurized at 725°C post-processed by DAQ of the measurement system. The contour map represents the result calculated for an instantaneous time at rupture.

CONCLUDING REMARKS

The applicability of stereo-imaging technique to evaluate the mechanical behaviors and response of a thin 3YSZ ceramic substrate as a function of temperature has been demonstrated. The deflection profiles obtained from the experimental measurements provides the basis for the determination of elastic moduli of 3YSZ at room temperatures and 725°C, respectively, when the measured data was coupled with finite element modeling of large deflection analysis of a membrane under pressure loading. The elastic moduli of 3YSZ specimens at room temperature and 725°C were found to be 215 GPa and 160 GPa, respectively. The rupture strengths were calculated to be 680 MPa and 225 MPa at room temperature and 725°C, respectively. The stereo-imaging technique also allows an *in-situ* observation and measurements at elevated temperatures where other behaviors of thin membrane, such as the maximum out-of-plane deflection is of interest for a given designed system.

ACKNOWLEDGEMENT

The authors would like to thank our colleagues at Corning Incorporated, namely to Dr. James E. Price for the nano-indentation measurement, Dr. John D. Helfinstine for the discussion on analysis and testing configuration, to Pamela Maurey and Mike Hobczuk for the specimen preparation and testing, respectively.

REFERENCES

[1]ARAMIS™ Optical Measurement System by GOM Optical Measuring Techniques, Braunschweig, Germany

[2]H. M. Berger, "A New Approach to the Analysis of Large Deflections of Plates," *J. Appl. Mech*, 465-472 (1955).

[3]H. Hencky, "On the Stress State in Circular Plates with Vanishing Bending Stiffness," *Zeitschrift fur Mathematik und Physik*, **63**, 311-317 (1915).

[4]W. B. Fichter, "Some Solutions for the Large Deflections of Uniformly Loaded Circular Membranes," *NASA Technical Paper 3658*, NASA Langley Research Center (July 1997).

[5]ANSYS™ ver. 5.7, Canonsburg, PA 15317

[6]P. Lemoine, J. F. Zhao, J. P. Quinn, J. A. McLaughlin, and P. Maquire, "Hardness Measurements at Shallow Depths on Ultra-thin Amorphous Carbon Films Deposited onto Silicon and Al_2O_3-TiC substrates," *Thin Solid Films*, **379**, 166-172 (2000).

[7]A. Atkinson and A. Selcuk, "Mechanical Behaviour of Ceramic Oxygen Ion-conducting Membranes," *Solid State Ionics*, **134**, 59-66 (2000).

Functional Nanomaterial Systems
Based on Ceramics

SYNTHESIS AND CHARACTERIZATION OF CUBIC SILICON CARBIDE (β-SiC) AND TRIGONAL SILICON NITRIDE (α-Si₃N₄) NANOWIRES

Karine Saulig-Wenger, Mikhael Bechelany, David Cornu*, Samuel Bernard, Fernand Chassagneux and Philippe Miele
Laboratoire des Multimatériaux et Interfaces UMR 5615 CNRS
University Claude Bernard - Lyon 1
43 Bd du 11 novembre 1918
F-69622 Villeurbanne Cedex, France
David.Cornu@univ-lyon1.fr

Thierry Epicier
GEMPPM UMR 5510 CNRS - INSA Lyon
20 Avenue Albert Einstein
F-69621 Villeurbanne Cedex, France

ABSTRACT

By varying the final heating temperature in the range 1050°C - 1300°C, cubic silicon carbide (β-SiC) and/or trigonal silicon nitride (α-Si₃N₄) nanowires (NWs) were prepared by direct thermal treatment under nitrogen, of commercial silicon powder and graphite. Long and highly curved β-SiC NWs were preferentially grown below 1200°C, while straight and short α-Si₃N₄ NWs were formed above 1300°C. Between these two temperatures, a mixture of both nanowires was obtained. The structure and chemical composition of these nanostructures have been investigated by SEM, HRTEM, EDX and EELS.

INTRODUCTION

Numerous studies have been recently devoted to ceramic and metallic nanowires (NWs) due to their outstanding properties which can be tailored by varying their chemical composition and also their crystalline structure. The possible applications of these NWs run from nanoelectronics to composite materials. Among the series, silicon-based NWs, made of cubic silicon carbide (β-SiC), silicon dioxide (SiO₂) or trigonal silicon nitride (α-Si₃N₄), are of particular interest due to their interesting mechanical[1,2], electrical[3,4] and/or optical[5] properties. However, industrial applications clearly need a cheap growth method for the large-scale fabrication of NWs. Moreover, works are also devoted to coaxial nanocables (NCs) due to their possible uses as reinforcement agents for mechanical applications, the outer sheet of the NC acting as an interface between the nanowire and the matrix.

Numerous routes have been reported for the fabrication of SiC NWs. The most promising ones for large scale production are (i) carbothermal reduction routes using carbon nanotubes (CNTs) as templates[6-8] or a mixture of SiO₂ nanoparticles and active carbon[9] and (ii) methods based on VLS (Vapour Liquid Solid) growth mechanism e.g. assisted by catalytic metallic nanoparticles[10-14]. Only few techniques have been described for the preparation of silicon nitride nanowires. They usually correspond to those reported for SiC NWs but they are driven at higher temperature (1400°C to 1650°C). The three main routes are the following: (i) carbothermal reduction of silica[15-19], (ii) the use of carbon nanotubes as templates[20] and (iii) a VLS growth technique[21]. All those synthetic methods require however, either expensive starting materials such as CNTs or expensive equipment.

We previously reported that the direct pyrolysis under nitrogen of commercial silicon powder in presence of graphite led to the formation of cubic silicon carbide (β-SiC) NWs[22,23]. These nanowires have diameters in the range 20 – 30 nm and micrometric lengths. We showed that when a mixture of argon and oxygen is used instead of nitrogen, amorphous silica nanowires were preferentially obtained[24,25]. As an extent to these results, we report in the present paper the influence of the final heating temperature on the structure and chemical composition of the resulting silicon-based nanostructures.

EXPERIMENTAL

All the experiments were conducted following the same experimental procedure. Silicon powder (Aldrich 99.999%, 60 mesh) was placed in an alumina boat containing a piece of graphite. This boat was then placed in the alumina tube of a horizontal tubular furnace, the tube being previously degassed *in vacuo* before filling with nitrogen (electronic grade). Under the gas flow (5 mL.min[-1]), the boat was heated up to the selected temperature (heating rate 200°C min[-1]), held for 1 hour then allowed to cool down to room temperature. Finally, powder was scraped from the alumina boat and analysed by scanning electron microscopy (SEM, Model N°S800, Hitachi), high-resolution transmission electron microscopy (HRTEM, Field Emission Gun microscope JEOL 2010F) and electron energy-loss spectroscopy (EELS, Digi-PEELS GATAN).

RESULTS AND DISCUSSION

In previous works, β-SiC nanowires have been obtained by heating silicon powder under nitrogen at 1200°C, this temperature being held during 1 hour before cooling down[22,23]. In order to examine the influence of the final heating temperature on the yield, the crystallographic structure and the chemical composition of the resulting nanowires, five independent experiments have been conducted up to 1050°C, 1150°C, 1200°C, 1250°C and 1300°C, respectively. In all these experiments, the final temperature was held during 1h before cooling down. Each crude product was first analysed by SEM and representative images of each sample are shown in fig.1.

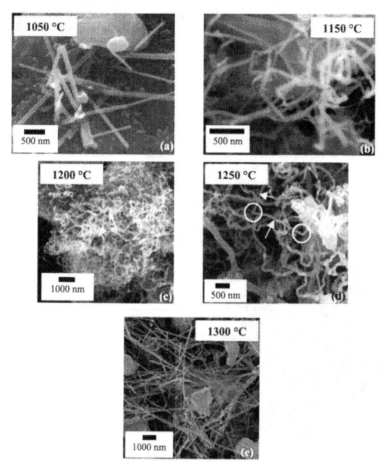

<u>Fig. 1:</u> SEM images of the crude products obtained at different final heating temperature: (a) 1050°C, (b) 1150°C, (c) 1200°C, (d) 1250°C and (e) 1300°C.

The crude sample heated up to 1050°C contained only few nanowires (Fig. 1a), located on the surface of the residual silicon particles. These NWs are straight with scattered diameters (from ~10 to ~300 nm) and short lengths (below ~3 μm). When the experiment was conducted up to 1150°C, it resulted in the formation of larger amount of nanowires (Fig. 1b). Their diameters are in the same range but they are longer with lengths estimated above ~8 μm. As illustrated by Fig. 1b, this result should be related with their highly curved shape. Nanowires exhibiting similar diameters and lengths were obtained at 1200°C (Fig. 1c). In that case, the yield, estimated from the SEM images, was however significantly improved. In contrast, when the thermal treatment was driven up to 1250°C, the yield was not significantly improved (Fig. 1d). At this temperature, two kinds of nanowires were observed: numerous long and highly curved nanowires similar to those obtained at 1200°C mixed with few straight nanowires (Fig. 1d, white arrows). The latter are shorter and exhibit well-defined angles which can be interpreted as changes in direction of the NW axis (Fig. 1d, white circles). The SEM image of the samples obtained after heating up to 1300°C shows a strong modification in the shape of the obtained NWs (Fig. 1e). No long and curved nanowires were detected but numerous straight NWs were observed. Their diameters are comparable to those previously obtained but their lengths are shorter and below ~8 µm.

HRTEM investigations coupled with EELS analysis have been conducted in order to determine if there is a difference in structure and/or chemical composition within the two kinds of NWs observed by SEM. Figure 2a shows a HRTEM image of a typical long and curved nanowire obtained at 1200°C. On the corresponding EELS spectrum (Fig. 2b), two main features are observed at ~100 eV and ~284 eV corresponding to Si-L and C-K edges, respectively. As expected, further selected area electron diffraction (SAED) analysis showed that cubic silicon carbide (β-SiC) nanowires have been formed. According to HRTEM investigation, a high density of stacking faults was observed (fig. 2a) which can be related to high growth rate, as previously mentioned[22,23].

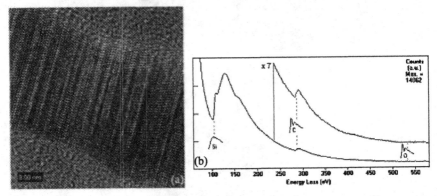

Fig. 2: HRTEM image of a β-SiC NW (a) with the corresponding EELS spectrum (b).

In contrast, figure 3a shows a typical HRTEM image of a straight and short nanowire obtained at 1300°C. The chemical composition of the observed NW has been determined by EELS analysis performed on its core (Fig. 3b). Two main features are observed at ~100 eV and ~400 eV corresponding to Si-L and N-K edges, respectively. Features at ~284 eV and ~532 eV, corresponding to C-K edge and O-K edges respectively, were not detected. A coating of ~2.5 nm thickness was observed around the analysed NW. Its EELS analysis performed using a 2 nm probe revealed that it is composed of carbon. As illustrated by figure 3c, fast Fourier transformation (FFT) shows that the core of the NW is composed of the trigonal polymorph of silicon nitride (α-Si_3N_4). Complementary structural analysis showed that the carbon coating is amorphous. The obtained nanostructures can be considered as carbon sheathed silicon nitride nanocables (α-Si$_3$N$_4$@C).

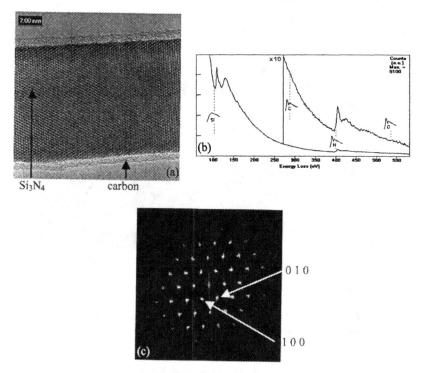

Fig. 3: (a) HRTEM image of a α-Si$_3$N$_4$ NW coated with amorphous carbon with corresponding (b) EELS spectrum and (c) fast Fourier transformation (FFT) of the core of the NW.

In previous works, we suggested that the silicon carbide nanowires were formed at 1200°C by a Vapour-Solid (VS) nucleation process, carbon being transported to silicon nanoparticles by nitrogen through the formation of CN-like derivatives[22,23]. For the formation of α-Si$_3$N$_4$ NWs e.g. at higher temperature (>1250°C), we can suggest a similar VS growth mechanism with a preferential nucleation of silicon nitride onto the surface of the silicon nanoparticles due to the higher temperature reached during the experiment. The formation of an amorphous coating of carbon onto the NWs can be interpreted by considering that carbon transportation by nitrogen was still effective at that temperature. Comparative analysis of nanowires obtained at 1300°C revealed that main α-Si$_3$N$_4$ NWs were free of carbon coating. Moreover, no SiC nanowires were detected in the sample obtained by heating up to 1300°C. This result is of primary importance for the determination of the growth mechanism of these nanowires because it indicates that their formation did not occurred when the temperature was increased, but either when the final temperature is reached or, less probably, during the cooling step.

While only α-Si$_3$N$_4$ NWs were formed at 1300°C, a mixture of β-SiC and α-Si$_3$N$_4$ NWs were obtained when the experiment was conducted up to 1250°C. This is evidenced by the bright field (Fig. 4a) and dark field (Fig. 4b) TEM images recorded on a whole mass of nanowires.

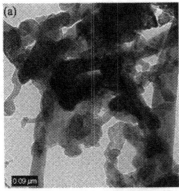 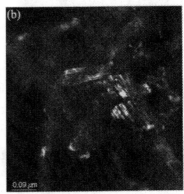

Fig. 4: bright field (a) and dark field (b) TEM images of a whole mass of β-SiC and α-Si$_3$N$_4$ NWs.

The image in bright field reveals the presence of Si$_3$N$_4$ nanowires, which appear in bright in Fig. 4a. The dark field image 4b, made with a β-SiC reflection, reveals that the Si$_3$N$_4$ NWs were mixed with SiC NWs which appear in bright in this image. These observations clearly point out that between 1200°C and 1300°C, a mixture of β-SiC NWs and α-Si$_3$N$_4$ NWs were obtained.

CONCLUSION

The direct thermal treatment of a silicon powder under nitrogen yielded the formation of cubic silicon carbide (β-SiC) in the range 1050°C – 1200°C. As we found, when the experiment was driven in the range 1200°C – 1300°C, a mixture of β-SiC and trigonal silicon nitride (α-Si$_3$N$_4$) nanowires (NWs) were obtained. Moreover, when the experiment was conducted up to 1300°C, only α-Si$_3$N$_4$ NWs were observed within the sample. The growth mechanism for these nanowires were presumed to be a VS process, starting with the nucleation of silicon carbide or silicon nitride nuclei onto the surface of the silicon nanoparticles.

REFERENCES

[1]E.W. Wong, P.E. Sheehan, C.M. Lieber, "Nanobeam mechanics: elasticity, strength, and toughness of nanorods and nanotubes", *Science*, **277**, 1971-1975 (1997).

[2]Y. Zhang, N. Wang, R. He, Q. Zhang, J. Zhu, Y. Yan, "Reversible bending of Si$_3$N$_4$ nanowire" *J. Mater. Res.*, **15**, 1048-1051. (2000)

[3]K. W. Wong, X. T. Zhou, F. C. K. Au, H. L. Lai, C. S. Lee, S. T. Lee, "Field-emission characteristics of SiC nanowires prepared by chemical-vapor deposition", *Appl. Phys. Lett.*, **75**, 2918-2920. (1999)

[4]Z. Pan, H.-L. Lai, F. C. K. Au, X. Duan, W. Zhou, W. Shi, N. Wang, C.-S. Lee, N.-B. Wong, S.-. Lee, S. Xie, "Oriented silicon carbide nanowires: synthesis and field emission properties", *Adv. Mater.*, **12**,1186-1190. (2000)

[5]D.P. Yu, L. Hang, Y. Ding, H.Z. Zhang, Z.G. Bai, J.J. Wang, Y.H. Zou, W. Qian, G.C. Xiong, S.Q. Feng, "Amorphous silica nanowires: intensive blue light emitters", *Appl. Phys. Lett.*, **73**, 3076-3078 (1998)

[6]W. Han, S. Fan, Q. Li, W. Liang, B. Gu, D. Yu, "Continuous synthesis and characterization of silicon carbide nanorods", *Chem. Phys. Lett.*, **265**, 374-378 (1997)

[7]C. C. Tang, S. S. Fan, H. Y. Dang, J. H. Zhao, C. Zhang, P. Li, Q. Gu, "Growth of SiC nanorods prepared by carbon nanotubes-confined reaction", *J. Cryst. Growth*, **210**, 595-599. (2000)

[8]J-M. Nhut, R. Vieira, L. Pesant, J-P. Tessonnier, N. Keller, G. Ehret, C. Pham-Huu, M. J. Ledoux, "Synthesis and catalytic uses of carbon and silicon carbide nanostructures", *Catal. Today*, **76**, 11-32. (2002)

[9]J. Pelous, M. Foret, R. Vacher, "Colloidal vs. polymeric aerogels: structure and vibrational modes", *J. Non-Cryst. Solids*, **145**, 63-70. (1992)

[10]X. T. Zhou, H. L. Lai, H. Y. Peng, F. C. K. Au, L. S. Liao, N. Wang, I. Bello, C. S. Lee, S. T. Lee, "Thin β-SiC nanorods and their field emission properties", *Chem. Phys. Lett.*, **318**, 58-62. (2000)

[11]B. Q. Wei, J. W. Ward, R. Vajtai, P. M. Azjayan, R. Ma, G. Ramanath, "Simultaneous growth of silicon carbide nanorods and carbon nanotubes by chemical vapor deposition" *Chem. Phys. Lett.*, **354**, 264-268. (2002)

[12]X. T. Zhou, N. Wang, H. L. Lai, H. Y. Peng, I. Bello, N. B. Bello, N. B. Wong, C. S. Lee, S. T. Lee, "β-SiC nanorods synthesized by hot filament chemical vapor deposition" *Appl. Phys. Lett.*, **74**, 3942-3944. (1999)

[13]H. L. Lai, N. B. Wong, X. T. Zhou, H. Y. Peng, F. C. K. Au, N. Wang, I. Bello, C. S. Lee, S. T. Lee, X. F. Duan, "Straight a-SiC nanorods synthesized by using C-Si-SiO₂", *Appl. Phys. Lett.*, **76**, 294-296. (2000)

[14]Y. Zhang, N. L. Wang, R. He, X. Chen, J. Zhu, "Synthesis of SiC nanorods using floating catalyst", *Solid State Comm.*, **118**, 595-598. (2001)

[15]L. D. Zhang, G. W. Meng, F. Phillipp, "Synthesis and characterization of nanowires and nanocables", *Mater. Sci. and Eng.* A, **286**, 34-38 (2000)

[16]X. C. Wu, W. H. Song, W. D. Huang, M. H. Pu, B. Zhao, Y. P. Sun, J. J. Du, "Simultaneous growth of α-Si₃N₄ and β-SiC nanorods" *Mater. Res. Bull.*, **36**, 847-852 (2001)

[17]X. C. Wu, W. H. Song,B. Zhao, W. D. Huang, M. H. Pu, Y. P. Sun, J. J. Du, "Synthesis of coaxial nanowires of silicon nitride sheathed with silicon and silicon oxide", *Solid State Comm.*, **115**, 683-686 (2000)

[18]Y. H. Gao, Y. Bando, K. Kurashima, T. Sato, "Synthesis and microstructural analysis of Si₃N₄ nanorods", *Microsc. Microanal.*, **8**, 5-10. (2002)

[19]Y. H. Gao, Y. Bando, K. Kurashima, T. Sato, "Si₃N₄/SiC interface structure in SiC-nanocrystal-embedded α-Si₃N₄ nanorods", *J. Appl. Phys.*, **91**, 1515-1519. (2002)

[20]W. Han, S. Fan, Q. Li, B. Gu, X. Zhang, D. Yu, "Synthesis of silicon nitride nanorods using carbon nanotube as a template", *Appl. Phys. Lett.*, **71**, 2271-2273 (1997)

[21]H. Young Kim, J. Park, H. Yang, "Synthesis of silicon nitride nanowires directly from the silicon substrates", *Chem. Phys. Lett.*, **372**, 269-274. (2003)

[22]K. Saulig-Wenger, D. Cornu, F. Chassagneux, G. Ferro, T. Epicier, P. Miele, "Direct synthesis of β-SiC and h-BN coated β-SiC nanowires", *Solid State Comm.*, **124,** 157-161 (2002)

[23]K. Saulig-Wenger, D. Cornu, F. Chassagneux, G. Ferro, P. Miele, T. Epicier, "Synthesis and characterization of β-SiC nanowires and h-BN sheathed β-SiC nanocables", *Ceramic Transactions, (Ceramic Nanomaterials and Nanotechnology)* **137,** 93-99. (2003)

[24]K. Saulig-Wenger, D. Cornu, F. Chassagneux, T. Epicier, P. Miele, "Direct synthesis of amorphous silicon dioxide nanowires and helical self-assembled nanostructures derived therefrom", *J. Mater Chem.*, **13,** 3058-3061. (2003)

[25]K. Saulig-Wenger, D. Cornu, P. Miele, F. Chassagneux, S. Parola, T. Epicier, "Synthesis of Si-based nanowires", *Ceramic Transactions, (Ceramic Nanomaterials and Nanotechnology II)* **148,** 131-137. (2004)

HIGH ENERGY MILLING BEHAVIOR OF ALPHA SILICON CARBIDE

Maria Aparecida Pinheiro dos Santos
Instituto de Pesquisas da Marinha - IPqM , Grupo de Materiais
Rua Ipirú n° 2 - Ilha do Governador
CEP 21931-090, Rio de Janeiro/RJ, Brasil

Célio Albano da.Costa Neto
Programa de Engenharia Metalúrgica e de Materiais, COPPE – UFRJ,
CP 68505, CEP 21945-970, Rio de Janeiro/RJ, Brasil

ABSTRACT

Alpha silicon carbide (α- SiC) powder was comminuted in a planetary mill during the periods of time of ½, 2, 4 and 6h. The rotation speed was 300rpm, the milling media used was isopropilic alcohol and the grinding bodies were spheres of zirconia estabilized with ceria. The milling powders were characterized concerning the particle size distribution, chemical composition and density. It was observed a great reduction in the particle size, from micrometric to submicrometric size, showing the capability of producing even nanometric particles.

INTRODUCTION

Processing of powder in the nanometer range increases the sinterability of ceramic materials, especially for those that are covalent bonded. If the grain size is kept in nanometer range, promising functional and structural properties may arise. Also, small grain size and large amount of grain boundaries area can lead to superplasticity and extreme hardness, since hardness of nanophase ceramics often increases with decreasing grain size [1].

The majority of problems found in processing particles of submicron size arises from the high surface area created in the powders, which often outcome in surface contamination, low packing density and formation of agglomerates [2,3].

The present study has the objective of refining a coarse α-SiC powder produced in Brazil, not used for sintering, up to submicron size via planetary mill.

EXPERIMENTAL PROCEDURES

The starting material was α-SiC powder (Alcoa SiC 1000) with mean grain size of 5,5 μm, Figure 1, and purity of 98,7% [4]. This material was comminuted in a planetary mill (300 rpm) during ½, 2, 4h and 6h. The milling vessel was specific design for this study, having the capacity of 500 ml, made of stainless steel and coated with WC-CO by HVOF. The milling media was composed of isopropylic alcohol and ZrO_2 (Zirconox ®, Netsch Brasil Ltda) balls with size in the range of 0,7 to 1,2 mm.

Figure 1 - SEM photomicrograph of the as received SiC powder. The average particle size (d_{50}) was 1.77 μm and morphology showed marked sharp edges.

After milling, the powders were analyzed by particle size distribution (granulometer CILAS 1064), surface area (BET - Model Gemini III 2375), particle density (He picnometer -ACCUPYC 1330 Micromeritics). Before analyze the milled powders in the granulometer, they were ultrasonic dispersed for 5 minutes in a sodium pyrophosphate solution.

RESULTS AND DISCUSSION

Table I shows the average particle size (d_{50}) as a function of the milling time in isopropilic alcohol, and Table II shows the average particle size (d_{50}), surface area (S), the amount of ZrO_2 added to the system by the milling process and the powder density, either measured by picnometry and calculated by X-ray spectroscopy.

Table I- Average particle size (d_{50}) as a function of the milling time in isopropyl alcohol [4].

Milling time (h)	Average particle size d_{50} (μm)
Original (0)	1,77
½	0,41
2	0,38
4	0,36
6	0,36

Table II– Characteristics of α-SiC as received and milled for 4 and 6h [4].

t_m (h)	$d_{50}(\mu m)$	$S(m^2/g)$	ZrO_2 (%)	$d^*(g/cm^3)$ Picnom. He
-	1.77	3.483	-	3.269
4	<0,5	10.089	12	3.906
6	<0,5	11.479	16	3.968

It can be noted that the powder density has increased from a value of 3.2 g/cm³, measured for the as received SiC, to approximately 4 g/cm³ for 4 and 6 h of milling time. The increase in density was attributed to a major contamination of ZrO_2, as shown in Figure 2, which was detected by x-ray fluorescence. It is worth mention that it increased to 12% and 16 %wt of ZrO_2 for 4 and 6 hours of milling, respectively. The density of ZrO_2 is about 6 g/cm³, and the milled powder had its density increased form 3.2 to 3.9 g/cm³, an increase of 20%.

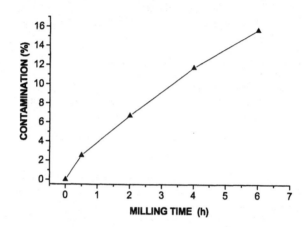

Figure 2 – The amount of zirconia, measured by X-ray fluorescence, added by the milling process as a function of the milling time.

Laser granulometry analysis showed that 100% of the particles have sizes bellow 1μm and the mean particle size was around 0.4 μm for the milling times of 4 and 6h. X-ray diffraction showed the presence of small broad peaks, and this type of feature is expected when particle size lies in the range of 0.1 to 0.5μm [5], which is confirmer in figure 3.

Figure 3 – High magnification of milled SiC powder showing the large amount of submicron particles, mainly SiC and also ZrO₂.

The present data showed that planetary mill can be used to produce submicron size powder in a very short time, since ½ h of milling resulted in mean particle size of 0.4 μm. The secondary additivation of the ZrO_2 was not observed to impart the sinterability of the material, since values of 99% density were obtained [4], and the fracture toughness was measured to be 5.8 MPa.m$^{1/2}$ [6]. This improves behavior was due to transformations of ZrO_2 into ZrC [4,7].

CONCLUSIONS

The planetary milling process was quite efficient in reducing the mean particle size of α-SiC ceramic. Techniques employed to evaluate particle size indicate that most of the particles lie in the nanometer range between 0.5 to 0.1 μm. Contamination of ZrO_2 was observed and it helped to density the material to 99% of the theoretical density. Further research has to be conducted in order to understand the effect of ZrO_2 during the sintering process.

REFERENCES

[1] H. Hahn, R. Averback , J.Am.Ceram.Soc. 74 (11), 2918 (1991).
[2] B.Matos, "Influência do meio na moagem ultrafina de carbeto de silício", Projeto de Formatura, Departamento de Engenharia Metalúrgica e de Materiais da Escola de Engenharia da Universidade Federal do Rio de Janeiro, Rio de Janeiro, Abril (2002).
[3] H.Tanaka , " Sintering of Silicon Carbide ", Silicon Carbide Ceramics –1 Fundamental and Solid Reaction – Edited by Shigeyuki Sŏmiya and Yoshizo Inomata – Ceramic Research and Development in Japan Series – Elsevier Applied Science (1988).
[4] M.A. P. Santos, "Processamento e sinterização de Carbeto de Silício Nacional", Tese de Doutorado, PEMM, COPPE, UFRJ(2003).

[5] K. Jiang, " On The Applicability of The X-ray Diffraction Line Profile Analysis in Extracting Grain Size and Microstrain in Nanocrystalline Materials",Journal of Materials Research, V.14(2),pp.549-559 (1999).

[6] E.Souza Lima, "Mechanical Behavior of alfa - SiC Based Nanocomposites", II SBPMat
Rio de Janeiro, RJ (2003)

[7] C.K. Narula, Ceramic Precursor Technology and its Application,Macel Dekker.,New.

ACKNOWLEDGEMENTS

Thanks to Dr. Francisco Cristovão de Melo (IAE/CTA) and Dr. Leonardo Ajdelsztajn (UC Irvine) for helping this research.

SYNTHESIS OF BORON NITRIDE NANOTUBES FOR ENGINEERING APPLICATIONS

Janet Hurst
David Hull
NASA Glenn Research Center
21000 Brookpark Rd
Cleveland, Ohio 44135

Daniel Gorican
QSS Group
21000 Brookpark Rd
Cleveland, Ohio 44135

ABSTRACT
Boron nitride nanotubes (BNNT) are of significant interest to the scientific and technical communities for many of the same reasons that carbon nanotubes (CNT) have attracted wide attention. Both materials have potentially unique and important properties for structural and electronic applications. However of even more consequence than their similarities may be the complementary differences between carbon and boron nitride nanotubes. While BNNT possess a very high modulus similar to CNT, they also possess superior chemical and thermal stability. Additionally, BNNT have more uniform electronic properties, with a uniform band gap of 5.5 eV while CNT vary from semi-conductive to highly conductive behavior.

Boron nitride nanotubes have been synthesized both in the literature and at NASA Glenn Research Center, by a variety of methods such as chemical vapor deposition, arc discharge and reactive milling. Consistent large scale production of a reliable product has proven to be difficult. Progress in the reproducible synthesis of 1-2 gram sized batches of boron nitride nanotubes will be discussed as well as potential uses for this unique material.

INTRODUCTION

In the last decade, significant attention from the scientific community has been focused on the area of nanotechnology and specifically upon nanotube synthesis. While carbon nanotubes have generated the bulk of interest to date, other compositions offer promise as well and may have advantages or complementary properties relative to carbon nanotubes for various applications. At NASA Glenn Research Center, where application interests are often focused on high temperature propulsion, both BN and SiC nanotube synthesis are currently under investigation for high temperature structural and electronic materials (1,2). The focus of the current effort is BNNT synthesis. While boron nitride nanotubes are known to be structurally similar to carbon nanotubes, inasmuch as both are formed from graphene sheets, much less is known about BNNT. In large part this is due to the difficulty in synthesizing this material rather than lack of interest. It has been found that BNNT have excellent mechanical properties with a measured Young's modulus of 1.22 +/- 0.24 TPa (3). BNNT also have a constant band gap of about 5.5 eV (4). In contrast,

CNT vary from semi-conducting to conducting behavior depending on chirality and diameter of the product. Little progress has been demonstrated in the control of the chiral angle and hence the electronic properties of CNT. On the other hand, BNNT preferentially forms the zig-zag structure rather than the armchair or chiral structures due to the polar nature of the B-N bond (5). Recently, it has also been shown that BNNT systems have excellent piezoelectric properties, superior to those of piezoelectric polymers (6). Additionally, the expected oxidation resistance of BNNT relative to CNT suggests BNNT may be suitable for high temperature structural applications. This stability may be an important safety consideration for some applications, such as hydrogen storage, as carbon nanotubes readily burn in air. Many synthesis approaches have been tried with varying degrees of success. Among these approaches are pyrolysis over Co (7), CVD methods (8), arc discharge (9) laser ablation (10), and reactive milling techniques (11,12). The reported approach developed at NASA Glenn Research Center produced BNNT of significant length and abundance.

EXPERIMENTAL PROCEDURE

BNNT were prepared by reacting amorphous boron powder in a flowing atmosphere of nitrogen with a small amount of NH_3. Prior to heat treatment, fine iron catalyst particles were added in the range of up to several weight percent and briefly milled in polyethylene bottles with a hydrocarbon solvent and ceramic grinding media. Batch sizes of 2 grams are typically produced but the process should be easily scaleable to larger sizes. Milled material was applied to various high temperature substrates such as alumina, silicon carbide, platinum and molybdenum. Nanotubes were formed during heat treatments to temperatures ranging from 1100 C to 1400 C for brief times, 20 minutes to 2 hours.

Nanotubes were imaged with a Hitachi S4700 field emission scanning electron microscope with a super thin window EDAX Genesis System energy dispersive spectrometer (EDS) or Phillips CM200 transmission electron microscope operated at 200 kV, with Gatan electron energy loss spectrometer (EELS). Thermogravimetric analysis (TGA) was done in air up to 1000 C.

RESULTS

BNNT synthesized by this method are shown in Figure 1. The low magnification photo in Figure 1a shows an as-produced flake of BNNT. The flake was removed by tweezers from a 2.5 mm x 5 mm substrate or crucible. It is robust and easily handled in this form. Also, as the nanotubes are anchored within a growth media, there is no respiration hazard. BN nanotubes grow extensively both from the "top" and "bottom" of the growth media, where "top" refers to the side exposed to the atmosphere. Both top and bottom layers can be observed in Figure 1a as the edges were slightly rolled during handling. The nanotubes are quite long, 100 microns being common as shown in Figures 1b and 1c. On the bottom of the growth media, a layer of shorter nanotubes develops between the film and substrate or crucible. EDS also showed that the BNNT grow upon a film composed primarily of B, N, O, Fe and some purities. Nanotube diameters can be very consistent throughout their length; however this is processing temperature dependent. Not surprisingly, variation in the heat treatment temperature resulted in somewhat different products. A lower processing temperature resulted in more fine, uniform nanotubes, as those shown in figure 1b. Diameters of 20nm to 50 nm were typical. Higher temperatures and/or longer times resulted in large "nanotube" growth as shown in Figure 2a. Growth originated from fine nuclei but with excessive temperature, large structures quickly developed, up to a few microns in diameter. As shown in Figures 2a and 2b, secondary nucleation of small diameter BNNT also occurred on these larger structures. Higher processing temperatures, as well as excess catalyst concentration, resulted in interesting structures, such as nanohorns or nanoflowers, as seen in figure 2d. These structures also generally had considerable amounts of oxygen found by EDS, up to 6 w/o.

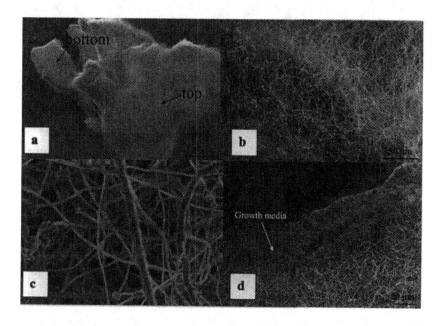

Figure 1 Field emission scanning electron microscope images of BNNT. (a) Typical flake peeled from substrate. (b)(c) Higher magnification photos of typical areas. (d) BNNT growing from media.

Transmission electron microscopy results are shown in Figures 3 and 4. TEM results showed the nanotubes to be nearly stoichiometric BN and a mixture of both straight walled nanotubes, as shown in Figure 3, and the "bamboo" structures, Figure 4. Predominately, the product from this method is multiwalled, commonly composed of 15-30 lattice layers, although this can be affected by processing conditions. Diameters of the

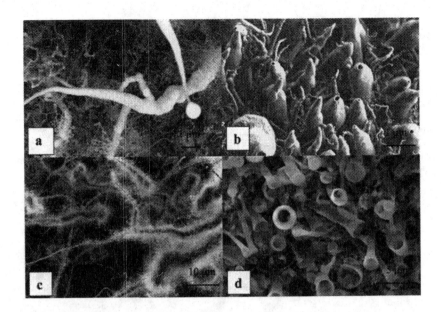

Figure 2. Examples of less typical structures synthesized at processing extremes.
a) Adjacent extremes in size. b) open ended nanopods c) fine BNNT nucleated on larger tubes
d) nanohorns.

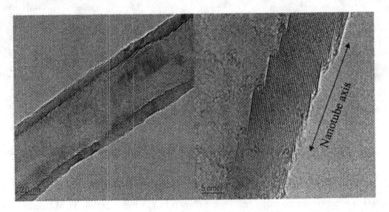

Figure 3 TEM photos of typical straight walled BNNT

multiwalled BNNT were often in the 20-40 nm range, again determined by processing
conditions and catalysts concentration. Figure 3 shows the atomic planes within the
straight walled nanotubes exhibited lattice fringes at an angle of 12.5° with respect to the

tube axis. This has also been noted elsewhere (5, 13) and may be an indication of rhomohedral stacking order (12). In Figure 4 the typically highly faulted lattice walls of the bamboo structure are evident. These faulted short walls, with their open edge layers, have been suggested to be superior for hydrogen storage (14,15). The structure of the BN layers are analogous to stacked paper cups with potential hydrogen storage sites on the surface and between lattice planes, and also perhaps within the isolated voids. Predominately bamboo structures can be consistently produced by this processing method. Figure 5 shows a region of exclusively bamboo BNNT.

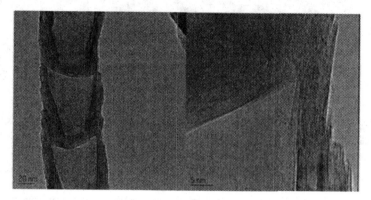

Figure 4 TEM photos of typical bamboo BNNT

The temperature stability of as-produced BN nanotubes was investigated by thermal gravimetric analysis. The results were compared to commercial as-produced carbon nanotubes*. Photos of the as-produced CNT and BNNT materials are shown in Figure 6 as well as those following heating in the TGA. The BNNT structure is clearly intact with the CNT decomposed, leaving behind the extensively oxidized iron catalyst. Figure 7 shows the TGA data in air, confirming that the carbon nanotubes have decomposed by 400 C. However, the BNNT are unaffected by the heat treatment with the exception of some slight weight gain from oxidation above 1000 C.

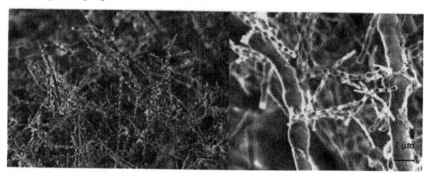

Figure 5.Regions of predominately bamboo nanotubes structures

Figure 6. As-processed BNNT and CNT samples were examined before and after exposure in air to 1000 C in a TGA. Photos on the left are as-processed samples. On the right are photos of material remaining following a TGA run to 1000C

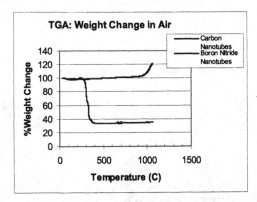

Figure 7. TGA results for BNNT and CNT samples showing the superior stability of BNNT relative to CNT

360

CONCLUSION

Boron nitride nanotubes were successfully and reproducibly grown by a NASA Glenn Research Center developed process. Currently 1-2 gram batches are being synthesized; however, the process is scaleable to much larger batch sizes. Sufficient amounts are now available so that boron nitride nanotubes are currently being incorporated into composites to provide high strength behavior at high temperature. It is also possible that this material may have many applications for sensors, electronics, piezoelectrics, among other applications of interest to NASA and the technical community.

A preponderance of bamboo structured nanotubes could be achieved by careful additions of catalyst materials and control of processing conditions. These bamboo structures are of interest for hydrogen storage applications. The boron nitride nanotubes were found to be much more stable at high temperatures than carbon nanotubes.

ACKNOWLEDGEMENTS

This work was sponsored by the Alternate Fuel Foundation Technologies (AFFT) Subproject of the Low Emissions Alternative Power (LEAP) Project at the NASA Glenn Research Center.

*Carbon Nanotechnologies, Inc.,16200 Park Row, Houston, TX

REFERENCES

1. .J. Hurst NASA R&T 2002, NASA TM – 2002-211333.

2. D. Larkin NASA R&T 2002, NASA TM – 2002-211333

3. N.G. Chopra, A. Zettl, "Measurement of the Elastic Modulus of a Multi-wall Boron Nitride Nanotube", Solid State Commun. 105 (1998) 297.

4. X. Blaise, A. Rubio, S.G. Louie and M.L. Cohen, Europhys. Lett. 28 (1994) 335.

5. L. Bourgeois, Y. Bando and T. Sato, "Tubes of Rhombohedral Boron Nitride", J. Phys. D: Appl. Phys. 33 (2000)1902.

S. M. Nakmanson, A. Cazolari, V. Meunier, J. Bernholc, and M. Buongiorno Nardelli, "Spontaneous Polarization and Piezoelectricity in Boron Nitride Nanotubes", Phy. Rev. B 67, 235406 (2003).

7. R. Sen, B.C. Satishkumar, A. Govindaraj, K.R. Harikumar, G. Raina, J.P. Zhang, A.K. Cheetham, C.N.R. Rao, "B-C-N, C-N and B-N Nanotubes Produced by the Pyrolysis of Precursor Molecules Over Co Catalysts", Chem. Phys. Lett. 287 (1998) 671.

8. P. Ma, Y. Bando, and T. Santo, "CVD Synthesis of Boron Nitride Nanotubes Without Metal Catalysts", Chem. Phys. Lett. 337 (2001) 61-64.

9. N.G. Chopra, R.J. Luyren, K. Cherry, V.H. Cresi, M.L. Cohen, S.G. Louie, A. Zettle, "Boron Nitride Nanotubes", Science, New Series, 269 (1995) 966.

10. D.P. Yu, X.S. Sun, C.S. Lee, I. Bello, S. T. Lee, H. D. Gu, K. M. Leung, G.W. Zhou, Z. F. Dong, Z. Zhang, Appl. Phys. Lett 72 (1998) 1966.

[11.] Y. Chen, L.T. Chadderton, J.S Williams, and J. Fitz Gerald, "Solid State Formation of Carbon and Boron Nitride Nanotubes", Mater. Sci. Forum 343, 63 (2000).

[12.] Y. Chen, M. Conway, J.S Williams, J. Zou, "Large-Quantity Production of High-Yield Boron Nitride Nanotubes", J. Mater. Res. Vol 17 No.8 (2002) 1896-1899.

[13.] R. Ma, Y. Bando, T. Sato and K. Kurashima, " Growth, Morphology, and Structure of Boron Nitride Nanotubes", Chem. Mater. (2001) 13 2965.

[14.] R. Chen, Y. Bando, H. Zhu, T. Sato, C. Xu, and D. Wu, "Hydrogen Uptake in Boron Nitride Nanotubes at Room Temperature", J. Am. Chem. Soc. 124, 7672 (2002).

[15.] T.Oku, M. Kuno, I. Narita, "Hydrogen Storage in Boron Nitride Nanomaterials Studied by TG/DTA and Cluster Calculations", J. Phys. and Chem. of Solids, 65, 549 (2004).

COMPARISON OF ELECTROMAGNETIC SHIELDING IN GFR-NANO COMPOSITES

Woo-Kyun Jung
Seoul National University of Korea
301B/D 1255-1, San56-1, Shinlim, Kwanak,
Seoul, Korea, 151-742

Sung-Hoon Ahn
Seoul National University of Korea
301B/D 1205, San56-1, Shinlim, Kwanak,
Seoul, Korea, 151-742

Myong-Shik Won
Agency for Defence Development
Yuseong P.O.Box 35,
Daejeon, Korea, 305-600

ABSTRACT

The research on electromagnetic shielding has been advanced for military applications as well as for commercial products. Utilizing the reflective properties and absorptive properties of shielding material, the replied signal measured at the rear surface, or at the signal source can be minimized. The shielding effect was obtained from such materials that have high absorptive properties and structural characteristics, for example stacking sequence. In this research {glass fiber}/ {epoxy} / {nano particle} composites (referred to GFR-Nano composites) was fabricated using various nano particles, and their properties in electromagnetic shielding were compared. For visual observation of the nano composite materials, SEM(Scanning Electron Microscope) and TEM(Transmission Electron Microscope) were used. For measurement of electromagnetic shielding, HP8719ES S-parameter Vector Network Analyser System was used on the frequency range from 8 GHz to 12GHz. Among the nano particles, carbon black and Multi-walled Carbon Nano-tube (MWCNT) revealed outstanding electromagnetic shielding. Although silver nano particles (flake and powder) were expected to have effective electromagnetic shielding due to their excellent electric conductivities, test results showed relatively little shielding effect.

INTRODUCTION

Electromagnetic (EM) shielding is defined as the protection of the propagation of electric and magnetic waves from one region to another by using conducting or magnetic materials. The shielding can be achieved by minimizing replied signal or the signal passing through the material using reflective properties and absorptive properties of the material. This "Stealth" technology has been especially advanced for military application such as RADAR protection as well as for commercial products.

The shielding effect has been reported for such materials having high absorptive properties as Ferrite or Carbon-black [1-5]. Another mechanism of EM shielding was obtained from structural characteristics such as stacking sequence of composites [1, 6, 7]. Recently researchers have studied the electromagnetic properties of nano size particles [8-10].

In this research, Glass fiber reinforced composite made of {Glass fiber} / {Epoxy} / {Nano particles} were fabricated to observe EM shielding by nano particles mixed with Glass / Epoxy. Four types of nano particles were tested, and their properties in EM shielding were compared on X-band frequency (8GHz~12GHz).

EFFICIENCY OF ELECTROMAGNETIC SHIELDING

Figure 1 shows a schematic structure of a typical multi-layered material. A part of the incident EM wave is reflected at the surface while rest of the wave penetrates to inner materials. The wave undergoes surface reflection and interference at each layers to be absorbed, and finally attenuated wave escapes from the back of the material (transmitted wave).

EM-wave shielding is expressed by the shielding efficiency (SE), which uses unit of decibel (dB). The shielding efficiency is defined as the ratio of the power of incident EM-wave (P_I) to the power of transmitted EM-wave (P_T). The power of EM-wave can be related to electric field, incident electric field (E_I), and transmitted electric field (E_T) as Eq. (1) [11].

$$\begin{aligned} \text{SE(Shielding Efficiency)} \quad &= 10 \log (P_I / P_T) \\ &= 20 \log |E_I / E_T| \end{aligned} \tag{1}$$

Appling the boundary conditions of transverse EM-wave, the relation among the electric fields of incident, reflected, and transmitted EM-wave of a multi-layered material is generally described as Eq. (2).

$$\begin{bmatrix} E_I \\ E_I{}' \end{bmatrix} = A_I^{-1} A_1 B_1^{-1} \cdots A_N B_N^{-1} \begin{bmatrix} 1 \\ k_T \end{bmatrix} E_T \tag{2}$$

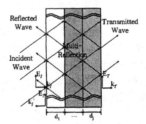

Figure 1. Schematic of electromagnetic shielding by a multilayer material.

From Figure 1, E_I' is the electric field of reflected electromagnetic-wave. Matrices A and B can be written as follows :

$$A_j = \begin{bmatrix} 1 & 1 \\ k_j & -k_j \end{bmatrix} \quad \text{and} \quad B_j = \begin{bmatrix} \exp(ik_j d_j) & \exp(-ik_j d_j) \\ k_j \exp(ik_j d_j) & -k_j \exp(-ik_j d_j) \end{bmatrix}$$

where, $i = \sqrt{-1}$, and k_j is the wave vector of the EM-wave in the j-th layer of multi-layered materials and d_j is thickness of the j-th layer. Minutely, k_j is the function of permittivity (ε_j), permeability (μ_j), and conductivity (σ_j) of the j-th layer at a given frequency ($f = \omega/2\pi$) such as

$$k_j^2 = \omega^2 \mu_j \varepsilon_j + i\omega \mu_j \sigma_j.$$

The EM-wave shielding obtained from Eqs. (1) and (2) is described as sum of the contributions due to reflection (SE_R), absorption (SE_A), and multi-reflections (SE_M) in the following Eqs. (3)~(6).

$$SE = SE_R + SE_A + SE_M \quad \text{(dB)} \tag{3}$$

$$SE_R = 20\log\left|\frac{(1+n)^2}{4n}\right| \quad \text{(dB)} \tag{4}$$

$$SE_A = 20\text{Im}(k)d \log e \quad \text{(dB)} \tag{5}$$

$$SE_M = 20\log\left|1 - \frac{(1-n)^2}{(1+n)^2}\exp(2ikd)\right| \quad \text{(dB)} \tag{6}$$

where, n is the index of refraction of the shielding material, and $\text{Im}(k)$ is the imaginary part of the wave vector in the shielding material.

According to the analysis of S parameter of two-port network system, transmittance (T), reflectance (R), and absorbance (A) can be described as :

$$T = |E_T/E_I|^2 = |S_{12}|^2 \ (=|S_{21}|^2) \tag{7}$$

$$R = |E_R/E_I|^2 = |S_{11}|^2 \ (=|S_{22}|^2) \tag{8}$$

$$A = 1 - R - T \tag{9}$$

From Eqs. (7)~(9), the effective absorbance (A_{eff}) can be described as

$$A_{eff} = (1-R-T)/(1-R)$$

with respect to the power of the effectively incident EM-wave inside the shielding material. It can be expressed as the form such as Eqs. (10) and (11).

$$SE_R = -10 \log (1 - R) \quad \text{(dB)} \tag{10}$$

$$SE_A = -10 \log (1 - A)$$
$$= -10 \log (T / (1 - R)) \quad \text{(dB)} \tag{11}$$

This analysis of the contribution of absorption and reflection to total EM-wave shielding is highly relevant for practical applications of the shielding materials.

EXPERIMENTAL

Specimen fabrication

Four types of nano particles (Table 1) were mixed with epoxy resin (YBD-500A80, *Br* series, Kukdo chemical Co.) and plain weave glass fabric (KN 1800, 300g/m², KPI Co.). The mixture of nano particles and epoxy resin were agitated by homogenizer to have 5wt% of particles in the mixture. After two hours of agitation, the mixture impregnated glass fabric. The impregnated three-phase composite was heated up to 120°C and held for one minute to become a prepreg. The GFR-nano composite parts were then fabricated by cutting into desired geometries, laying up, and curing these prepregs.

Specimens to measure EM-wave shielding were fabricated by curing process shown in Figure 2. The prepreg material was cured in a hot plate for two hours at 120℃ and 5.5 atm.

In order to see the effect of stacking sequence, each specimen was made either in cross-ply or in quasi-isotropic laminate. To see the effect of thickness on EM-wave shielding, 8 plies and 16 plies were laid for each type of specimens

For the visual observation of the nano composites and nano particles, Scanning Electron Microscope (SEM, JSM-5600) and Transmission Electron Microscope (TEM, JEM-3000F) were used.

Table 1. Nano particles used in this study

Nano Particle	Average Diameter (nm)
Multi-Walled Carbon Nano-Tube(MWCNT)	15
Carbon Black(CB)	40
Silver Nano Flake(SNF)	40
Silver Nano Powder (SNP)	40

366

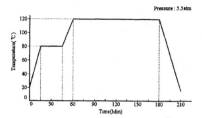

Figure 2. Cure cycle of GFR-nano composites.

Measurement of electromagnetic-wave shielding

The measurement of EM-wave shielding was performed using S parameter Vector Network Analyzer System (HP8719ES). The frequency range from 8GHz to 12GHz was scanned. From the measured data of reflected signal S_{11} and transmitted signal S_{21}, transmission coefficient and reflection coefficient were obtained. Using these coefficients, EM-wave shielding by absorption and reflection were calculated. Figure 3 shows the test setup for EM-wave shielding measurement.

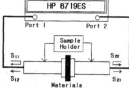

Figure 3. Test setup for electromagnetic shielding measurement.

RESULTS AND DISCUSSION

Morphology

Figure 4 shows TEM pictures of nano particles used in this study. The nano particles had 15nm~40nm average diameter. MWCNT showed 15nm average diameter and aspect ratio over 1,000.

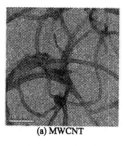   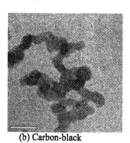

(a) MWCNT   (b) Carbon-black

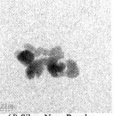

(c) Silver Nano Flake   (d) Silver Nano Powder
Figure 4. TEM pictures of nano particles used in this study.

The SNF had thin shell shapes while SNP had deformed ellipsoidal shapes.

Using SEM, structures of GFR-nano composites was observed. Although the cured epoxy and glass fiber of the composite material were observed with magnitudes between 1,000 and 2,000, the mingled nano particles, however, was not observable. Figure 5 (a) shows surface of the composite after grinding along fiber direction with 1,000 magnification. Figure 5 (b) is the cross-sectional view of the GFR-nano composite with 2,000 magnification. Even with higher magnification of 10,000, only the MWCNT was visually observed separately from the epoxy base (Figure 5 (c)). Lumped with epoxy, CB particles were not distinguishable, nor SNF and SNP particles.

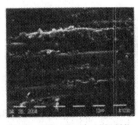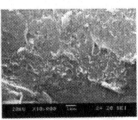

(a) Along fiber direction (X1,000)   (b) Cross section (X2,000)   (c) MWCNT in GFR- nano composites
(X10,000)

Figure 5. SEM pictures of GFR-nano composites.

Evaluation of electromagnetic-wave shielding

The measured EM-shielding efficiency was evaluated by characteristics of nano particles, thickness of specimens, and directions of fiber.

Figure 6 is an experimental data (GFR-MWCNT, 16C) of EM shielding as a function of frequency ranged from 8GHz to 12GHz.

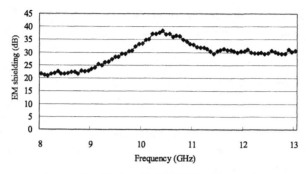

Figure 6. EM shielding of GFR-MWCNT (16C) composites.

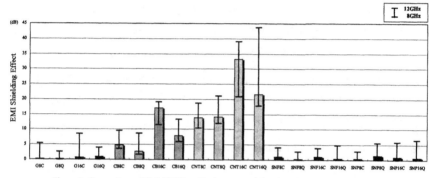

Figure 7. Comparison of EM shielding : Each nano particles, Stacking sequence, Thickness

Figure 7 shows the results of EM shielding of GFR-nano composites with various parameters. The raw GFR composites (G8C, G8Q, G16C, G16Q) had low EM shielding effect while GFR-nano composites containing MWCNT and CB (CB and CNT series) showed excellent shielding efficiency. Especially, GFR-MWCNT composites were three times higher efficiencies than those of CB series. The silver composites (SNF and SNP series) had higher efficiency than raw GFR-composites. The specimens of 8 ply laminate were 1.30mm thick in average, and specimens of 16 ply laminate were 2.50mm thick in average. Except for SNF and SNP series, as composite becomes thicker, shielding efficiency was increased. However, silver composites showed no change in shielding efficiency.

To investigate the effect of fiber orientation, prepergs were laid up with two conditions. One was cross-ply [0_8], and the other was quasi-isotropic [$(0 / 45)_2$]$_s$. In case of raw GFR composites and GFR-CB composites, cross-ply lay-up showed superior shielding efficiency to quasi-isotropic lay-up. For thin (8 plies) GFR-MWCNT composites, cross-ply lay-up had higher

369

shielding efficiency than quasi-isotropic lay-up. Thick (16 plies) GFR-MWCNT composites showed different distribution as a function of frequency range. At lower frequency range (under 10GHz), cross-ply lay-up shielded EM-wave more efficiently. At higher frequency range (over 10GHz), to the contrary, quasi-isotropic lay-up had higner shielding efficiency. The SNF and SNP series had no special effects on the shielding efficiency.

CONCLUSIONS

Through a series of experiments, EM-wave shielding of GFR nano composites was observed. Four nano particles added in GFR composites were investigated by measuring EM-wave shielding of the composite materials. CB and MWCNT showed outstanding shielding effect of 17dB~43dB on 10GHz frequency range, but silver nano particles did not show significant effect. Besides the electromagnetic properties, the addition of nano particles may improve the structural properties of composite, for which further study is required.

REFERENCES

[1]J. H. Oh, K. S. Oh, C. G. Kim, and C. S. Hong, "Design of radar absorbing structures using Glass/epoxy composite containing carbon black in X-band frequency ranges," *Composites Part B*, **35**, 49 - 56 (2004).

[2]M. S. Pinho, M. L. Gregori, R. C. R. Nunes and B. G. Soares, "Performance of radar absorbing materials by waveguide measurements for X and Ku-band frequencies," *European Polymer Journal*, **38**, 2321 - 2327 (2002).

[3]D. D. L. Chung, "Electromagnetic interference shielding effectiveness of carbon materials," *Carbon*, **39**, 279 - 285 (2001).

[4]G. Li, G. G. Hu, H. D. Zhou, X. J. Fan and X. G. Li, "Absorption of microwaves in $La_{1-x}Sr_xMnO_3$ manganese powders over a wide bandwidth," *Journal of Applied Physics*, **90**, 5512 - 5514 (2001).

[5]K. B. Cheng, S. Ramakrishna and K. C. Lee, "Electro magnetic shielding effectiveness of copper/glass fiber knitted fabric reinforced poly-propylene composites," *Composites Part A*, **31**, 1039 - 1045 (2000).

[6]S. A. Tretyakov and S. I. Maslovski, "Thin Absorbing Structure for all incidence angles based on the use of a high-impedance surface," *Microwave and Optical Technology Letters*, **38**, 175 - 178 (2003).

[7]K. Matous and G. J. Dvorak, "Optimization of Electromagnetic Absorption in Laminated Composite Plates," *Transactions on Magnetics*, **39**, 1827 - 1835 (2003).

[8] A. Kajima, R. Nakayama, T. Fujii and M. Inoue, "Variation of dielectric permeability by applying magnetic field in nano-composite Bi_2O_3 - Fe_2O_3 - $PbTiO_3$ sputtered films," *Journal of magnetism and magnetic materials*, **258**, 597 - 599 (2003).

[9] L. I. Trakhtenberg, E. Axelrod, G. N. Gerasimov, E. V. Nikolaeva and E. I. Smirnova, "New nano-composite metal-polymer materials : dielectric behaviour," *Non-Crystalline Solids*, **305**, 190 - 196 (2002).

[10] P. Talbot, A. M. Konn and C. Brosseau, "Electromagnetic characterization of fine-scale particulate composite materials," *Journal of magnetism and magnetic materials*, **249**, 481 - 485 (2002).

[11] Y. K. Hong, C. Y. Lee, C. K. Jeong, D. E. Lee, K. Kim and J. Joo, "Method and apparatus to measure electromagnetic interference shielding efficiency and its shielding characteristics in broadband frequency ranges," *Review of Scientific Instruments*, **74**, 1098 - 1102 (2003).

DENSIFICATION BEHAVIOR OF ZIRCONIA CERAMICS SINTERED USING HIGH-FREQUENCY MICROWAVES

M. Wolff, G. Falk, R. Clasen
Saarland University, Department of Powder Technology
Building 43, D-66123 Saarbrücken, Germany

G. Link, S. Takayama, M. Thumm
Forschungszentrum Karlsruhe GmbH, IHM
Hermann-von-Helmholtz-Platz 1, D-76344 Eggenstein-Leopoldshafen, Germany

ABSTRACT

The aim of the investigation was the formation of zirconia ceramics with a fine-grained microstructure and low porosity. The potential of high frequency (30 GHz) microwave sintering in this field was studied. The application of microwaves allowed high heating rates and short processing times because of volumetric heating and enhanced sintering kinetics and thereby a better control of the microstructure. Green bodies were made of tetragonal (8 mol-% Y_2O_3) and cubic (10 mol-% Y_2O_3) nanoscale and submicron zirconia powders, dispersed in aqueous suspensions. They were prepared by electrophoretic deposition (EPD) and had relatively high green densities and a homogenous pore size distribution.

Green bodies and ceramics were characterized by means of XRD, SEM, density measurements by means of the Archimedes method and Hg-porosimetry. The main focus of this work is on the investigation of grain size evolution as a function of density.

INTRODUCTION

In this work, the application of advanced green body shaping techniques, in combination with sintering using high-frequency microwaves was investigated to determine the possibilities of grain size and density control. The microstructure of ceramic materials has an important effect on their mechanical, electrical and optical properties. Porosity and grain size affect especially hardness, optical transmittance, thermal and ionic conductivity. The production of highly dense ceramics with small grain sizes makes high demands on shaping as well as on sintering methods. The use of fine powders as raw materials is required to obtain fine-grained microstructures in ceramics. But shaping of homogeneous ceramic green bodies with high densities, in particular, from submicron or nanosized powders, is a challenge. Dense and homogenous green bodies with small pores are a precondition for homogenous densification and low porosity after sintering. Removing large pores during sintering is much more difficult than elimination of small ones, sometimes even impossible and requires higher temperatures, which may result in strong grain growth. Furthermore, nanosized ceramics show slow grain growth only up to a certain critical temperature [1]. Thus substantial increase of density without grain growth cannot be achieved with long dwelling times at high temperatures.

While the primary particle size of the powders is determined by their physical characteristics, the degree of particle agglomeration and the density of the green bodies depend on the shaping process. Dry pressing methods are not suitable to shape nanosized and submicron powders, because the compaction of very fine particles is difficult, due to their high degree of aggregation. Dry pressing of such powders results in non-homogenous pore size distributions and mechanical stresses within the green bodies. Electrophoretic deposition (EPD) from aqueous sus-

373

pensions is a simple and fast method to produce homogeneous and highly dense structures and was used to shape the green bodies for the investigations in this work [2]. For optimal shaping with EPD a deaggregation of the powders is required. The dispersion of the powders into stable primary particles and thus the supply of an agglomerate-free suspension, as well as high solids loadings, are fundamental conditions to produce green bodies with high green densities.

The heating process for sintering is an important step in the processing of engineering ceramics. Different sintering techniques such as conventional sintering (CS), microwave sintering (MWS), spark plasma sintering, sinter forging, hot pressing and hot isostatic pressing are applied in order to control the microstructure. The use of microwaves allows transfer of energy directly into the volume of dielectric materials. Through absorption mechanisms such as ionic conduction, dipole relaxation and photon–phonon interaction, the energy is converted to heat. Using MWS, volumetric heating can be realized, which allows the application of high heating rates and significantly shortens the processing time. Furthermore, the densification process of ceramic bodies seems to be enhanced by sintering in a microwave field, allowing a reduction of sintering temperatures and/or dwelling times compared to the processing in a standard sintering furnace [3, 4].

Heating of ceramics under microwave radiation depends on the dielectric properties. The absorbed power P inside the material is proportional to frequency ω and dielectric loss ε'', which itself is a function not only of frequency but also of temperature T (Eq. 1). This loss factor increases with temperature as well as with frequency in the microwave range for many low-loss ceramics.

$$P = \frac{1}{2}\omega\varepsilon_0\varepsilon''(\omega,T)|E|^2 \qquad \text{(Eq. 1)}$$

Even though microwaves with a frequency of 2.45 GHz and 915 MHz have the broadest application, high power, millimeter-wave sources (gyrotrons) are an interesting alternative [5, 6]. In the case that the installed microwave power is not sufficient to reach sintering temperatures for heating experiments with low-loss ceramics, it is possible to preheat the ceramic by infrared radiation to temperatures where coupling of the ceramic samples is sufficiently high for microwave heating. Another possibility is to change to heating with higher frequencies such as 30 GHz, since power absorption is by a factor of 10–100 higher compared to the absorption at magnetron frequencies.

The sintering behavior of tetragonal stabilized zirconia (TZP) in a conventional and a microwave furnace with heating rates of 12 K/min was investigated in [7]. Maximal densities of 99.5 %TD were obtained at 1250 °C in the MW furnace whereas same densities were determined at 1350 °C in a conventional furnace. Important differences in the densification level were noticed between 1100 and 1300 °C. This was attributed to accelerated diffusion especially of O^{2-}-ions. No significant differences in grain structure were determined. An investigation in "ultrafast" sintering with 100 K/min in the MW furnace resulted in slightly lower sintered densities and crack generation.

Nightingale [8] compared the density-grain size relationship of conventionally and microwave sintered TZP and cubic stabilized zirconia (CSZ). The activation energy for grain growth is significantly lower in cubic than in tetragonal zirconia, which results in much stronger grain growth of cubic zirconia. For microwave sintered TZP samples, significantly smaller grain sizes were found at densities < 96 %TD and a slightly smaller grain sizes at higher densities, suggesting that microwaves tend to accelerate lattice diffusion more than they do surface and grain boundary diffusion. Nevertheless no differences were determined for CSZ, indicating a balance between the different diffusion processes. Accelerated grain growth after ageing at 1500 °C for 15 h was determined in [4], comparing MWS and conventional sintering. The grain size of the TZP samples was 1.5 times greater for MWS than for conventionally sintered ones.

EXPERIMENTAL

The green bodies were manufactured from nanosized powders of 8 mol-% yttria stabilized zirconia NA8Y (8Y Zirconium Oxide, Nanostructured and Amorphous Materials, USA) and powders of 10 mol-% yttria fully stabilized zirconia (TZ10Y, Tosoh Corp., Japan). The powders were characterized by XRD and TEM. The NA8Y powder had a mean grain size of 50 nm and consisted of monoclinic and tetragonal phase. TZ10Y had a mean aggregate size of 0.5 μm and consisted of cubic phase.

Electrophoretic deposition (EPD) was applied to manufacture the green bodies. As preparation for EPD, the powders were dispersed in water and the suspensions were electrosterically stabilized by tetramethylammoniumhydroxide (TMAH) and a dispersing aid. The dispersion process was carried out by means of a Dispermat N1–SIP (VMA-Getzmann GmbH, Germany), using a serrated disc. Electrophoretic deposition (EPD) was carried out in a horizontal cell (10 x 40 x 40 mm) under constant voltage, applying the membrane process [9]. The green bodies were first dried at room temperature for 24 h, then in a drying chamber at 120 °C for 24 h. Both drying processes were performed in air. The pore sizes were determined by Hg-porosimetry (Pascal 440, CE Instruments, Italy). The green bodies made of NA8Y had a green density of approximately 56 % of the theoretical density (TD) and those of TZ10Y, 54 %TD. The pore size distribution of the green bodies was homogenous and monomodal. NA8Y and TZ10Y had a mean pore size of 12 nm and 60 nm, respectively.

The conventional sintering experiments were carried out in air at 1300-1700 °C. The microwave sintering experiments were performed in a compact gyrotron system operating at 30 GHz in air with sintering temperatures starting from 1300 °C up to 1600 °C and a heating rate of 50 K/min. To avoid the evolution of an inverse temperature profile, which is a specific feature of microwave volumetric heating, the samples were sintered within a ZrO_2 crucible, placed in a box of mullite ceramic fiber boards, for thermal insulation.

Densities of the sintered samples was measured using the Archimedes principle, assuming a theoretical density of 6.07 g/cm^3 for tetragonal Y-ZrO$_2$ and 5.95 g/cm^3 for cubic Y-ZrO$_2$. To determine the grain sizes, the samples were thermally etched at 1240 °C for 30 min. Characterization of grain structure was performed by high resolution scanning electron microscopy (HRSEM). Grain sizes were measured applying the linear intercept technique, using a4i software (Aquinto AG, Germany).

RESULTS AND DISCUSSION

The conventional sintering experiments were carried out at a heating rate of 10 K/min, while microwave sintering (MWS) was conducted at 50 K/min. After heating, the samples were cooled down directly without a dwelling time. Thermal insulation and thus cooling behavior of the conventional furnace was different from the gyrotron system. The gyrotron system allowed much higher cooling rates than did the conventional furnace, since the thermal mass of the sample and thermal insulation was much smaller

In Fig. 1, the sintering behavior of NA8Y at different sintering temperatures is shown. At 1300 °C, CS resulted in a higher density than did MWS. This may be attributed to the lower heat-

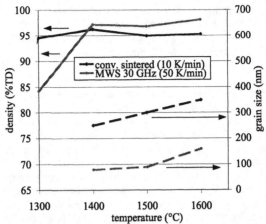

Fig.1: Density and estimated grain sizes after sintering of NA8Y
(tetragonal ZrO_2) in a conventional furnace and in a gyrotron system

ing rate in the case of CS. For higher sintering temperatures, densities of the microwave sintered samples were higher than for conventionally sintered ones, resulting in 1 %TD higher densities at 1400 °C, 2 % at 1500 °C and 3 % at 1600 °C.

Furthermore, the evolution of the grain sizes depended on the sintering techniques. MWS resulted in smaller grain sizes. Already at 1400 °C, in an earlier sintering stage, the conventionally sintered samples clearly had larger grain sizes. This is a possible explanation for the higher densification reached with MWS at higher temperatures. As the grain sizes remained smaller, a better elimination of porosity and thus a more homogenous densification resulted in higher densities.

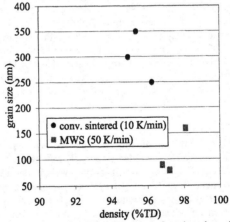

Fig. 2: Grain size as function of density after sintering of NA8Y
(tetragonal ZrO_2) in a conventional furnace and in a gyrotron system

a)   b)

Fig. 3: Microstructure of NA8Y: a) sintered with microwaves in the gyrotron system (50 K/min),
b) conventionally sintered (10 K/min) at 1500 °C

It is remarkable that even at higher sinter temperatures and clearly higher sintered densities, respectively, the grain size of the microwave sintered, tetragonal ZrO_2 remained smaller than
the conventionally sintered ones. In Fig. 2, the grain sizes are presented as function of the sintered density. At a density of around 95 %TD, the conventionally sintered samples had grain
sizes of about 300 nm. The maximum grain sizes of the MWS samples were about 150 nm, having a sintered density of 98 %TD. For sintered densities of 97 %TD, the grains were smaller than
100 nm. SEM pictures of the fracture surface of the samples, sintered at 1500 °C, are presented in
Fig. 3 a) and 3 b). As it can be seen, the morphology of the conventionally sintered samples was
clearly coarser than that of the microwave sintered samples.

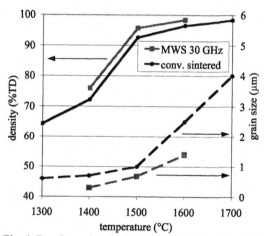

Fig. 4: Density and grain sizes after sintering of TZ10Y
(cubic ZrO_2) in a conventional furnace and in a gyrotron system

The sintering behavior of cubic zirconia (TZ10Y) is presented in Fig. 4. Densities and
grain sizes after heating up to 1600 °C of the samples are shown. The samples were cooled down
directly after heating. A difference in sintering behavior for conventional and microwave sinter-

ing can be determined for densification as well as for grain sizes. At 1400 °C, conventionally sintered samples had a density of 73 %TD whereas sintering with the gyrotron system results in densities of 76 %TD. Similar differences in density can be observed after sintering at 1500 °C (92 and 95 %TD) and 1600 °C (96 and 98 %TD). Simultaneously, MWS resulted in smaller grain sizes than CS, even though they had a higher sintered density. At 1400 °C, the average grain size of the MWS samples was about 300 nm whilst the conventionally sintered ceramics had grain sizes of about 700 nm.

A difference in grain growth can also be observed after sintering at 1500 °C. The average grain sizes of conventionally sintered samples reach 1 μm and 680 nm after sintering with microwaves. A stronger difference is determined after sintering at 1600 °C. The microstructure of the conventionally sintered samples was clearly coarser than the microwave sintered ones, having grain sizes of 2.5 and 1.4 μm, respectively.

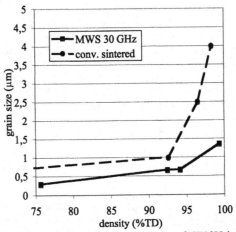

Fig. 5: Grain size as function of sintered density of TZ10Y (cubic ZrO₂), sintered in a conventional furnace and in a gyrotron system

The interaction of densification and grain growth is presented in Fig. 5, where grain size is shown as function of density. It can be observed, that at comparable sintered densities, conventional sintering resulted in explicitly coarser morphology. As can be seen in Fig. 4, in the conventional furnace, it was necessary to sinter at higher temperatures to reach the same densities as in the gyrotron system. At those higher sintering temperatures, strong grain growth occurred. Thus with a density of 92 %TD, the microwave sintered samples had average grain sizes of 600 nm. In order to reach comparable densities to conventional sintering, the samples already had grain sizes of 1 μm. Strong grain growth occurred mainly for densities higher than 92 %TD. Having densities of 97 %TD, conventionally sintered TZ10Y had grain sizes of 4 μm. Note that using CS, temperatures of 1700 °C were necessary to reach those densities. At a density of 98 %TD, the microwave sintered samples still had grain sizes < 1.5 μm. SEM images of polished and thermally etched samples of microwave and conventionally sintered TZ10Y are shown in Fig. 6 a) and 6 b). Distinct differences in morphology can be observed.

a)   b)

Fig.6: SEM images of TZ10Y, sintered a) in a gyrotron system (density 95 %TD, average grain size 700 nm) and b) in a conventional furnace (92 %TD, 1 μm) at 1500 °C

For a reliable comparison of process temperatures from different sintering techniques as presented in this work, the different heating mechanisms between conventional and microwave sintering and the resulting differences in temperature distribution within the ceramic body must be considered. In standard sintering furnaces, while heating, temperature gradients are induced in ceramics due to the low penetration depth of infrared radiation. This results in thermal gradients with a hot surface and a colder interior, which is compensated by a thermal conduction process depending on the thermal conductivity of the sintered material. The thermal conductivity of zirconia is low: tetragonal stabilized zirconia as well as cubic stabilized zirconia exhibit $\lambda \sim 2 \ Wm^{-1}K^{-1}$, which is even much smaller in a powder compact. Therefore, it is necessary to apply an optimized time-temperature program with relatively low heating rates to avoid crack formation or even destruction of the samples due to thermal stresses.

Using MWS, a volumetric heating can be realized which allows the application of high heating rates. On the other hand, a reverse problem to the conventional heating, namely temperature gradients from the hotter inside of the ceramic body to the surface, can exist. Within the gyrotron system, this effect is avoided by placing the samples within a zirconia crucible surrounded by mullite fiber boards for thermal insulation.

Furthermore, measurement of temperature was different for the sintering methods presented which may lead to systematic errors and therefore to wrong conclusions if the measured temperature is compared from different systems. The thermocouple of conventional furnaces usually is placed on the thermal insulation of the furnace wall. Thus the temperature is not detected directly next to the sample but it is assumed that the furnace has a homogenous temperature distribution. In the gyrotron system, the thermocouple is in contact with the sample surface. Nevertheless, especially for high heating rates, higher temperatures inside the sample may occur.

As exact comparison of temperature is difficult for the different sintering methods, the relationship between density and grain size shows reliable differences in densification behavior. Here the temperature parameter is not needed to discuss microwave specific effects on sintering. In Fig. 2 and Fig. 5 it can clearly be seen that the different sintering methods resulted in different sintering behavior. Using microwave sintering, higher sintered densities could be reached and the grains were significantly smaller than after conventional sintering. Furthermore it was possible to densify the ceramic with slower grain growth. As the grains remained small at an advanced sintering stage, higher maximum final sintered densities could be reached. The advanced sintering behavior can be attributed to higher sintering rates and an enhanced densification behavior induced by the microwave field.

CONCLUSION

The application of high frequency microwave sintering techniques in comparison with conventional sintering on tetragonal and cubic zirconia was investigated. A combination of advanced shaping and advanced sintering methods was carried out. The green bodies were formed by means of electrophoretic deposition (EPD) to produce homogenous and dense structures. Sintering experiments were carried out in a compact gyrotron system at 30 GHz. For high frequency microwave and conventional sintering, different densification behavior was found. As microwave sintering allows higher heating rates and enhanced densification, smaller grain sizes could be achieved for tetragonal as well as for cubic zirconia during intermediate and final sintering stage. Furthermore, higher final sintered densities were reached by applying microwave sintering. This is related to the smaller grain sizes during densification and thus a better elimination of porosity. Distinct differences in grain growth in relation to the densification state of the zirconia ceramics were observed. At their maximum sintered densities, microwave sintered tetragonal zirconia had nearly half as big grain sizes as conventional sintered ones. An even stronger effect was detected for cubic zirconia. Conventional sintering resulted in 2.5 times coarser grains.

ACKNOWLEDGEMENTS

The financial support from the Deutsche Forschungsgemeinschaft (DFG, Graduiertenkolleg 232) and the Helmholtz-Gemeinschaft (VH-FZ-024) is gratefully acknowledged.

REFERENCES

[1] M. J. Mayo, D. C. Hague and D.-J. Chen, "Processing Nanocrystalline Ceramics for Applications in Superplasticity," *Mater. Sci. Eng. A*, **166** 145-159 (1993).

[2] J. Tabellion and R. Clasen, "Electrophoretic Deposition from Aqueous Suspensions for Near-shape Manufacturing of Advanced Ceramics and Glasses - Applications," *J. Mater. Sci.*, **39** 803-811 (2004).

[3] M. A. Janney, H. D. Kimrey, W. R. Allen and J. O. Kiggans, "Enhanced Diffusion in Sapphire During Microwave Heating," *J. Mater. Sci.*, **32** 1347-1355 (1997).

[4] S. A. Nightingale, D. P. Dunne and H. K. Worner, "Sintering and Grain Growth of 3 mol% Yttria Zirconia in a Microwave Field," *J. Mater. Sci.*, **31** 5039-5043 (1996).

[5] H. D. Kimrey, M. A. Janney and P. F. Becker, "Techniques for Ceramic Sintering Using Microwave Energy"; pp. 136-137 in *Int. Conference on Infrared and Millimeter Waves*, ed. Edited by Orlando, 1987.

[6] G. Link, L. Feher, M. Thumm, H. J. Ritzhaupt-Kleissl, R. Böhme and A. Weisenburger, "Sintering of Advanced Ceramics Using a 30-GHz, 10-kW, CW Industrial Gyrotron," *IEEE Trans. Plasma Sci.*, **27** [2] 547-555 (1999).

[7] A. Goldstein, N. Travitzky, A. Singurindy and M. Kravchik, "Direct Microwave Sintering of Yttria-stabilized Zirconia at 2.45 GHz," *J. Europ. Ceram. Soc.*, **19** 2067-2072 (1999).

[8] S. A. Nightingale, H. K. Worner and D. P. Dunne, "Microstructural Development during the Microwave Sintering of Yttria-Zirconia Ceramics," *J. Am. Ceram. Soc.*, **80** [2] 394-400 (1997).

[9] R. Clasen, S. Janes, C. Oswald and D. Ranker, "Electrophoretic Deposition of Nanosized Ceramic Powders"; pp. 481-486 in *Ceram. Transactions*, ed. Edited by H. Hausner, G. L. Messing and S. Hirano. Am. Ceram. Soc., Westerville (USA), 1995.

MANUFACTURING OF DOPED GLASSES USING REACTIVE ELECTROPHORETIC DEPOSITION (REPD)

Dirk Jung, Jan Tabellion and Rolf Clasen
Saarland University, Department of Powder Technology
Building. 43, D-66123 Saarbrucken, Germany

ABSTRACT

Doped glasses can be manufactured by means of gas infiltration, soaking of green bodies, by using powder mixtures or by melting. The melting point of silica glass is 2100 °C and most dopants evaporate at temperatures in this range. Because the sintering temperature of silica glass is about 700 °C lower then melting point, dopants, that would evaporate during the melting process, can be used. But it is difficult to achieve homogeneous green bodies by using powder mixtures because separation occurs as the particles have different sizes and densities. In case of the soaking method, during the drying process a surface segregation of the salt ions leads to samples with an inhomogeneous distribution. A promising method to manufacture homogeneous silica green bodies is the electrophoretic deposition (EPD) [1]. In the first approximation the mobility of the particles is independent of their size [2].

A modification of the EPD, by adding salts to the suspension, leads to reactive electrophoretic deposition (REPD). By varying the amount of added salts to the suspension, the dissolved ions modify the occupancy of particle surfaces and the composition of the cloud of ions. The adsorbed ions can be co-deposited with the particles leading to a very homogeneously doped green body. It is tested first for a suspension of SiO_2 that contained different amounts of boric acid and cobalt chloride. It is shown that the green bodies doped with boric acid can be sintered at lower temperatures compared to undoped ones. However, the sintering temperature depends on the amount of boric acid added to the suspension before.

INDRODUCTION

By doping of silica glass with different types and amounts of additives important properties like thermal expansion coefficient, refractive index or color can be tailored. Conventional doped glasses are fabricated via melting-routes. The melting point of silica glass is 2100 °C. In this case most of dopants cannot be used because they evaporate at 2100 °C.

Nano-sized, fumed silica powders (DEGUSSA OX50, A380) can be sintered at 1300 °C [3] because of their large sintering activity. Due to the sintering temperature being lower than the melting point, dopants can be used that normally evaporate at 2100 °C. Using powders to manufacture glasses offers alternative methods like gas infiltration, soaking of green bodies and employing powder mixtures to dope them. Gas infiltration is limited by the small variety of possible species and by the pore size of the green body. The pore size has to be larger than the mean free length of path of the gas molecules. In the case of green body soaking, ion-movement to the green body surface during the drying process leads to inhomogeneously-doped glasses [4]. Using powder mixtures to fabricate green bodies the different particle densities and sizes lead to decomposition and inhomogeneous green bodies are achieved. Another suspension-based method to fabricate green bodies using powders is electrophoretic deposition (EPD) [5]. In the case of aqueous slips the decomposition of water at applied DC>1.5 V leads to gas bubble formation at the electrodes, which are assembled into the green body. Using an ion-permeable membrane [6], which divides the electrophoresis cell into a suspension chamber and a second chamber, the deposition and recombination of ions are separated in space. Figure 1 illustrates the principle of the membrane method.

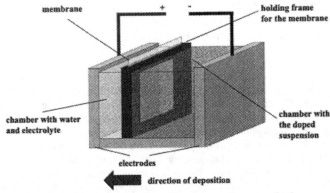

Figure 1: Schematic of the electrophoretic deposition by membrane method

Via EPD near-netshape manufacturing is possible since the surface tension of suspensions of silica particles ($\approx$ 20-70 mN/m) is lower than that of silica glass melts (340 mN/m) [7]. Moreover, it is possible to shape at room temperature. Via EPD it is possible to fabricate silica green bodies with a relative density of 45 %. In ref. [8] silica green bodies were fabricated by means of EPD from aqueous suspensions with a bimodal mixture of powders with a mean particle-size of 15 µm (SE15) and 50 nm (OX50). Maximum green density of 84 %TD was reached by a SE15 to OX50 ratio of 10:90. At the same time, the linear shrinkage was lowered to 4.7 %.

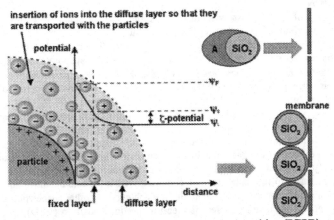

Figure 2: Functional principle of the reactive electrophoretric deposition (REPD)

Due to the dissociation of water there are ions in suspension. These ions arrange around the dispersed particles to compensate their surface charge. They form a cloud, which consists of anions and cations [9]. By shaping for EPD, a part of these ions will be co-deposited with the particles. The difference of reactive electrophoretic deposition (REPD) is the addition of water-soluble salt to the suspension to modify the cloud of ions. In Figure 2, the functional principle is schematically shown.

EXPERIMENTAL

Boric acid and cobalt chloride were used as a model system to test if REPD works with anions and cations. The experiments were carried out with a fused silica powder, which has a mean particle-size of 40 nm (DEGUSSA OX50) and a broad particle size distribution. Furthermore, boric acid (Roth) with a purity of 99.9 % and cobalt chloride (Fluka) with a purity of 98 % were used as dopants. The pH-value was adjusted using tetramethylamoniumhydroxide (TMAH), which is a strong base. The suspensions were prepared by dispersing nanosized OX50 in bidistilled water by means of a dissolver (PC-Laborsystem, Typ LDV1). Boric acid (anion doping) and cobalt chloride (cation doping) respectively were added to the suspension and the pH-value was adjusted to 11 by using TMAH.

The doped silica green bodies were shaped by REPD under constant electrical field by means of the membrane method [6]. The second chamber was filled with bidistilled water containing different amounts of TMAH. The conductivity was adjusted by adding TMAH to the 5-, 10- and 20-fold of the suspension's conductivity. In a modified experiment, boric acid or cobalt chloride were added to the liquid in the second chamber, to analyze if these ions pass the membrane to be deposited in the green body. After shaping, the green bodies were dried at room temperature for 12 hours. The content of ions was measured by inductively coupled plasma emission spectrometry (ICP-ES) (Jobin Yvon, JY 24). Furthermore, the green bodies were sintered at 900, 1000, 1100, 1200, 1300, 1350 and 1400 °C in a zone-sintering furnace.

The optical properties of the sintered samples were studied with an UV/VIS-spectrometer (Brucker, IFS 66v). Sintering kinetics were measured with a vertical dilatometer (Linseis, L75).

RESULTS AND DISCUSSION

The variation of the sintering temperature is an appropriate method to determine qualitatively the amount of boron oxide in the green bodies, because the concentration of boron oxide decreases the sintering temperature. In Figure 3, the linear shrinkage behavior of different doped green bodies is shown. The deposition was carried out from suspensions containing 0, 1, 5, 10 and 15 wt.% of boric acid. It is obvious, that the sintering temperature decreased with an increasing amount of boric acid in the suspension. Sintering of green bodies without boron oxide started at a temperatures of 1100 °C. 15 wt.% of boric acid decreased the first sintering stage to 900 °C. The zeta-potential of silicon dioxide at pH 11 was about -70mV. Due to the negative zeta-potential it was expected that only cations can co-deposit, because a negative zeta potential results in adsorption of cations. This is reasonable since a different electrical charge leads to an attractive force. Boron acid generates the anion $B(OH)_4^-$. Because of the negative zeta-potential it is remarkable that doping of green bodies via electrophoretic deposition using boron acid is successful. A co-deposition of anions and particles proves that anions as well as cations are part of the cloud of ions around the particles. To determine the incorporated amount of ions in the green body, an ICP analysis was performed to determine the final quantity of elements so that the efficiency could be calculated.

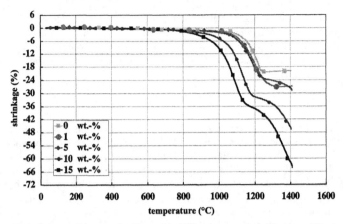

Figure 3: Sintering diagram for green bodies with different contents of boric acid

In Figure 4, the content of boron oxide in the green body as a function of the amount of boric acid in the suspension is shown. There is a linear correlation between the quantity of ions in the suspension and that in the green body. As shown in the diagram, less than half of the quantity in the suspension can be determined in the green bodies. As a matter of fact, only ions are co-deposited that are in the cloud around the particles. The quantity of ions in green bodies has to be lower than in suspension. Moreover, it is shown that an addition of boric acid in the second chamber also increases the concentration of boron oxide in the green bodies. Hereby, the suspension contained 5 wt.% and the water in the second chamber 0, 1, 3 and 5 wt.% of boric acid. In this case the amount of boric acid in green bodies was enhanced from 1.75 wt.% (0 wt.% in second chamber) to 3.6 wt.% (3 wt.% in second chamber).

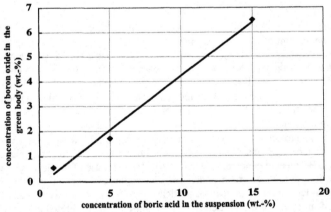

Figure 4: Concentration of boric oxide in green bodies increases with increasing amount of boric acid in suspension

Due to the negative zeta-potential a co-deposition of cations and particles is conceivable, due to the adsorption of cations needed to compensate the surface charge of particles. In Figure 5, the correlation between the concentration of cobalt chloride in the suspension and that in the green body is shown. The quantity of cobalt chloride in the green bodies is a linear function of cobalt chloride content in the suspension. Obviously, a higher deposition efficiency compared to the use of boric acid can be reached.

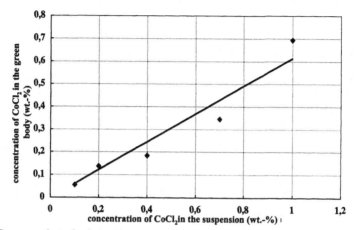

Figure 5: Concentration of cobalt chloride in the green bodies increases with increasing amount of cobalt chloride in suspension

Adding cobalt chloride to the liquid of the second chamber resulted in green bodies with an inhomogeneous structure can be observed in a gradient of color. Deposition proceeded from the cathode to anode. As the second chamber is almost saturated with the cations, the movement of those ions from suspension to this chamber is low. Anions from the second chamber can pass the membrane and displace the cations out of the green body, resulting in a gradient In Figure 6, two sintered glasses doped with cobalt chloride are shown.

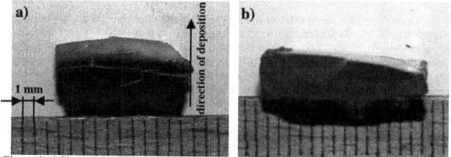

Figure 6: a) Sintered sample, deposited with cobalt chloride in both chambers of the electrophoresis cell, b) cobalt chloride only in the suspension chamber

The optical micrograph on the left hand side shows a glass that is produced with cobalt chloride in the second chamber. Decomposition can be clearly noted. This effect is not visible in the picture on the right-hand side. This glass was fabricated using cobalt chloride in only the suspension chamber. Boric acid-doped green bodies could also have a gradient in the concentration of ions but it could not be analyzed. Analyzing via ICP-ES needs dissolved samples using HF. That is why one is not able to observe gradients in the distribution of ions. The measurement of a gradient by means of sintering kinetics is also impossible because the total shrinkage was measured.

Furthermore, the transparency of the cobalt colored glasses was analyzed. Green bodies have been manufactured using suspensions with 0, 0.1, 0.2, 0.4, 0.7 and 1 wt.% cobalt chloride. After sintering the optical properties of the glasses produced were studied using UV/VIS-spectroscopy. Hereby, only glasses from suspensions with an amount of 0, 0.1, 0.2 and 0.4 wt.% cobalt chloride could be analyzed. Glasses obtained from suspensions with 0.7 and 1 wt.% were deep blue so that measurements did not lead to reasonable results. In Figure 7, the transmittance of sintered silica glass co-deposited from suspensions with different amounts of cobalt chloride is shown. The spectra are compared to that of a glass commercial fabricated.

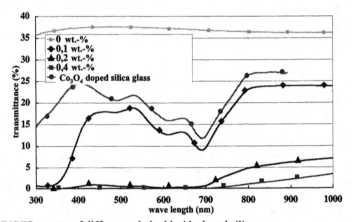

Figure 7: UV/VIS-spectra of different cobalt chloride doped silica

A parameter that limits the maximum amount of boric acid or cobalt chloride in the suspension is the electrical conductivity. Electrophoretic deposition of SiO_2 from suspensions with a value of σ higher than 20 mS/cm was not possible. If conductivity is too high the suspension will be heated to boiling and bubbles were deposited resulting in pores in the green body. Beside suspensions conductivity, the conductivity of the water with TMAH in the second chamber is also very important. It was determined that the conductivity in the second chamber had to be 10–fold of the suspensions conductivity. Due to the co-deposition of the ions, we achieved a doped green body that was more homogeneous compared to other methods such as gas infiltration, soaking of green bodies or by means of powder mixtures. Because of using nano-sized powders there were short paths of diffusion.

CONCLUSIONS

It was shown that manufacturing of doped of silica glass in a single-step process using EPD was possible. Boric acid and cobalt chloride were used as a model system to analyze if anions as well as cations could be co-deposited with nano-sized particles. Since boron oxide decreased the sintering temperature, the quantity was analyzed qualitatively by dilatometry combined by measuring the sintering temperature. The quantity of cobalt chloride in the glasses was appraised by the coloration. The amount of dopants in the green bodies was determined by ICP whereas it was shown that the quantity in the green bodies was twice larger than that in the suspension. The electrical conductivity in the suspensions must be smaller than 20 mS/cm in order to make deposition possible. Deposition with a ratio in conductivity between suspension and the second chamber about 1:10 resulted in better green density and green strength then for other ratios.

LITERATURE

[1] J. Tabellion and R. Clasen, "Advanced Ceramic or Glass Components and Composites by Electrophoretic Deposition/ Dmpregnation using Nanosized Particles", pp. 617-627, H.-T. Lin, M. Singh, Eds., 26th Annual Conference on Composites, Advanced Ceramics, Materials and Structures (The American Ceramic Society, Cocoa Beach, Florida, USA, 2002).

[2] E. Hückel, "Die Kataphorese der Kugel", *Physik. Z.*, **25** 204-210 (1924).

[3] J. Tabellion, R. Clasen, J. Reinshagen, R. Oberacker and M. J. Hoffmann, "Correlation between Structure and Rheological Properties of Suspension of Nanosized Powders", *Key Engineering Materials*, **206-213** 139-142 (2002).

[4] R. Clasen, "Verfahren zur Herstellung von dotierten Gläsern", *Fortschrittsber. Dtsch. Keram. Ges.*, **17** [1] 118-126 (2002).

[5] R. Moreno and B. Ferrari, "Advanced Ceramics via EPD of Aqueous Slurries", *Am. Ceram. Soc. Bull.*, **Jan 2000** 44-48 (2000).

[6] R. Clasen, "Forming of Compacts of Submicron Silica Particles by Electrophoretic Deposition", pp. 633-640, H. Hausner, G. L. Messing, S. Hiranos, Eds., 2nd Int. Conf. on Powder Processing Science (Deutsche Keramische Gesellschaft, Köln, Berchtesgaden, Okt. 12.-14. 1988, (1988).

[7] J. Tabellion and R. Clasen, "Near-Shape Manufacturing of Complex Silica Glasses by Electrophoretic Deposition of Mixtures of Nanosized and Coarser Particles", 28th International Cocoa Beach Conference on Advanced Ceramics and Composites (The American Ceramic Society, Cocoa Beach, Florida, (2004).

[8] J. Tabellion and R. Clasen, "Electrophoretic Deposition from Aqueous Suspensions for Near-shape Manufacturing of Advanced Ceramics and Glasses - Applications", *J. Mater. Sci.*, **39** 803-811 (2004).

[9] O. Stern, "Zur Theorie der elektrolytischen Doppelschicht", *Z. Elektrochem.*, **30** 508-516 (1924).

SHAPING OF BULK GLASSES AND CERAMICS WITH NANOSIZED PARTICLES

Jan Tabellion
Institute of Microsystem Technology
Laboratory for Materials Process Technology
University of Freiburg
Georges-Köhler-Allee 102
D 79110 Freiburg, Germany

Rolf Clasen
Department of Powder Technology
Saarland University, Geb. 43
D 66123 Saarbrücken, Germany

ABSTRACT

Manufacturing of functional ceramics and high-performance glasses by shaping of nano-particles and subsequent sintering combines significant advantages. Due to the high specific surface area of nanoparticles significantly increased sintering activity is achieved, which results in much lower sintering temperature. However, most of the common shaping techniques are not adapted to the intrinsic properties of nano particles. Due to their high specific surface area and low bulk density dry pressing can not provide an economic alternative. Suspension-based techniques seem to be much more promising to achieve green bodies with high density and good homogeneity. Nevertheless, with slip or pressure casting only comparably low compaction rates can be achieved, decreasing with particle size. In contrast, deposition rate is independent of particle size in case of electrophoretic deposition (EPD). Thus, EPD from aqueous suspensions is a fast and economic shaping technique for nanosized particles. A deposition rate of up to 3.5 mm/min was achieved, controlled mainly by the applied electric field strength (typically 1 to 10 V/cm). However, due to the high specific surface area of nano-particles, the density of the electrophoretically deposited green bodies was limited to about 50 %TD. By combining nanosized with larger particles, significantly higher green densities could be reached. From a suspension with optimized properties and adjusted parameters during EPD a green density of up to 81 %TD could be reached, resulting in a strong decrease in sintering shrinkage down to app. 6 %. Samples are shown for silica glasses and zirconia.

INTRODUCTION

In recent years, an increasing interest exists in using nanosized particles not only as fillers or functional isolated particles within a structural matrix but also for the manufacturing of mono-lithic glass and ceramic components. Due to their comparably very high specific surface area nano-particles bring a strongly enhanced sintering activity about. Thus sintering temperatures are reduced significantly compared to conventional microsized powders. Fully dense ceramics can be achieved avoiding high temperature phase transformations e.g. in case of zirconia[1]. Furthermore, changed mechanical and physical properties can be achieved due to the high ratio of grain boundary to grain bulk[2]. In case of powder manufacturing of glasses, the use nano-particles is mandatory to avoid crystallization during sintering[3].

Manufacturing of large glass or ceramic components with high wall thicknesses from nanosized particles is accompanied by some severe disadvantages. First of all, preparation of nano-particles, which are appropriate to the manufacturing of ceramics and glasses on an industrial scale, is still cost-intensive, resulting in high cost of raw material. Furthermore, shaping of nano-particles is difficult, because most of the common techniques for advanced ceramics are not adaptable to the intrinsic requirements arising from the use of nano-particles. Both bulk density of nano-particles and green density after dry pressing are very low. The application of suspension-based methods seems therefore much more promising. However, for most of these techniques deposition rates depend strongly on particle size. In case of slip or pressure casting e.g., deposition rate decreases significantly with decreasing pore radius of the generated green body, which in turn is mainly governed by particle size of the powder used.

In contrast, deposition rate is independent of particle size in case of electrophoretic deposition (EPD)[4]. Thus a fast compaction of nano-particles can be achieved. Electrophoretic deposition is a shaping process where the driving force arises due to an externally applied electric field. Charged particles are forced to move through a dispersing medium towards an oppositely charged electrode by the electric field. On an oppositely charged electrode or on a porous dielectric mould arranged perpendicular to the moving direction of the particles between the electrodes the particles coagulate and form s stable deposit. Reviews about EPD as a shaping technique for advanced ceramics and glasses are given in references[5, 6]. Bulk ceramics shaped from nano-particles by EPD include alumina[7] and zirconia[8]. Furthermore, EPD has been used for the manufacture of transparent silica glasses. Thus, silica green bodies with a thickness of 10 mm could be shaped within 5 minutes and sintered to transparent glass at 1450 °C from nanosized Aerosil OX50[9]. Higher green densities were achieved from identical suspensions by means of EPD compared to slip casting or pressure casting[10]. Even for comparable green densities, a lower sintering temperature was observed for electrophoretically deposited silica green bodies, which is related to the higher microstructural homogeneity[11].

The aim of this work was to highlight the potential of electrophoretic deposition of both nanosized particles and powder mixtures of nanosized and conventional larger particles for the manufacture of bulk glass and ceramic components.

EXPERIMENTAL

Two kinds of both silica and zirconia particles were used, one nanosized powder and another one with a mean particle size of several microns. The silica powders used were Aerosil OX50 (Degussa) with a mean particle size of 40 nm and Excelica SE15 (Tokuyama, $D_{50} = 15 \mu m$). Furthermore, tetragonally stabilized zirconia powder with a D_{50} of 0.6 μm (Tosoh, TZ3YS) and a nanosized zirconia powder (Degussa Zirconia-3YSZ, mean particle size 30 nm) were used. Aqueous suspensions of OX50, mixtures of OX50 and SE15 and a mixture of 3YSZ and TZ3YS were prepared by dispersing the particles gradually in bidistilled water by means of a dissolver (LDV1, PC Laborsysteme), adding different amounts of tetramethylammoniumhydroxide (TMAH). Vacuum was applied to avoid incorporation of air bubbles. TMAH was used to adjust pH and thus the ζ-potential of the silica particles as well as the viscosity of the suspension, both of which are important factors concerning EPD. The solids content of the suspensions was varied between 10 and 60 vol. %, depending on the powders used.

Electrophoretic deposition was carried out under constant applied voltage by the membrane method[12]. A porous ion-permeable membrane was used as deposition surface. The electrophoresis cell was subdivided by this polymer mould in two chambers, so that deposition of particles (onto the porous mould) and recombination of ions (at the electrodes) were separated. Thus, no gas bubbles were found within the deposits. The cathode-sided chamber was filled with the suspension, the other with bidistilled water containing different amounts of TMAH. After determining the effective electric field strength within the electrophoresis cell, the width of the different chambers was adjusted, to allow for optimum deposition conditions[13]. A simple set-up with planar parallel electrodes and a planar polymer membrane was used to determine the influence of process parameters on deposition rate and green density. For more complex shaped components the set-up had to be re-adjusted concerning the shape of the porous mould as well as the electrode design. The applied voltage was varied between 1 and 15 V/cm. Deposition time was varied between 1 and 3 minutes.

After shaping, the green bodies were dried in air under ambient humidity. No cracking was observed. Sintering of the compacts was carried out either in vacuum (SiO_2 powder mixtures, 1480 °C), in air (zirconia, 1400 °C) or in a zone furnace (OX50, 1320 °C). Green density was measured by Archimedes method. The high depth of focus on the transmission micrograph was due to the software module EFI (Extended Focal Imaging by analySIS).

RESULT AND DISCUSSION

Concerning the manufacture of large bulk ceramics and glasses from nanoparticles, two key factors have to be considered. First of all, a deposition rate as high as possible is worthwhile and, secondly, a high green density in combination with good microstructural homogeneity is favorable in order to avoid high shrinkage during drying and sintering.

First of all the influence of the processing parameters on deposition rate was investigated. One important result is the fact, that shaping of green bodies from nanoparticles is outstandingly fast by means of EPD. As shown in Figure 1, for a suspension containing 31 vol.% OX50, which is about the highest solids content processable by EPD, a deposition rate of up to 30 g/cm²·min (dark, dotted line, y-axis on the left hand side) can be reached for an applied electric field strength of only 10 V/cm (in this case corresponding to an applied DC voltage of 30 V). With the resulting density of app. 40 % of the theoretical value (%TD) of 2,20 g/cm³, this corresponds to an growth rate of the green body of 3.5 mm/min. Furthermore, the deposition rate can be controlled by adjusting the electric field strength, allowing for a reproducible tailoring of the wall thickness of the green body. Thus, shaping of nanoparticles can be carried out very fast by means of EPD. The influence of other process parameters of the electrophoretic deposition on deposition rate was also investigated in detail for nanosized OX50 and is described elsewhere[14].

In comparison, the deposition rate for a suspension with 58 vol.% OX50 and SE15 (mixing ratio 10/90) is shown in Figure 1 (dark line). A slightly higher deposition rate was observed for the powder mixture, which was due to the higher solids content. However, in general, no influence of particle size on deposition rate was observed. Due to the significantly higher density (≈ 81 %TD) of the green body consisting of the mixture of SE15 and OX50 a distinctly lower growth rate of 2 mm/min was observed in this case.

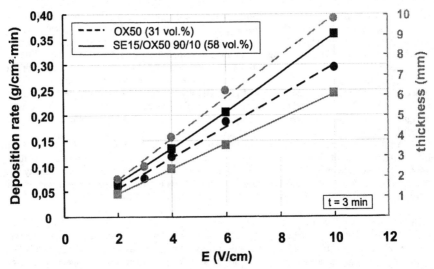

Fig.1: Deposition rate for EPD of nanosized fumed silica OX50 and a bimodal mixture of OX50 and SE15

In Fig. 2 the linear sintering shrinkage of numerous silica green bodies shaped EPD under different parameters and from suspensions with different solids contents is shown as a function of their green density. The dark dots represent samples shaped solely from OX50. As can be seen in Fig. 2, the green density of the OX50-samples is limited to values below app. 50 %TD. In comparison with other techniques this is still high[10]. The highest green density of 49.8 %TD was found for the sample deposited from the suspension with the highest solids loading (31 vol.-%). A further increase in solids content was not possible due to a strong increase in viscosity[15]., which made shaping by electrophoretic deposition impossible. As a result, linear shrinkage of more than 20 % occurs during sintering the OX50 green bodies at 1320 °C, to fully dense and transparent silica glasses. Apart from the typically high cost of raw material, this is the most significant disadvantage of using nanoparticles for the manufacturing of bulk ceramics and glasses. High shrinkage appears along with the risk of distortion, and near-shaping of complex geometries is impossible. Furthermore, the necessary oversize of the sintering furnace adversely affects the process cost.

Nevertheless, to benefit from the advantageous properties of nanomaterials, like the high sintering activity or the small resulting grain size after sintering, these disadvantages can be avoided by combining nano-particles with conventional, larger particles. As a result of an optimized ratio of particle sizes and an adjusted mixing ratio of nanosized and bigger particles, green bodies with very high green densities could be shaped. A very good microstructural homogeneity of the green bodies could be observed after drying. No size-dependent particle separation occurred. The linear sintering shrinkage of numerous silica green bodies shaped under different conditions by EPD from mixtures of SE15 and OX50 is also shown in Fig. 2 as function of green density (light rhombi). Much higher green densities could be achieved with these powder mixtures with a maximum of 81.7 %TD. This value was achieved for a green body shaped from a suspension with 58 vol.% OX50 and SE15 in a mixing ratio of 10/90. The resulting linear shrinkage during sintering to a fully dense and transparent glass at 1480°C was observed to be only 6 %.

Thus, two significant disadvantages of the use of nanoparticles for the manufacturing of bulk components can be overcome by using mixtures of nanosized and larger conventional microsized particles. Only ten percent of nanoparticles are necessary to achieve such high green density. Thus the cost of raw material can be reduced significantly. Furthermore, near-shape manufacturing of complex-shaped silica glass components is possible.

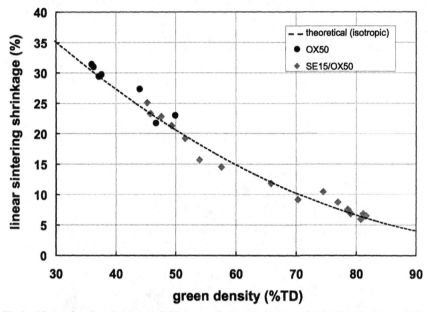

Fig.2: Linear sintering shrinkage of silica green bodies electrophoretically deposited from different OX50-suspensions (dark dots) and different suspensions containing bimodal mixtures of OX50 and SE15 (light dots)

Figure 3 shows a sintered silica glass triple mirror array as received after electrophoretic deposition and sintering. The corresponding green body was deposited from a suspension containing 58 vol.% OX50 and SE15 in a mixing ratio of 10/90. Sintering was carried out at 1480 °C. Due to the high green density only very low shrinkage occurred during sintering as shown above (cp. Figure 2). Since the shrinkage occurred isotropically, no deformation of the structures was achieved, as can be seen on the transmission micrograph on the right-hand side of Fig. 3. Sharp edges and defined contours were achieved without any post-processing, which is not possible in case of silica glass by conventional manufacturing via the melting route.

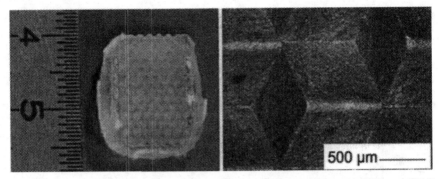

Fig.3: Silica glass microstructure as received after EPD (bimodal mixture of OX50 and SE15) and sintering (right hand side: transmission micrograph with software module EFI)

A similar optimization process was carried out in case of zirconia, but will not be described in detail here. Figure 4 shows a sintered zirconia tube with an inner diameter of 7 mm and a wall thickness of 1.3 mm after sintering at 1600 °C. Isotropic shrinkage of only 12 % (linear) occurred during sintering.

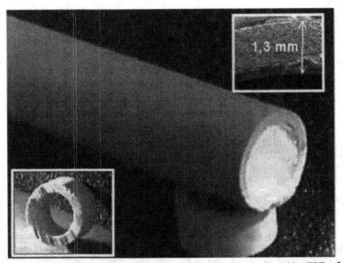

Fig.4: Zirconia tube (inner diameter 7 mm, wall thickness 1.3 mm) shaped by EPD of a bimodal powder mixture, as received after sintering (1600°C)

CONCLUSIONS

Using nanoparticles for the manufacturing of bulk ceramic and glass components can result in components with enhanced properties. However, still the high cost of raw materials as well as the typically low density of green bodies from nanoparticles are getting in the way of a large-scale production of bulk components from nanosized particles. Although very high deposition rates can be reached by electrophoretic deposition of nano-particles (up to 3.5 mm/min for nanosized fumed silica OX50) even for a very low energy input (E < 10 V/cm, t <3 min), the green density is limited to about 50 %TD. Combining nanosized particles with conventional particles, having a mean size of several microns, can result in green bodies with significantly enhanced density. Due to the fact that the deposition rate is independent of particle size in case of EPD, very homogeneous green bodies were achieved, showing no size-dependent particle separation. A maximum green density of app. 81 %TD was reached. Correspondingly, the sintering shrinkage could be reduced significantly (app. 6 % linear). Thus near-shape manufacturing of complex shaped silica glasses or zirconia components is possible, allowing at the same time for a much more cost effective production of bulk ceramics and glasses from nano-powders, because much less nano-particles are necessary to achieve similar favorable materials properties.

REFERENCES

[1]M. J. Mayo, "Processing of Nanocrystalline Ceramics from Ultrafine Particles," *Int. Mater. Rev.,* **41,** 85-115 (1996).

[2]H. Hahn and K. A. Padmanabhan, "Mechanical Response of Nanostructured Materials," *Nanostructured Materials,* **6,** 191-200 (1995).

[3]E. Rabinovich, "Review: Preparation of Glass by Sintering," *J. Mater. Sci.,* **20,** 4259-4297 (1985).

[4]C. Hamaker, "Formation of a Deposit by Electrophoresis", *Trans. Faraday Soc.,* **36,** 279-287 (1940).

[5]M. S. J. Gani, "Electrophoretic Deposition - A Review," *Industrial Ceramics,* **14 [4],** 163-174 (1994).

[6]P. Sarkar and P. S. Nicholson, "Electrophoretic Deposition (EPD): Mechanisms, Kinetics, and Application to Ceramics," *J. Am. Ceram. Soc.,* **79 [8],** 1987-2001 (1996).

[7]B. Ferrari and R. Moreno, "Electrophoretic Deposition of Aqueous Alumina Slips," *J. Eur. Ceram. Soc.,* **17,** 549-556 (1996).

[8]K. Moritz, R. Thauer, E. Müller, "Electrophoretic Deposition of Nano-scaled Zirconia Powders Prepared by Laser Evaporation," *cfi/Ber,* **77 [8],** E8-E14 (2000).

[9]R. Clasen, "Preparation and Sintering of High-Density Green Bodies to High Purity Silica Glasses," *J. Non-Cryst. Solids,* **89,** 335-344 (1987).

[10]J. Tabellion; R. Clasen, "Electrophoretic Deposition from Aqueous Suspensions for Near-shape Manufacturing of Advanced Ceramics and Glasses—Applications", *J. Mater. Sci.,* **39 [3],** 803-811 (2004).

[11]J. Tabellion, E. Jungblut and R. Clasen, "Near-Shape Manufacturing of Ceramics and Glasses by Electrophoretic Deposition Using Nanosized Powders"; in *Ceramic Engineering and Science Proceedings,* **24 [3],** AcerS (Westerville), 75-80 (2003).

[12]R. Clasen, "Forming of Compacts of Submicron Silica Particles by Electrophoretic Deposition", in *2nd Int. Conf. on Powder Processing Science,* ed. H. Hausner, G. L. Messing and S. Hirano, Berchtesgaden,: DKG,. 633-640 (1988).

[13]J. Tabellion and R. Clasen, "In-Situ Characterization of the Electrophoretic Deposition process"; in *Innovative Processing and Synthesis of Ceramics, Glasses, and Composites IV.*, ed. by N. P. Bansal and J. P. Singh. Am. Ceram. Soc., Westerville, 197-208 (2000).

[14]J. Tabellion and R. Clasen, "Controlling of Green Density and Pore Size Distribution of Electrophoretically Deposited Green Bodies"; in *Innovative Processing and Synthesis of Ceramics, Glasses, and Composites IV.*, Edited by N. P. Bansal and J. P. Singh. Am. Ceram. Soc., Westerville, 185-196 (2000).

[15]J. Tabellion, R. Clasen, J. Reinshagen, R. Oberacker and M. J. Hoffmann, "Correlation Between Suspension Structure and Rheological Properties of Suspensions of Nanosized Fumed Silica Powders", in *Improved Ceramics through New Measurements, Processing, and Standards*, ed. M. Matsui, S. Jahanmir, H. Mostaghaci, M. Naito, K. Uematsu, R. Wäsche and R. Morell, ACerS, Westerville (USA), 183-188 (2002).

Author Index

9 781574 982619